PREFAB PROTOTYPES

PREFAB PROTOTYPES

SITE-SPECIFIC DESIGN FOR OFFSITE CONSTRUCTION

Mark Anderson and Peter Anderson

PRINCETON ARCHITECTURAL PRESS / NEW YORK

Published by
Princeton Architectural Press
37 East 7th Street
New York, New York 10003

For a free catalog of books, call 1.800.722.6657.
Visit our web site at www.papress.com.

Printed and bound in the United States
10 09 08 07 06 5 4 3 2 1 First edition

Editor: Scott Tennent
Design: Paul Wagner

Special thanks to: Nettie Aljian, Sara Bader,
Dorothy Ball, Nicola Bednarek, Janet Behning,
Penny (Yuen Pik) Chu, Russell Fernandez,
Pete Fitzpatrick, Sara Hart, Jan Haux, Clare Jacobson,
John King, Mark Lamster, Nancy Eklund Later,
Linda Lee, Katharine Myers, Lauren Nelson Packard,
Jennifer Thompson, Joseph Weston, and Deb Wood
of Princeton Architectural Press
–Kevin C. Lippert, publisher

Library of Congress Cataloging-in-Publication Data
Anderson, Mark (Mark S. T.)
Prefab prototypes : site-specific design for offsite
construction / Mark Anderson and Peter Anderson.
p. cm.
Includes bibliographical references and index.
ISBN-13: 978-1-56898-560-2 (alk. paper)
ISBN-10: 1-56898-560-6 (alk. paper)
1. Prefabricated houses–Design and construction.
I. Anderson, Peter (Peter C. O.) II. Title.
TH4819.P7A584 2006
690'.837–dc22
2006001193

Photo Credits

Amerikaya Construction, p. 33
Charles Benton, p. 231
Dan Brondyk, p. 156
John Clark, p. 115
International Hi-Tech Industries, pp. 84, 85
Jeffrey Jordan, pp. 48, 49, 148
Karen Moskowitz, pp. 27, 30, 31
Michael Scarbrough, pp. 47
Cameron Schoepp, pp. 93, 95
Scott Stafne, p. 120
Timbercraft, p. 55
Anthony Vizzari, jacket, pp. 51, 96, 97, 104–07, 116, 131,
132, 135, 138–40, 150, 151, 153, 158–61, 163, 169,
170, 184, 191, 207, 237, 241, 245, 246
Madeline Williamson, p. 174
All other images by Anderson Anderson Architecture

Contents

Introduction

Designing Construction

We have often said that our main interest in construction is tools and trucks.[1] Architecture as a creative endeavor certainly trumps this fascination, but as much as we enjoy and respect fine craft, our own interests tend toward more abstract concepts at one level and machinery and rough process at another. Quality and logic at all levels are important to us, ultimately more so than the finer points of hand craft. Unless we are doing the work ourselves or collaborating directly with excellent craftspeople—in which case we get as caught up in the nuances of material and precision as much as anybody else—designing for a handcrafted approach to building construction is generally a recipe for exclusivity and unaffordable expense. From a practical point of view, mechanization and speed bring far greater access to quality of design and construction when considering the overall context of society and the built environment.

Prefabrication is an important step in the construction of every building, referring to the making of parts in an offsite workshop or factory prior to installation at the site. The primary purpose of prefabrication is to produce building components in an efficient work environment with access to specialized skills and equipment in order to reduce cost and time expenditures on the site while enhancing quality and consistency. Since very early in the history of building, some degree of offsite prefabrication has contributed to the majority of construction projects in all parts of the world. Today, with rare exceptions, industrial prefabrication of building elements plays a major role in nearly every project, ranging in degree from wholly prefabricated doors and windows to structural framing elements such as steel beams and engineered wood I-joists with highly standardized, consistent dimensions and industrially produced connection hardware, all frequently shipped to the site as small pieces to be assembled in place.

Since in this sense most every building is already prefabricated to some extent, the recent renewed interest in prefab housing reflects a primary concern for advancing the efficiency that can be brought to bear in reducing housing costs while simultaneously enhancing design and construction quality. Despite the reasonable and historically prevalent concern that increased standardization and industrialization will lead to homogenization and reduction of choice and variety, in the case of most contemporary homebuilding, increased prefabrication may very well lead to higher quality and a wider array of choices in housing types and products. To understand why this should be requires a grasp of the current construction industry in terms of both economic and technical production issues. The following chapters will address these topics through a discussion of the various systems of structural prefabrication already on the market, further illustrated by case studies of built and prototype designs by our firm, Anderson Anderson Architecture.

The positive payoff for increased prefabrication includes reduced building costs; a more stable and fairly remunerated construction industry with improved safety and working conditions; greater investment in research, design creativity, and product development; reduced consumption of energy and materials; and the generally increased availability of better designed, higher quality built environments. If the focus of prefabrication is, on the one hand, primarily in the area of cost savings alone—whether from the point of view of producers or consumers—or, on the other hand, if one views prefabrication as a superficial design style rather than a comprehensive systems approach, then the full promise of increasingly industrialized building construction will be much diminished, even to the point of opening further opportunities for exploitative consolidation within the building industry, leading to increased land consumption, lowered wages, higher prices, and reduced competition, choice, and quality. Prefabrication is not a clearly separable design issue in itself, but in relation to a much larger design, production, and economic context, it may again be a popularizing concept capable of catalyzing substantial change and progress in the construction industry, restoring the dream of high-quality, imaginatively conceived, affordable houses and buildings of all types.

Prefabrication from Many Points of View

Since starting our firm as a design/build construction company in Seattle more than twenty years ago, we have been working with prototype applications of building prefabrication systems through many built and unbuilt projects. Sometimes we work with panelized stud-wall or structural insulated panel (SIPs) systems, prefab steel structural frames, or precast concrete systems; sometimes we work with high-tech CAD/CAM laminated timber frame systems or even low-tech, hand-cut log systems. Almost always we work with sustainably harvested engineered wood structural components and other resource-efficient, low energy consumption products and building techniques. At times we experiment with new or unusual materials and processes, but more often we look for ways to creatively adapt more common materials, products, and techniques in order to best engage with the more mainstream building industry—and to push it, gently, forward.

Over the last two decades, we have continually developed and refined our own modular approach to the design, manufacture, and implementation of affordable, high-quality prefabrication systems, which we have come to regard as a unique way of making buildings composed not just of physical building components but also a set of ideas about design and construction that can be thoughtfully adapted, as required by individualized sites and client programs, for houses or for other buildings and urban structures of all types and sizes.

A most important thread within all of our work and throughout this book concerns site-specificity and the ability to individualize architectural objectives within prefabricated systems. It is important to assert individuality and a unique point of view at all stages in the production, reading, and inhabitation of architecture. Although we are dealing here with issues of mass production, we consider our work always to be quite personal, as all work should be, and therefore we are interested in introducing important influences on our point of view and in suggesting potentials for the always unique, personal, and site-specific within economical prefabricated construction.

To appreciate the past richness of ideas as well as the future potential of rationalized construction systems, prefabrication must be considered within the context of a long history of works by many architects, engineers, builders, government officials, entrepreneurs, and social theorists who have addressed these issues since the dawn of the industrial age, and all of this must be considered within a realistic understanding of the contemporary construction industry.[2] This is an immense task, and we cannot attempt to be comprehensive in this book. The construction industry is the single largest economic sector in the world, of which prefabrication is an enormous sub-industry. The housing and other building needs of both developed and developing societies are great, and many national and international governmental and non-profit organizations continue to devote significant resources to projects for construction system rationalization, resource efficiency, and industrialized prefabrication.[3]

Acknowledging this inability to be comprehensive, the point of view informing this book is that of practicing architects engaged in building construction as a creative process and concerned with making tangible ideas for human experience and inhabitation. As with any such creative endeavor, the effective realization of building projects relies on large amounts of technical, social, and historical research, as well as direct, responsible design involvement at all levels of the production process. Our own primary interest is not in research for its own sake, in pure systems logic, or in abstract engineering. We have a practical interest in the means of producing meaningful architectural projects: harnessing tools and techniques to conceptualize the positive experience of living in the constructed world. The case studies we have selected for this book highlight a variety of ideas, systems, and processes, as well as the broad implications each entails. Using both technical arguments and project-specific anecdotes, we are interested in both the difficulties and the potentials involved in the prefabrication and construction of ambitious architecture.

Prefabrication is an issue interwoven with the central threads of modernism. Stylistically, many people associate prefabrication with the simple forms, flat roofs, and minimalist detailing of modern buildings. Technically, prefabrication has also been much used in building traditional, anti-modern, or stylistically unconcerned or misguided buildings, in which there is no particular formal or spatial outcome that is necessarily intrinsic to prefabrication methods. This is unfortunate, because buildings always come out best when they emerge logically and honestly from their immediate material, process, and cultural contexts.

The interest here is not in issues of style, however, but in fundamental concepts and imperatives in modern thought and theory. Standardization, systemization, rationalization, material honesty, realism, and technical innovation in support of social progress are all central concepts of the modern period, and they are vital to the development of contemporary production processes and distribution networks as much as to particular architectural developments in space and form. Architecture has in the past been instrumental in the theoretical and practical development of modernist ideas and practices, but it has lost momentum, particularly in the realm of practical construction, during the conservative architectural retrenchment of recent decades that has distanced architecture from its modernist tradition of social and political responsibility and abandoned its most powerful tools of engagement in the economics and technology of construction. From this weakened position, architecture has been slow to embrace the most far-reaching and potentially advantageous technological advances of the current age, in which the vanguard of progressive inquiry has passed into other fields. The revitalized interest of architects in new and emerging technologies intertwined with natural and societal systems has begun to reinvigorate this stalled engagement between architecture and the advancing frontiers of modern society. Initially this interest has been primarily driven by a regard for new formal, spatial, and conceptual potentials offered by new technology, which has been wonderful experimentation in itself. Now there is beginning to be a deeper and richer engagement of digital tools and their attendant conceptual concerns with more complex human and natural fields of design and inquiry. The old frontier of practical, responsible, economical, prefabricated building construction now seems almost nostalgic and retrograde in present form. Yet the issues inherent in the slow progress of construction innovation remain the same, and while sharply piercing experiments beneficially push the outer frontiers of technological and social experimentation in housing and construction, a broad, mass-market push—within the everyday worlds of construction and business that most impact the day-to-day lives of most people—still encounters many of the same hurdles that have thwarted remarkable efforts in prefabrication for the past hundred years.[4] The big issues are not primarily technological, nor even resulting from any widespread social or economic resistance to the simple logic of rationalized production that has transformed most every other human endeavor during modern times. Prefabrication is simple to understand conceptually and not so difficult to achieve technically. It is primarily an issue of investment and organization, which is a disappointing recognition for most architects, who most typically lack capacity for either, and are more interested in pursuing the concept, the space, the form, the innovative details.

The last two centuries have seen many determined attempts at large-scale prefabrication projects, with many important architects and entrepreneurs championing off-site building approaches through their work on significant advances in building systems and innovative designs for prefabricated structures.[5] Early in the nineteenth century there were successful prefabrication enterprises focused on export to Britain's expanding international colonies. Packaged houses, churches, commercial buildings, and elaborate provincial government structures in wood or in cast-iron were shipped to Australia, India, and the far corners of the empire.[6] At the turn of the twentieth century, prefabricated cast-iron commercial building systems became common in American cities, while Sears, Roebuck and Company offered comprehensive kit-house catalogs offering plans, pre-cut pieces, and all the fixtures and furniture as a single, easy-to-order package, shipped to your site on the great American prairie.[7] Many of these houses still exist today and are typically sturdy and well-designed examples of the common houses of their time.

By contrast, with the experiments of modern architecture emerging in parallel with such wholesale change and cultural upheaval as the century progressed, all assumptions were off the table and every aspect of architecture and construction was brought up for re-examination and redesign. In this context, prefabrication and mass-production were central to the logic and objectives of modern architecture, formally, socially, and politically. The tremendous period of cultural creativity between the World Wars, particularly in Europe, brought forth prefabricated housing experiments as diverse and differently directed as those of Buckminster Fuller, Le Corbusier, Walter Gropius, Hannes Meyer, and many others.[8] With the outbreak of World War II, manufacturing production underwent radical re-engineering and expansion and drew most every other aspect of culture and society into entirely new relationships with industry. During and after the war, many European architects now in America, along with a new generation of American architects, opened a new chapter in prefabrication, creating high expectations and great experimental buildings. California was at the center of these great hopes, and held particular promise as a new frontier, with a limited history free of preconceptions, an accommodatingly benign climate, and, most importantly, a vast new population, recently migrated to the war-era industrial plants on the coast.[9] In this moment was a great flowering of ideas, high expectations, and wide-open potential for modernist experiments in factory-built housing. Charles and Ray Eames produced the most compelling work and brilliant, all-encompassing engagement with modern ideas for living.[10] The Case Study program of experimental houses commissioned and published by *Arts and Architecture* magazine under the leadership of John Entenza produced a legacy of great, highly influential prototypes.[11]

The experience of the postwar prefabricated housing experiment of Walter Gropius, Konrad Wachsmann, and the General Panel Corporation is emblematic of the difficulties of making a significant impact on the housing industry, at least in North America, in such a way and at such a time. Gropius and Wachsmann began working together at the Bauhaus in pre-war Germany on prototypes for panelized, prefabricated houses then pursued larger-scale factory-based development of their concepts in the U.S. after the war. This later enterprise was the most extraordinary and promising venture in the history of prefabricated housing and enlisted the substantial partnership and financial underwriting of the U.S. government and the world's largest manufacturer, General Motors. A large and sophisticated plant was constructed in the heart of the California market, and extensive design and prototyping was pursued. For a great variety of reasons, however, the General Panel Corporation soon collapsed without producing any significant output.[12] Yet the particular approach of General Panel continues to be the model for many ambitions in the current revival of broad interest in prefabrication, and thus it remains a cautionary tale of pre-war architecture and industrial assumptions outstripped by a society and economy modernizing more rapidly and in a far more decentralized and individualized structure than could be engaged by rigid, centralizing industrial organization. Since that time the construction industry has slowly begun to consolidate; trade organizations such as the American Plywood Association and the National Association of Homebuilders have made sporadic attempts at research-based innovation across the market; and developments in product manufacturing theory, in particular with respect to just-in-time process flow and concepts of mass-customization, all point to more flexible models for fabricating building components at the factory scale and integrating them more fully into the construction market as a whole.[13] Nevertheless, many market factors remain much the same, and most design-focused prefabrication proposals put forward by architects continue even now to repeat the same mistakes of the immediate postwar era, focusing too narrowly on specific structural approaches little adapted to the more comprehensive issues of production and marketing.

Another example of modern housing experimentation that met with considerably greater success—due at least in part to the limited number of production process variables engaged—emerged in the large-scale, middle-class housing developments built by Joseph Eichler in Northern California during the 1950s, '60s, and '70s. Influenced by the Case Study homes, Eichler hired many of California's best modern architects to design high-quality houses well integrated with the casual, indoor-outdoor lifestyle ambitions of the California dream. While designs were repeated in significant quantities, these development projects were well-conceived in planning, landscape, and quality-of-life home design, and they remain highly desirable living environments to this day. While these projects adopted many of the architectural and social aspirations shared with the proponents of industrialized house production, Eichler was quite practical in adopting the prevalent mass-housing production method of the time, harnessing the booming, decentralized, low-technology, low cost, 2x4 framing sector as his industrial production engine rather than the far more capital-intensive path of a dedicated manufacturing plant.[14] The success of the Eichler approach, which was dependent on the ready availability of inexpensive skilled carpentry labor as well as the optimistic, progressive, and lifestyle-focused mood of the homebuying public of the time, may have been one of the significant nails in the coffin of the postwar prefabricated homebuilding industry, particularly on the west coast, where a tight, machine-focused and somewhat bloodless image with high initial production costs could hardly compete in the public imagination. Even while capitalizing on highly efficient and innovative project delivery systems, which can be seen as a kind of modularization of the process, the Eichler houses demonstrated that the relationship of the building to its site was ultimately a more significant factor in establishing a successful market for homes than was the realization of production efficiencies in the factory and during assembly at the site.

The pragmatic innovation of Joseph Eichler and his projects has long been an interest of ours, sparked at an early age, perhaps, by the fact that our parents chose an Eichler as their first home purchase when beginning their young family in Palo Alto. Although their first home exists for us primarily as a collection of images from photographs and glowing recollections from our parents, it has always been the standard against which all our other living situations have been judged, and the plausibility of an affordable, well-designed, stylish, happy, and site-integrated house has always been a part of our consciousness, influencing our own design, construction, and development work. Although idealists in most every sense, we have of necessity always been very practical about our design opportunities and our abilities to achieve radical construction system change within any one project. We approach many commissioned building projects as opportunities for the development of prototype prefabrication and project-delivery systems that focus on the adaptation and implementation of readily available materials and building techniques to the needs of specific sites and programs rather than as a platform for entirely experimental technologies. Our experience in both custom and prefabricated affordable construction suggests that most often these goals can only be achieved with a highly realistic approach to selecting and adapting already existing components while organizing these elements into a system that will point the way toward significantly greater cost and time efficiencies once applied to higher-volume production.

One of the lessons that can be learned from the many previous attempts at prefabricated housing production is that uniquely proprietary systems of single-source components are too costly to develop and have almost always ended in economic failure, even when excellent in design, detailing, and production concept. We have come to believe that the most effective path to achieving the benefits of prefabrication comes from an incremental transition from site-based craft and assembly to offsite componentization of building elements, accompanied by a deeper analysis and understanding of existing social and economic forces outside of design and mechanics. During the past thirty years in particular, building components have increasingly transitioned toward standardized, manufactured sub-assemblies ready to be deployed on the job site as complete modular components.[15] During this period of industry consolidation and entrepreneurial advances in production, marketing, and delivery methods, an important missing link in the widespread implementation of prefabrication into the homebuilding industry has been the lack of sophisticated architects and designers interested in mass-market construction who are capable of integrating the availability of existing products and systems into appealing and affordable buildings. The development of architectural design interest in a viable, prefabricated housing market that is responsive to a demand for high-quality design and construction will open a vast world of new opportunities for future exploration of experimental approaches to material use and new production techniques within a significantly more advanced industrial economy that is now ripe for such cooperative advance.

People, Landscape, and Quality of Life

Buildings are built in specific places by and for specific people for specific reasons. Such a statement may seem contrary to the typical, generalized notion of prefabrication, but there are many reasons why prefabrication need not necessarily become wholly generalizing, some of these dealing with new, computer-based potentials for mass customization of products available within endlessly overlapping and intertwining orders of digitally networked individualities, and others dealing with the myriad complexities of everyday life that seem always to override even the best intentions toward order and consistency. As artists, certainly we are concerned about the specter of too much standardization. As experienced builders, we can safely say that this will be the least of our worries for the foreseeable future. Today, there is a good deal of wasteful construction system chaos, directly contributing to homogenized landscapes and wearisomely redundant buildings and products that can be described as "custom" in name only.

A number of approaches to positive prefabrication reasoning will be discussed throughout this book. Before launching into details of either theory or technology, however, there are certain big-picture issues that should guide any thoughtful discussion of architecture, buildings, housing, and human life itself—whether in natural, constructed, or conceptual environments. The big picture for prefabrication, as with any other issue of architecture and building—and indeed all matters of human endeavor generally—must focus on its contribution to quality of life. This is a pragmatic approach, requiring that architectural thought and products functionally contribute to a larger social purpose. This may seem either aesthetically conservative or politically leftist, depending upon one's point of view, but in fact human purpose and quality of life certainly must transcend any argument limited to prosaic conceptions of practical or political functionality. Functionality and fitness for purpose must include aesthetic and spiritual ambitions far beyond the satisfaction of basic economic, work, and living requirements. Without venturing too far into philosophical questions of human purpose, the basic criteria behind each decision at every step in the architect's, builder's, and manufacturer's work—from determining what degree of sunlight should fall on the breakfast table to what degree of toxic off-gassing may be acceptable from a phenolic resin product—should consider fundamental issues affecting human life and experience.

Addressing the functionality of a building in respect to its impact on and contribution to the people and environments it affects is further complicated in a for-profit economy, where one primary purpose of building construction is the production of income. The profit function in real estate development and construction can be a strong force for innovation, but it can also skew processes and concepts of functionality and suitability toward limited sets of criteria for judging and developing construction options. There has always been a good deal of innovative business interest in prefabrication, but this is set against even larger entrenched forces established within existing systems. Aligning profit motives with social and aesthetic imperatives offers considerable potential for prefabrication and rationalized construction processes to significantly contribute to positive growth and progress in the built environment.

An underlying fear with any form of industrialized production is the suspicion that individuality, craft, and specificity to local sites and cultural conditions will be forsaken in the drive toward efficiency and the maximization of profit. There is the suspicion among architects and skilled builders that in an already debased construction craft environment, prefabrication will simply be the straw that breaks the camel's back, and that originality and handcraft will be thrown out the window altogether. Based on substantial experience in the construction world, a strong case could be made for quite the opposite projection.

There are several important components to the question of individuality in design, craft, and living. First of all, two great joys in life—perhaps the greatest joys beside love itself—reside in the recognition and cherishing of difference and character among people and landscapes. This infinity of distinctions among people and places is manifest in the endless variety found in nature, cities, language, custom, cuisine, art, and intellectual insight. The great differences that we enjoy in life are particularly admired in architecture and in thinking about architecture. How does prefabrication fit into the valuing of difference in life and architecture?

We have subtitled this book *Site-Specific Design for Offsite Construction*. This subtitle reflects our interest in prefabrication as a tool, means, method, or process—but not an end unto itself—and links it to other issues that are at the core of our concerns as architects. All of our design and construction work draws its form, space, and inspiration from the particularities of a specific site and its inhabitants and not, as one might expect from people interested in prefabrication, from technical concerns and production methods. *Inhabitants* is a clumsy, anthropological-sounding term that we intend here to encompass all people concerned with and affected by the site and its construction. This includes, of course, the client, who has hired the architects and builders to produce a building for a specific purpose, and whose involvement in a project's creative process invariably embeds his or her individuality within the design. But client and architect also recognize a larger group of people inhabiting the affected space of the project, and these people, if not individually known, still exert strong particularizing forces on the design: the construction workers who will spend much time on site, the surrounding populace, visitors, and future generations of occupants, owners, and neighbors. Many of these people exert direct influences in the form of legal requirements embodied in zoning and building codes, laws such as the Americans with Disabilities Act, environmental regulations, health and safety codes, union rules, neighborhood covenants, easements, and title restrictions.

Whenever such legal issues arise in discussion with our design students, we always try to stress the larger issue of responsibility to the safety and quality of life of individuals and the general public, rather than allowing a focus on the arcane and seemingly inhuman details of the restrictions themselves. For example, in discussing an issue like fire stairs with our students, we have found it quite telling to ask the question, "Even though it may sometimes interfere with a competing design concern, why is it important to always make sure that you have proper fire exits?" Invariably, the first response refers to a concern for liability. With considerable coaxing we eventually manage to get someone to state the obvious—that we don't, as architects and as human beings, want anybody to burn up in a fire. Of course we all share these same fundamental human concerns when we are not lost in the details of the rule books, but we mention this anecdote here as a reminder of the obligation to maintain a strong focus on the greater responsibilities to people, environment, art, and culture, and the absolute imperative to avoid becoming lost in the endless arcana of detail and regulation.

How quickly our discussion of the site and its people shifted from a lofty point of view—even mentioning love and cherishing—before quickly descending into reference to laws, regulations, and liabilities. With each passing year, design and construction is increasingly affected by endlessly detailed and specific rules and laws. Typically these rule sets are divided into national—and increasingly international—technical, engineering, and life safety issues dealing with the building construction itself, and progressively more site-specific local zoning and micro-habitat environmental regulations governing what and where and how buildings are placed. At their roots, nearly all of these enactments are very well intentioned and specifically meant to protect both people and their environment. In the complex of issues concerning globalization and international trade that increasingly involves and affects the construction industry, there are more and more opportunities and discussions revolving around the protection of economies, cultures, and production traditions, as well. The growing global network has allowed for incredible access to new products, processes, skills, and opportunities, while at the same time—partly due to the attendant problems resulting from this new interconnectivity—is increasingly constrained and arcanely regulated, as well. All of this new complexity in the world of construction, far beyond the mere introduction of new materials or technical processes, demands accordingly sophisticated methods of organization and administration. Physical production process alone is no longer the sole driving logic in prefabrication. There are significant external pressures increasing the need for further modularization and rationalization of planning, permitting, financing, and management.

Landscape Pressured by Time, Money, and Zoning

Among the most important issues related to prefabrication, and especially to modular design and construction, is the relationship of the building unit to the landscape. By landscape we are referring to the immediate context of the natural and adjacent built environment—topography, vegetation, neighboring structures, streets, animals, and people. While the building itself may be quite distinct as a unique, offsite-manufactured object or aggregation of modular objects, this in no way requires that the relationship of these objects to their new context need be insensitive or lacking in specificity. Even in our most uniquely one-off buildings on pristine natural sites, for example, we have frequently followed two divergent approaches. In one approach, we bend and twist and drape the building logic to conform to the complex rolling forms and orientations of the site. In other cases, we stand up a strong geometric form floating above or sharply cut into the landscape as a distinct counterpoint to the natural site. The reasoning behind individual decisions along these lines is subtle and deeply considered while standing within these landscapes, and also in careful mapping and analytical diagramming of various conditions of the site and the relationship of these to the building purpose and the aspirations and concerns of the people involved. There is no singularly appropriate response to any specific urban or natural context. One cannot say that it is always best to hide a building within nature, for example, or that it is always best to emulate the immediate architectural context within a city. The relationship of any new construction to its immediate context involves subtle and complex decision-making that cannot be reduced to simple rules, despite the misguided efforts of many planners and regulators who seek to enforce specific and simplistic design guidelines. As a tour through most design-guidelined cities and suburbs will quickly prove, the well-intentioned prefabrication of architectural style and process guidelines does vastly more harm than good and typically precludes genuine architectural quality while encouraging the most banal and formulaic responses of developers and project owners who may now thoughtlessly check off minimal rules in lieu of thoughtful engagement with architecture. Increasingly, the spirit of architecture and of unique and meaningful places is bargained away in favor of development checklist accounting administered by the burgeoning planning professions and their increasingly complex rule-making. While the end result of unimaginatively restrictive planning is quite frequently endless homogenization and sameness, most often the arcana of overly rigid and cumbersome rules makes most difficult the thoughtful incorporation of rational modularization and prefabrication. If a fraction of the construction expense now committed to poorly conceived, inefficiently administered, and uncreative planning compliance were instead committed to architecture and construction systems design, along with a spirited community interest in establishing higher critical and market standards for the built environment, far greater individuality and quality would result than might ever be achieved within circumscribing rules. From this point of view, the increasingly sophisticated design of rationally constructed, well-designed, enjoyable-to-live-within buildings and landscapes that will eventually emerge from increasingly modularized and prefabricated building units will equally yield more rational land use and even greater variety, quality, and affordable site-specificity than is possible now, with the essentially mandated, lowest-common-denominator sameness of the current mass-market.

Introduction

Central issues of prefabrication and modularity come into particular focus when considering the overall complexity of the construction economy. Houses in particular are an important engine of economic expansion in many parts of the world and respond to and initiate forces far outside the immediate realm of design or even the desires of homebuyers, much less the best interests of society as a whole in imagining how to make good places for people to live. There are many ways to research and also to critique and lament this particularly strong relationship between macro-economic forces and the most basic needs and desires of individuals and whole populations, but failure to recognize the complexities of this relationship within the design process leads back to overly simplistic prefabrication approaches overly focused on detailing and manufacturing process without acknowledging the bigger picture. While architects and manufacturers may concentrate on simplifying and streamlining the physical facts, the increasingly larger hurdles in regulation, finance, insurance, and liability work backward toward increasing complexity, particularity, and uniqueness in individual conditions. For example, during the first years of the twenty-first century, when interest rates have been historically low and the market for houses has been at a historic high, the process of construction has been increasingly slowed by usually well-intentioned new levels of review and required documentation for financing and zoning approval that are not well coordinated or integrated across the market, governmental jurisdictions, or the landscape itself. Unlike building codes, which have become increasingly rational and well-integrated at national and international levels, zoning and planning review rules have gone in the opposite direction, toward intentionally dilatory and frustratingly complex adversarial agendas. As a direct result of these complexities, developers will generally seek to minimize all other variables and repeat common designs and production practices, rather than improve or change them and incur renewed regulatory review. We are in no way opposed to logical and beneficial regulation and oversight that will protect and improve the environment and quality of life for all people, but the design of these regulatory and finance systems must coordinate with the overall logic of design and production systems, so that progress and improvement can be a powerful and fast-reacting system, investing money, creativity, and effort in real progress toward beneficial goals rather than in the intended slowing of perceived negative potentials, when the net effect quickly becomes only the enshrinement of the status quo more slowly and expensively delivered. Interest in rational physical processes of construction must become more inclusive of interest in accompanying systemic processes as well, since the frustrating fact related to affordable construction in general and the efficiencies of prefabrication in particular is that less and less of a building's expense is primarily attributable to labor, materials, manufacturing, or anything physical at all, but instead to the vapor of overhead.

Design, Prototyping, and Manufacture

Central to the potential of prefabrication is the concept of iterative design and prototyping processes. A significant problem with one-off custom buildings is the inability to prototype and refine details prior to final construction. With manufactured products there is the opportunity to invest in research and testing prior to production, amortizing these development costs over a significant production run. This is the typical design and prototyping process used in manufactured items from automobiles to appliances, computers to cosmetics. In large-scale building projects, particularly those with a significant number of repetitive parts, there sometimes are full-scale mock-ups and testing trials on vital components such as curtain wall assemblies and building cladding. On most projects, however, there is no time or budget for prototyping or testing full assemblies. Increasingly, buildings are constructed from many manufactured subassemblies that have themselves been extensively tested and refined, and often graded or approved by regulatory agencies. Doors, windows, mechanical systems, lighting fixtures, and engineered structural elements such as trusses, beams, and floor joists are all likely to have been manufactured, tested, and approved as consistently performing products. There are testing standards and approval methods established both by government agencies and by industry laboratories and trade associations that objectively rate criteria such as fire resistance, sound transmission, UV resistance, thermal insulation, waterproofing, electrical safety, and hundreds of other construction issues and qualities. The testing of individual components is relatively well-developed for manufactured items, but the behavior of systems working together in larger assemblies or in whole buildings is far more difficult to assess, involves exponential variables, and typically has far less industry backing than that offered by the more consolidated industries manufacturing individual components. The complexity of building construction involves the coordination and fitting together of all these subassemblies. It is the integration of many tested, standardized parts with many one-off custom parts that costs a great deal of time, money, frustration, liability, and uncertainty. More and more, the role of the architect is focused on the design of the interface between pre-engineered manufactured building components, while the attention of the builder is focused on methods of assembly rather than onsite craft fabrication.

Although whole buildings are rarely prototyped (with the exception of small modular units), architects work with their engineering and construction consultants in an iterative design process that helps to break down design issues into manageable portions, cycling through numerous presentations, reviews, and coordinated redrawing. This is a time-consuming and expensive process, and yet the more time and money that can be spent on more iterations at this stage, the fewer questions and uncertainties will arise during construction.

Many architects work in scale models and repetitive rounds of drawings and sketch modeling to prototype and refine building designs, testing the architecture spatially, technically, and conceptually in multiple rounds before preparing final construction document sets. With increasingly complex regulation, there are also typically a number of rounds of presentation, testing, and reworking relative to zoning boards, environmental agencies, and building departments. The advantages of working in prefabricated design and building systems increase with the need for repeated prototyping and proposals. It has become so complex to design and obtain permits and financing for new construction projects that some form of standardization has become essential. In most cases, standardization comes in the form of repetitive banality rather than creative research-based, iterative advancement of modular knowns.

While a typical contemporary development and design approach will maximize cost efficiency by applying a creative design veneer to standardized, unexamined assumptions about substantive market, regulatory, and construction processes, a far more powerful design method may involve acceptance of less apparent surface and singular system creativity in exchange for more deeply applied creativity and steady advancement in the design and implementation of comprehensive systems encompassing the entire process of building. While there is an assumption of greater variety possible in a non-modularized construction economy, the contrast with the automotive industry is very instructive. Although there has been an unfortunate, century-long reduction in the number of automakers worldwide, there is a relatively great variety of excellent choices available in these mass-produced products. By contrast, while the construction industry itself is far less consolidated, and there are far more individual builders making ostensibly individualized buildings, there is far less variety even in the surface of typical building construction, much less in the actual mechanics and performance. By contrast again, the putatively more standardized automobile offerings also advance far more rapidly technologically and in terms of human satisfaction due to the modular efficiency of research and production methods. While the auto industry is in no way a model for the highest ambitions of architecture, it does offer a disconcerting comparison with the monotony and conceptual illogic of the not-quite-skin-deep creativity in contemporary architecture and mass-market construction.

Comprehensive Imagination and the Responsible Architect

As the design and detailing of buildings and production systems is recognized as only a small piece of the puzzle in rationally implementing modernized construction methods, and as it becomes increasingly clear that the architect's interest and imagination must extend beyond these issues, the traditional role of the architect as a comprehensive master of construction is most diminished at the moment when the concept is again most relevant. With the critical recognition in the 1960s and 1970s that architecture had reached a point of particularly narrow, mechanically focused thinking, new specialties in architectural theory and practice developed to fill the gap.[16] As a result, both in universities and in practice, there are ever-narrowing divisions of responsibilities in architecture, and alongside this increasing specialization is an increasing loss of faith in the comprehensive responsibility of an overarching, integrative architectural design practice itself. In recent decades a similar condition had emerged and become recognized in American medicine, when the field had become so compartmentalized, with so little respect allotted to the general practitioner, that there was an increasing danger from the loss of anyone specializing in the comprehensive overview. Medicine has reacted to begin reversing this trend of over-specialization; other fields have begun to recognize this problem as well, and it can be hoped that the longest remaining true generalists—the practicing architect and the general building contractor—will maintain position before they become the last to fall and then the last to reform themselves and return to the big picture.

Traditionally, and as a holdover in many languages still, architects have been considered master builders, experts in the craft of construction. This is now rarely considered the case, as the compounding complexity of building options have made it impossible for individuals to be experts in such a broad range of construction technologies as well as in all of the attendant societal issues and responsibilities accompanying every building project. One solution for maintaining substantive expertise requires each individual architect's focus to narrow sufficiently to maintain a level of mastery: an architect becoming a specialist in curtain wall design, for instance, or in building code compliance, solar technology, or social impact studies. Such specialization among architects is becoming increasingly common, particularly within research universities and in some large architecture firms, where the immediate rewards of perceived mastery in detail increasingly obscure and subvert architecture's more traditional responsibility for structuring the conceptual and physical logic of the whole.

Although frequently discussed by architects in somewhat naïve terms when considered from a construction industry perspective, there is a great deal of current interest, theoretical discussion, and research directed toward the possibility of architects becoming once again more directly involved in the construction process, particularly through the mechanism of new computer-aided production technologies. Many suggest that with the increasingly comprehensive direction of machines and tools by computers rather than by skilled craftspeople, architects should be able to directly draw and control computer files that will now pass from architect to machine to reality without intervening layers of translation.[17] This is an area of great creative potential, and it has generated a renewal of architectural interest in the intrinsic creative potential that has always developed from the architect's direct engagement with material, technique, and process. This renewed alignment of theory and practice has initiated an exciting moment in architecture, with architects again moving beyond the surface and creatively engaging the technology and rapidly shifting cultural and economic assumptions central to this moment in history. While the technological and theoretical implications of digital production are clearly relevant to imagining a stronger and more comprehensive role for the architect in contemporary production, there is a far more important traditional mastery that architects must defend and maintain—mastery of the broad, creative overview.

The training and ambitions of the architect must retain the master builder approach, maintaining the generalist's overview even when the craft and detail cannot all be uniformly mastered. This is an important issue now commonly discussed in many engineering and technical fields, recognizing that as depth of knowledge increases in individual specialties, there is accompanying loss of expertise in more broadly integrative design and in the comprehensive, ethical overview. With its broad responsibilities, architecture has long been at risk of being viewed as a technologically irrelevant profession in the modern world. On the other hand, architecture's uniquely stubborn resistance to technical specialization in favor of breadth and comprehensive design logic offers a particularly powerful role for architecture as an integrating, generalist discipline at this moment when advancing science particularly recognizes the value of such an overview. Nevertheless, the traditional breadth of architecture is currently in danger, with increasing pressure to forgo the broad, integrative view that is increasingly essential to significant involvement in the complex realities of contemporary society and the increasingly specialized construction industry. Narrowly focused academic research and narrowly confined, poorly integrated design, development, and regulatory responsibilities within increasingly specialized design professions all contribute to the worst blights in contemporary design, land-use, and construction. With all of the spiraling complexity of contemporary construction, one great advantage of working within an environment of prefabrication and of pre-engineered modules and components is that an architect's attention may balance orderly technological innovation in collaboration with manufacturers, engineers, and technicians, with a more comprehensive commitment at the integrative, environmental coordinating level, where all of the broadest cultural responsibilities of architecture remain ever more relevant and are increasingly challenged.

The Ecology of Prefabrication and Standardization

Prefabrication is potentially efficient in terms far beyond simple cost savings. In essence, standardization and rationalized system integration at all levels of the design, permitting, and construction processes offer the potential for a well-integrated ecology of the built environment. Just as the conceptualizing of intertwining ecological systems revolutionized understanding of the natural world and has subsequently found broad application in relation to socio-cultural systems and urban structures, a similar understanding of the interconnected complexities of construction, architecture, society, and economics will greatly advance the cause of increasingly rationalized, high-quality, affordable, and systemically sustainable construction. A thoughtfully integrated ecology of construction can logically lead toward significant reductions in energy and transportation costs; reductions in materials waste and redundant warehousing; the reusability and recyclability of building components; and massive savings of time, frustration, injury, and redundancy on the job site. Finance costs and accounting and administrative overhead could be reduced; permitting and code compliance could be greatly streamlined; and increased predictability would lead to lower insurance and contingency costs. Architects and engineers could concentrate on design innovation within systems—and sometimes on radical proposals to depart from existing systems—with their efforts spent on creative production rather than on the navigation of poorly integrated half-systems of mundane and clumsily coordinated redundancy empty of creative design appreciation and lacking even in any particularly large target for effective attack with a new idea.

To a certain degree, there are already many underlying sub-systems within construction that create remarkable efficiencies and produce quite regular small innovations. Manufacturers are becoming more and more sophisticated and interested in good design, and American and much of Asian industry is advancing toward the levels of design and production quality that have long been the expectation in Europe and Japan. While the bulk of construction still occurs primarily on the job site, there have been many innovations in delivery, mobile fabrication, recycling, and waste removal systems that have tremendously rationalized onsite construction during the last decades of the twentieth century. Manufacturers increasingly prefabricate and pre-assemble ever-larger modules of construction, from doors and windows to heating and ventilation systems. Within basic materials industries such as those making steel and wood components, there are increasingly many and more sophisticated products that are wholly integrated as comprehensive sub-systems within construction. A prominent example of this progress is in the wood industry, once one of the least progressive construction material supply industries, and now among the most progressive. During recent decades in particular, with the widespread introduction of processed wood fiber, oriented-strand sheet, and lumber materials, the wood industry has rapidly evolved very well-integrated, well-cataloged, and easy-to-use pre-engineered wood construction systems that simultaneously address multiple construction problems and share systems across a variety of competitive suppliers and coordinate vertically with also competitively structured hardware and related parts industries. This rationalization of wood construction uses open-source research and development to minimize costs, improve standardization and quality control, reduce the cutting of large trees, and encourage sustainable rapid-growth fiber harvesting in place of clearcutting and the logging of older growth forests. Although to this point these major technological and process innovations have had little impact on structural systems, component shapes, and design applications themselves, the impact at business and production levels has been profound. Once design-based construction logic and imagination is further applied in these industries to harness the base potential of new materials, processes, and system integrations, radical new fiber-based architecture proposals may emerge as well.

There are many other recent innovations in mobile job-site networking and systemization that ameliorate the often inherent waste of job-site manufacture. There are many out-standing architecture, engineering, and construction administration innovations and experiments that could be discussed as well, but to a large extent the most significant rationalizations of the job site have been instituted through entrepreneurial commercial innovations servicing the essential margins of construction, rather than large conceptual shifts affecting the very core processes of design and construction strategy. Ultimately, the success of these small improvements makes the argument for ever greater degrees of prefabrication under ideal conditions prior to arrival on site, applying new digital tools better suited to the factory than the outdoor work site. There is a logic to incremental background change, of course, as it is always easiest to move and to innovate at edges, while to make wholesale changes in long-established systems that are intricately interwoven with legal systems, labor, insurance, finance, permitting, and contract structures is like swimming upstream through a river of molasses.

While on the one hand it can be said that relatively little progress toward factory-based construction has been made during the past fifty years, and that the long-promised benefits of prefabrication have periodically raised hopes only to disappoint with slow progress during this same time period, there are many underlying structural changes in the construction industry and in societal understandings and expectations that will increasingly create a better climate for rapid evolution of rationalized, manufacturing-based construction. For better and worse, materials manufacturing and building construction is becoming more consolidated, with fewer and larger companies becoming increasingly sophisticated and capable of funding extensive research and production development costs. While this has potential negative consequences, the positive potential of larger industries should lead architects, builders, and the public to demand and lead toward better research, more production investment, better design, and higher quality. Increasing awareness of the limited supply of quality raw materials and energy should also lead to greater demand for innovation, efficiency, and responsibility to the future and the larger environment, all of which can be better achieved within increasing factory production.

Based on past experience, successful progress in construction process will not come from singular innovations or proprietary systems. Architects and prefabrication entrepreneurs must first research and understand the construction industry as it now exists and has historically evolved and then focus imagination not just on changes in material and assembly mechanics, but also on looking deeply beyond this into the processes, networks, and system logic of the industry as an integrated complexity, to be unraveled piece by piece, understood, renegotiated, and rationally restructured as an effective and realizable whole. A part of this understanding must include a recognition that not all of construction can happen in a factory, that there must always be a role for and an increasing respect for the most complicated part of the process—both in design conceptualization and in physical construction—the attachment of the idea and the building to the ground. Not only is there a necessity for some of the messy handwork that will always be with us, but there will always also be an important role for creative hand craft, local building traditions, peculiarities of natural and urban sites, and anomalous departures in program, materials, and design, all of which are essential to a rich and meaningful environment. The most important part of any increasingly comprehensive system logic must be the welcome inclusion of the unexpected, novel, unplanned, fortuitous, and revolutionary. Finally, the design and production focus must include objectives beyond efficiency, affordability, logic, and systemic sustainability, and must include the larger functionality of human desire for culturally rich, environmentally responsible, and enjoyable inhabitation for all people.

NOTES

1 For earlier writings by Mark and Peter Anderson about their design work in the context of construction and prefabrication, see especially *Anderson Anderson: Architecture and Construction* (New York: Princeton Architectural Press, 2001), and "Playing in Traffic," *Offramp* 7 (Los Angeles: Southern California Institute of Architecture, 2000).

2 Please refer to the bibliography at the end of this book for a survey of original source documents and overview critiques on the history of prefabrication. See especially A. G. H. Dietz and L. S. Cutler, eds., *Industrialized Building Systems for Housing* (Cambridge, MA: MIT Press, 1971); B. Kelly, *The Prefabrication of Houses* (Cambridge, MA, and New York: The Technology Press of the Massachusetts Institute of Technology and John Wiley and Sons, joint pubs., 1951), T. R. B. White, *Prefabrication: A History of its Development in Britain* (London: H.M.S.O., 1965).

3 See especially British documents on prefabricated social housing initiatives during World War II and the postwar years, including B. Vale, *Prefabs: A History of the UK Temporary Housing Programme* (London and New York: E & FN Spon, 1995), and D. D. Harrison, J. M. Albery, et al. *A Survey of Prefabrication* (London: Ministry of Works Directorate of Post War Building, 1945).

4 There have been many ambitious and ultimately frustrating false starts in prefabrication. A particularly interesting example is the Lustron Corporation project. T. T. Fetters and V. Kohler, *The Lustron Home: The History of a Postwar Prefabricated Housing Experiment* (Jefferson, NC: McFarland & Company, 2006), and D. Knerr, *Suburban Steel: The Magnificent Failure of the Lustron Corporation, 1945–1951* (Columbus: Ohio State University Press, 2004).

5 Walter Gropius was a leading champion of industrialized construction who developed a coherent and consistent theoretical foundation for prefabrication. G. Herbert, *The Synthetic Vision of Walter Gropius* (Johannesburg: Witwatersrand University Press, 1959); Herbert, "House in 'Industry': A System for the Manufacture of Industrialized Building Elements," *Arts and Architecture* 644 (Nov. 1947): 28ff; and Gropius, "True Architectural Goals Yet to be Realized," *Architectural Record* 129 (June 1961): 147–52.

6 See G. Herbert, *Pioneers of Prefabrication: The British Contribution in the Nineteenth Century* (Baltimore: Johns Hopkins University Press, 1978).

7 J. Bogardus, D. D. Badger, et al., *The Origins of Cast Iron Architecture in America* (New York: Da Capo Press, 1970); Sears, Roebuck, *Sears, Roebuck Home Builder's Catalogue: The Complete Illustrated 1910 Edition* (New York: Dover Publications, 1990), Sears, Roebuck, *Small Houses of the Twenties: The Sears, Roebuck 1926 House Catalogue* (New York: Dover Publications, 1991).

8 For an overview of architects' views and debates on industrialized construction and prefabrication during this period, see Tim and Charlotte Benton, *Form and Function: A Source Book of the History of Architecture and Design 1890–1939* (London: Crosby Lockwood Staples, 1975).

9 For a historical overview of the social and economic context of postwar California, see K. Starr, *Embattled Dreams: California in War and Peace, 1940–1950* (Oxford: Oxford University Press, 2003), and J. T. Patterson, *Grand Expectations: The United States, 1945–1974* (Oxford: Oxford University Press, 1997). For a detailed story of a creative architect's work on construction rationalization and prefabrication within the context of this era, see W. Wagener, *Raphael Soriano* (London: Phaidon, 2002).

10 For an overview of Charles and Ray Eames' work in the context of their own prefabricated house, see M. and J. Neuhart, *Eames House* (Berlin: Ernst & Sohn, 1994).

11 See E. Smith and P. Goessel, *Case Study Houses* (Koln: Taschen, 2002); E. McCoy, *Case Study Houses: 1945–1962* (Santa Monica: Hennessey & Ingalls, 1977); and E. Smith, ed., *Blueprints for Modern Living: History and Legacy of the Case Study Houses* (Cambridge, MA: MIT Press, 1999).

12 For the complete story of Gropius's and Wachmann's work on the General Panel Corporation Project, see G. Herbert, *The Dream of the Factory-Made House* (Cambridge, MA: MIT Press, 1984).

13 For a readable perspective on the burgeoning area of industrial, business, and design theory, see M. Kratochvil and C. Carson, *Growing Modular: Mass Customization of Complex Products, Services and Software* (Berlin: Springer-Verlag, 2005).

14 See P. Adamson and M. Arbunich, *Eichler: Modernism Rebuilds the American Dream* (Salt Lake City: Gibbs-Smith, 2002), and J. Ditto, et al., *Eichler Homes: Design for Living* (San Francisco: Chronicle Books, 1995).

15 For a broad overview of many innovative construction ideas and systems described at the point of their introduction to the mass market in the 1950s, and yet only slowly becoming more commonplace today, see N. Cherner, *Fabricating Houses from Component Parts* (New York: Reinhold, 1957).

16 See K. M. Hays, ed., *Architecture Theory since 1968* (Cambridge, MA: MIT Press, 1998), and K. Nesbit, ed. *Theorizing a New Agenda for Architecture: An Anthology of Architectural Theory 1965–1995* (New York: Princeton Architectural Press, 1996).

17 See S. Kieran and J. Timberlake, *Refabricating Architecture: How Manufacturing Methodologies are Poised to Transform Building Construction* (New York: McGraw Hill Professional, 2003), and B. Kolarevic, ed., *Architecture in the Digital Age: Design and Manufacturing* (Oxford: Routledge, 2005).

Chapter 1

Panelized 2x4

Panelized building systems based on traditional wood-framing techniques are not new or revolutionary—indeed, they are just a simple extension of the onsite wood-framing system that has been the primary method of American residential construction since the late nineteenth century. They are among the most readily available and cost-effective ways to build. Because of low initial investment requirements, there are now a variety of shops throughout the world—from flat tables in a shed to large factories with highly automated production lines—that build prefabricated wood frame panels for 2x4 and 2x6 construction.

Even most site-built wood-frame structures actually use a panelized system, since wall sections are typically laid out and assembled flat on the ground, then tipped up into place as a unit, and finally connected to other sections to form the standing walls of the building. This is the simplest way of defining the term *prefabrication*, where some form of assembly occurs in one location because of some advantage or efficiency before being moved into its final location. In the case of this type of onsite wood framing, the advantage of laying out the wall panels on a flat surface is quite obvious: the parts that make up the wall can be relatively easily aligned and fastened to each other on the floor and would be nearly impossible to assemble piece by piece in a standing configuration. Usually these wall sections are built with their bottom plates exactly aligned with their eventual resting places, so that they can simply be pivoted up into position without requiring any other movement of the panel; they are assembled in sections that can be lifted using the available size of the carpentry crew without machinery. This standard method of platform framing seems quite common and unremarkable to contemporary American builders but was in fact a major technological shift in affordable housing construction that revolutionized American housing production during the twentieth century.

Understanding the clear efficiencies of this concept of assembly of components, it is easy to imagine greater degrees of dislocation of the panels, with attendant advantages and disadvantages. For instance, it may be necessary or easier to lay out the wall panels on a wider area of floor a few feet away, then carry or slide the assembly into place. Once panels are being moved from one location to another, the difference between framing a building onsite versus offsite panelization becomes a matter of degree and of definition.

In considering this method of construction, one must weigh the benefits of assembling panels off site against the possible drawbacks of transportation. There is a level of inefficiency in transporting pre-assembled panels, as opposed to neat bundles of

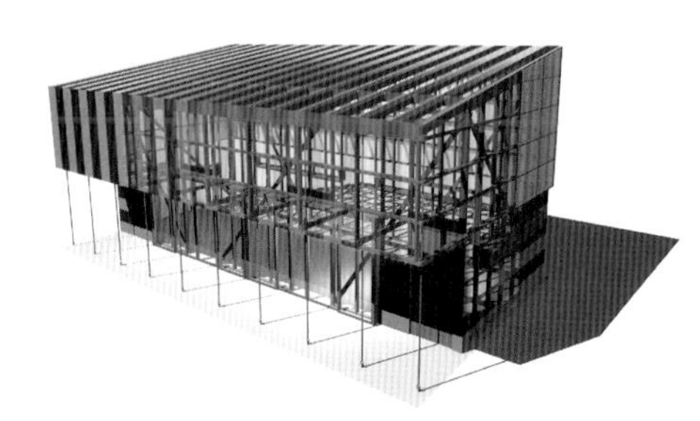

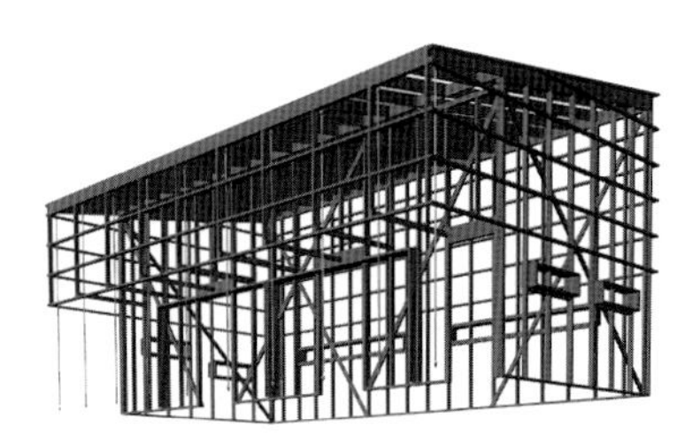

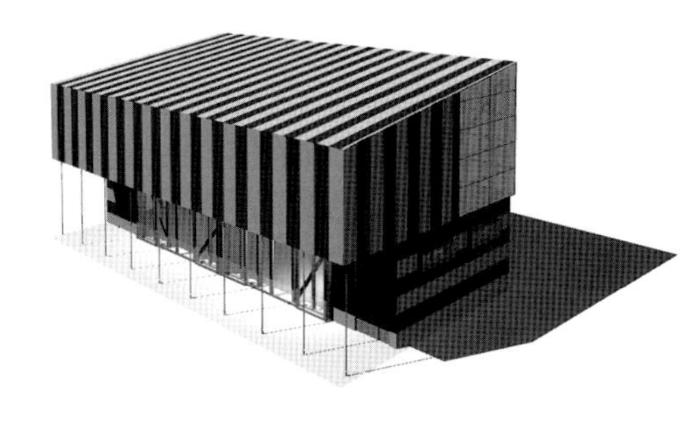

Zebra showroom project, Washizu, Japan

densely stacked lumber. Additionally, because the panels must be moved greater distances than site-built framing, they may require extra reinforcing and structure to allow for handling. These concerns can more than be offset, however, by the greater speed and consistency of the panel-making operation if done in the controlled environment of a factory, with the aid of layout tables, CAD/CAM control, gang nailing, and other factors. The greatest efficiency of factory panelization comes when the maximum amount of work on the panels can be completed offsite. In addition to the studs, headers, and plywood or OSB sheathing of the wall panels themselves, it is possible to install doors, windows, siding, trim, and other elements to create a more finished assembly. This in turn creates a heavier and perhaps more fragile panel, which further increases the cost and complexity of transport, but also significantly reduces the onsite labor and coordinating requirements.

Panelized wood stud wall systems do present certain design limitations, as they only achieve their best efficiencies when the wall panels readily fit the requirements of the panelizing equipment, which can be quite narrow. Having parallel top and bottom plates, for instance, is far easier to automate in a factory than producing walls with sloping tops or other irregular dimensions, which may have to be "stick-built" in the factory without gaining the benefits of the gang-nailing machinery. This is also true of going beyond the standardized dimensions anticipated by the production facility, such as working with walls that are taller than a certain height. Some panelizers cannot easily make panels taller than 9 or 10 feet (approximately 3 m). And although some can, trucking and transport limitations usually make it much more expensive to work with panels over 8.5 feet (2.6 m), the U.S. federal standard width limitation of the truck that will transport them. The truck width regulations are essentially the same in many parts of the world where prefabricated building systems are common, including the E.U., Australia, and Japan. Canada allows trucks up to 3m wide in most provinces. In some areas it may be possible to transport panels on a truck in a vertical or diagonal position to allow larger panels, but with much less load efficiency.

The overall length of the panels is more flexible, typically curtailed only by transport limitations, which in most areas allow as much as 45 feet (13.7 m) or longer, depending primarily on local road conditions near the job site. As a practical matter, from an assembly point of view it isn't likely that panels this long would be advantageous in any case, but the panel height limitation should be considered a critical design criteria when planning for optimal efficiency in panelized building. There are other transportation issues that may also play a defining role in selecting a prefabricated building system and sizing its components. Truck restrictions are important to consider even when transporting by rail or ship since trucks are likely to be used at some point in the panels' journey from factory to job site even if it is just to and from the port or rail yard. For ocean shipping, it is almost essential to plan for the dimensions of the standard shipping containers, which typically have nominal dimensions of 8 feet (2.44 m) wide, 8.5 or 9.5 feet (2.6 m or 2.9 m) high, and 20 or 40 feet (6.1 m or 12.2 m) long. Unlike on trucks, where the panels are usually carried horizontally, it is typical to load panels vertically in shipping containers since the usable interior width is only 7.5 feet (2.28 m). A standard container has a usable interior height of 7.8 feet (2.38 m), and the taller Hi-Cube container, which is more expensive to ship, can still not quite accept a panel the full width allowable on a truck, with its usable

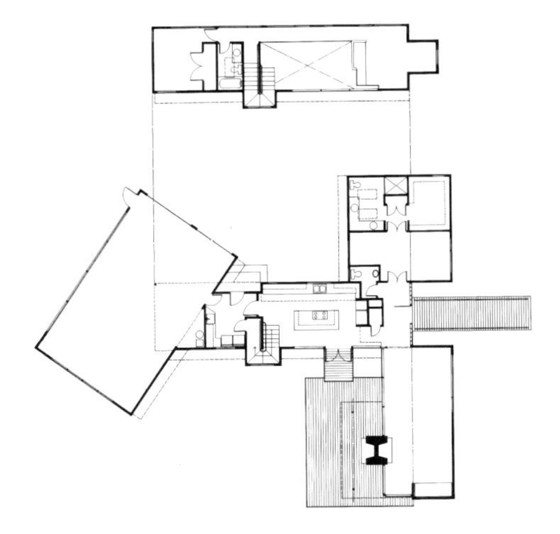

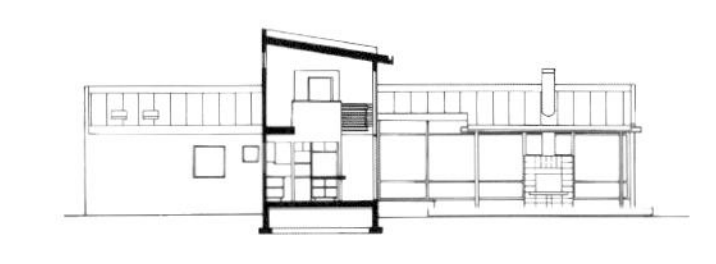

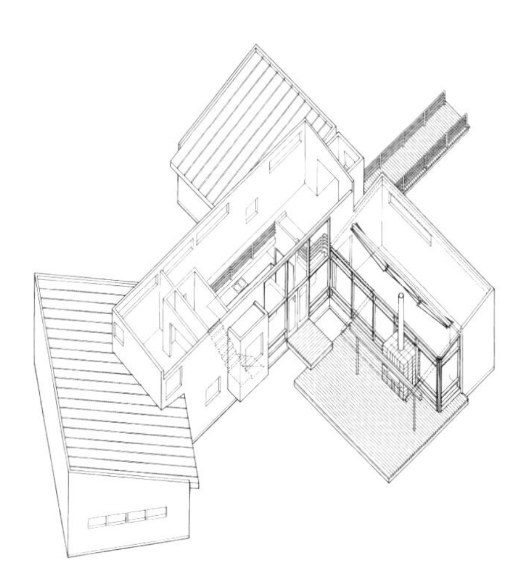

3-Part House plan, section, elevation, and axonometric views. Panelized modules can be assembled in different configurations for site adaptation

interior height of 8.4 feet (2.58 m). Building panels sized to maximum height of 8 feet (2.44 m) are compatible with any of the common transportation systems.

The limitations on panel configurations should not be considered a handicap to a building design's potential creativity or quality, for it equally makes little sense to choose a system of construction for its potential benefits in efficiency and cost, and then to not optimize the design to take the fullest advantage of those benefits. We have used this kind of panelized system on a wide variety of building types, including single-family houses, multi-family and mixed-use structures, and commercial office and showroom applications, and we have found that it is one of the easiest prefabrication systems to integrate into the standard processes of the construction industry. In its simplest form, it can directly follow the configuration of conventional framing techniques, yet it can also be used to significantly extend the design opportunities for wood-frame construction. For the architect, there is a short learning curve to begin designing for panelized construction, and for the builder, panelized stud framing is similar to onsite framing in terms of construction sequence, connection types, and integration with wiring, plumbing, heating, and other building systems. Perhaps most significantly, panelized stud framing is the easiest kind of prefabricated building system to mesh with planning approvals, building codes, lender financing, real estate appraisals and marketing, and neighborhood acceptance. In most situations, there is no need to make any distinction between onsite and offsite stud framing in these contexts, and no reason to raise anyone's fears about unconventional approaches to building. Unlike most other prefabricated building systems, the design and approval processes for many designs using stud wall construction allow the choice of whether or not to prefabricate to be kept as an open option long into the project schedule, even after the sitework and foundations are completed.

Fox Island House

Gig Harbor, Washington / 1992

View of house and recycled tire retaining wall from southwest

The Fox Island House was the first in a series of structures designed to take advantage of the cost-saving efficiencies of panelized framing systems without giving up the benefits of adaptation to specific requirements of individual building sites. The project series was initiated with the goal of producing a small, affordable, prefabricated house that would remain easily adaptable to a wide variety of site, light, and view conditions, particularly to the typically non-flat site conditions common to the western United States and Japan. The basic premise of the approach to site-adaptability was to use offsite fabricated 8-foot-wide vertical panels that remain standardized on the main floor and above, but are lengthened or shortened at their lower ends to adapt to varying slopes and lower-floor configurations.

The house is intended as a highly rational, mass-producible structure that is flexibly susceptible to the unique forces of each individual site and builder and does not reduce these primary influences to the lowest common denominator of bland abstractions. The flexibility of the design has been proven through several subsequent adaptations of the basic premise in a number of implementations in widely varying site and programmatic conditions, including houses in New York and Hawaii and commercial and multi-family structures in Japan. The structural approach is based on the idea of balloon framing proportioned to accommodate standard-dimension lumber framing spans in transportable panels. To maximize the strength of the panels for transport, create maximum flexibility in window layout, and expose the richness of this standard structure, the framing is continuous through most openings. The exposed 2x6 wall framing sets the aesthetic standard for the entire structure and finish materials that consist entirely of inexpensive framing grade materials and quick, framing-style carpentry methods for easy completion by homeowners or contractors.

The exterior of the house is clad in inexpensive fireproof materials: corrugated, galvanized steel roofing and granulated asphalt roll roofing used as siding. The siding pattern derives from strips of the largest panels of roll roofing that could be applied without the threat of drooping in hot weather. The siding expansion joints are galvanized sheet metal shingles. The windows are inexpensive standard-size aluminum units mulled together on site with standard aluminum square tubing and placed outside the 2x6 wall framing that continues uninterrupted through the window openings. This method eliminates the need for trimmers and headers and allows easy flexibility in window placement without altering the panel layouts.

Alongside the goal of rationality and affordable mass production is the equally emphasized ambition to create a house of material and spatial richness and experiential enjoyment. The arching wood ceiling is visible from every room in the house, uninterrupted except for clear glass panels to allow the eye a continuous, unifying view of the vast wooden ceiling curving out of sight beyond each room. The suspended stair and bedroom loft float in the space, allowing multiple views and filtered light throughout one large area and providing a sense of spatial limitlessness as well as of mystery and privacy.

The house is reduced to only the simplest, most practical elements, but every piece attempts to provide an intensified, meaningful experience of its presence and role in the occupants' physical and spiritual life in the house: the bookcase-lined bedroom loft floats like a raft in the air beneath the warm arch of a wooden sky. The hanging deck exposes an explicitly straining structure of tension and compression to reinforce a sense of hanging out on the face of the house, an experience similar to that on the suspended, open staircase built of 2x4s in a scissorlike cantilever that inexpensively reinforces the lightness and separation of the floating raft above. The continuous, strong 2x6 balloon frame is exposed before the window openings that float independent of the structure. These window openings can be interpreted either as carefully placed or as arbitrary and in motion, like sunlight filtering through clouds or a windswept forest. The ambition is for each piece of the house, and for the experience of the house as a whole, to be solid, real, primal, meaningful.

In addition to emphasizing economic affordability, the house also seeks to reduce environmental impact. The materials used in the balloon framing are readily available small-dimension lumber products from fast-growing managed forests, as opposed to large-dimension timbers and beams from older growth trees. The open and efficient interior space gives the feeling and livability of a larger home, allowing the inhabitants to live happily in a smaller and more economical structure. Materials throughout have been chosen for their low energy requirements during manufacturing and processing, and also for their long life span and low maintenance requirements.

Fox Island House

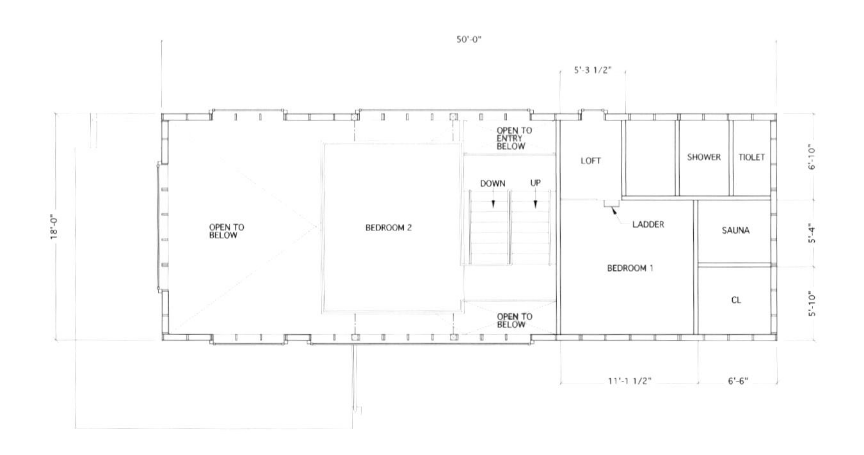

Top floor

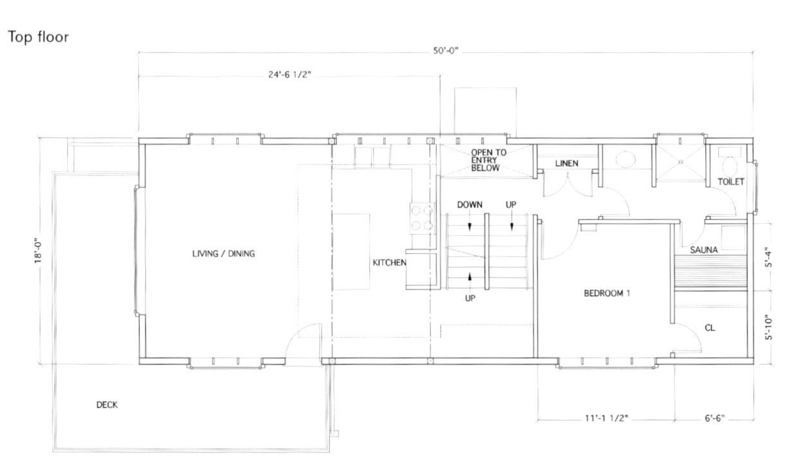

Main floor

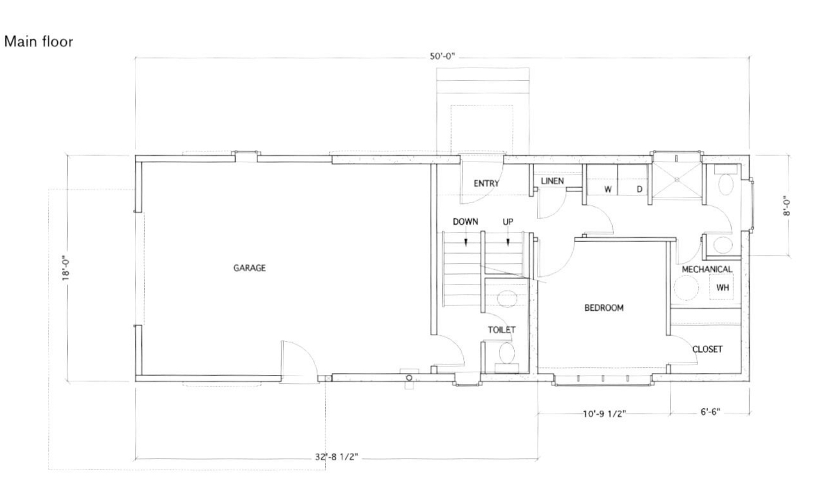

Ground floor

Left: Plans
Below: Axonometric drawing with exploded wall panels and curved roof structure

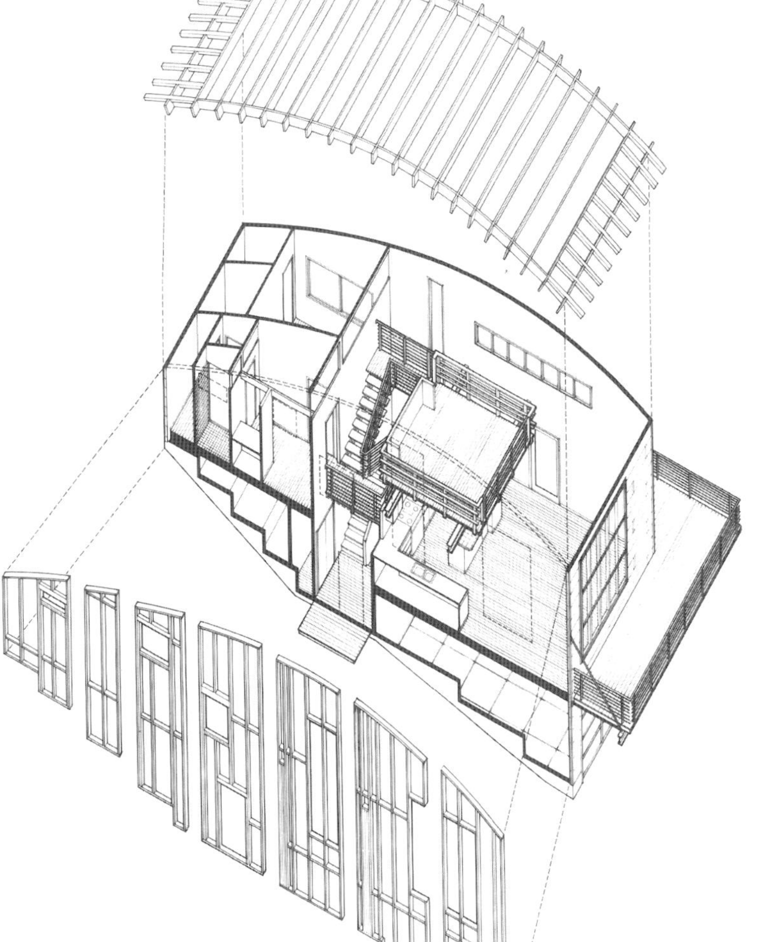

Panelized 2x4

Exterior wall panels are lifted into place by light crane and bolted into position. The building shell was completed in about six hours.

Exterior detail view of siding, windows, and balcony. Note continuous framing behind glass

Top left: Floating interior loft and curved ceiling
Bottom left: Dressing area looking out to loft. Note interior glass allowing curved ceiling to integrate all interior spaces
Right: Prefabricated steel and wood stair assembly

Amerikaya & Garden Pacific Prototypes

Fukui and Shizuoka Prefectures / Japan / 1992–2003

Building with wood has been the primary construction method in Japan since historical times, and finely crafted timber joinery has long been a cultural tradition that is fundamental to the refined aesthetic of traditional Japanese architecture. The opening of Japan to trade and contact with the West beginning in the 1850s brought an interest in Japanese wood craft that has strongly influenced international approaches in design and construction. This can particularly be seen in America through the work of Greene and Greene, whose Craftsman style buildings are heavily influenced by their exposure to Japanese wood craft, and through projects by Frank Lloyd Wright, Rudolph Schindler, and many others. In contrast to this tradition and perception of Japanese wood craft, contemporary wood construction in Japan has undergone a significant change in the last thirty years, where diminished sources of timber and skilled carpentry labor, together with an increased focus on structural integrity and seismic resistance, have led to a situation where the traditional materials and techniques are used for fewer and fewer projects, mostly in temples and other cultural buildings, and in very expensive residential architecture. The rest of the wood construction market has shifted to a variety of building systems that rely primarily on imported materials and less craft-oriented assembly techniques, most of which is concealed beyond surface finishes.

With the import of materials from other parts of the world have come a variety of building systems, and a logical progression toward increasing use of prefabrication techniques. With a substantial source of imported timber coming to Japan from Canada and the United States, the prevalent 2x4 and 2x6 stud wall wood framing systems in those countries began to displace the Japanese-dimension post-and-beam building systems with increasing frequency during the 1990s. Other North American building components and materials, such as windows, doors, siding, cabinets, and flooring, have been imported along with the primary wood structural components.

During this period, our firm was involved with these transitions in the Japanese building methods on several levels, beginning with supplying carpentry labor and training supervisors for Japanese construction firms new to these building techniques. We were among a group of American architects, builders, manufacturers, and distributors asked by the U.S. Department of Agriculture and Washington State trade offices to develop a teaching curriculum to introduce American building methods to the Japanese construction industry, and we taught seminar courses to architects and builders in the Tokyo and Kobe/Osaka areas for several years while doing our own design, construction, consulting, and materials supply work with Japanese companies.

The primary interest in American building techniques was the potential for increased efficiency, based on the statistics that showed that building construction costs in North America were far lower than in Japan, with the same building often costing three times as much if built in Japan. Although the perception on both sides of the Pacific was that this was due in large part to differences between wood framing systems, it became increasingly clear to us during this time period that the answer was more complex and involved all stages of the project delivery systems, including labor issues, materials sourcing and distribution, subcontractor structures, and supervisory and administrative procedures at the job site and in the office. Although we were teaching that 2x4 building systems were more efficient than Japanese timber systems, we became increasingly aware of and interested in the shortcomings of all of these approaches and in the opportunities for improvement in standardization, modularization, and prefabrication techniques at all levels of the project delivery process.

Given the shortage of carpenters in Japan, particularly of those with experience working with American-style 2x4 construction, the potential efficiencies of the system could not be achieved without finding ways to decrease the reliance on the site-assembly that was the standard in North America, where panelization of walls was not common because the distributed availability of relatively inexpensive carpentry labor made it less necessary economically. Our hands-on experience with the Fox Island prototype led us to believe, however, that there were numerous advantages to be realized in offsite panelization beyond carpentry labor efficiencies, particularly in allowing expanded design opportunities and in shortened building schedules and improved project management.

Our first chance to apply what we learned from the Fox Island Prototype to a project in Japan was with a Japanese commercial construction company, used to working in concrete and steel, that was seeking a way to enter the currently stronger market sector of wood-frame residential building. When we first met Kenji Shinohara, an architect himself and owner of Amerikaya Construction, he was searching for products from the North American building industry that could be imported and assembled to create houses in his area of western Japan. He soon discovered that he could directly purchase lumber, doors, windows, and even whole prefabricated sunrooms from North America, thus bypassing the much more expensive Japanese manufacturing and distribution sources, but he did not know how to determine the exact specifications and quantities for these items in order to put them all together. Since there was little experience with this kind of framing in his region, and none at all in his staff, he asked us to work with him to develop a series of small home designs that could be easily componentized in the U.S. or Canada, then shipped to his site and rapidly assembled by a mixed crew of American and Japanese construction workers. The designs rely primarily on panelized 2x6 framing systems, used in conjunction with engineered wood roof structures.

Considering the first projects to be demonstration houses and prototypes for larger-scale development, Mr. Shinohara requested as much variety in forms and materials as possible. The series has curved roof structures

Above: Completed Amerikaya Model Home, Tsuruga, Japan
Right: Prefabricated window gang units are installed into panelized structure

made with custom-laminated wood beams as well as long-span shed roofs, walkable flat roof sections, and other hip and dormer configurations. Considerable attention was given to creating open interiors with large volumes of ceiling height in order to help the small houses gain a feeling of airy spaciousness.

The smallest and simplest of the prototypes is a rectangular two-story balloon-framed box that can be assembled in a variety of configurations and includes a series of separate elements such as greenhouses, trellises, fences, and landscape structures to allow for flexible site development and definition of outdoor living spaces through the many possible assembly configurations. A larger version of this basic system was adapted to the specific needs of the builder's own family and the land on which he wished to erect a model home he would live in himself. The site is a terraced hillside overlooking the port city of Tsuruga, surrounding mountains, and the Sea of Japan. On the other sides are close neighboring houses, and we chose to create privacy on the narrow site by linking the building modules in a "U" shape around a central courtyard. This configuration allows an integration of interior and exterior spaces to make optimal use of the entire site. Spaces and windows are arranged to provide maximum benefit from sunlight and shade as well as to open the house to the best views while providing spaces for privacy, separation, and seclusion.

The walls of the house were panelized at a plant in Portland, Oregon, using a typical 2x6 stud wall assembly; and the curved, flat, and shed open-beam structural roof assemblies were fabricated to our specifications by a custom wood laminator in Tacoma, Washington. All of these parts, along with windows, sunroom, doors, cabinets, hardware, and the central heating system, were consolidated by an ocean freight company and shipped in four containers directly from the west coast of the United States to the port in Tsuruga, then offloaded to trucks for the short drive to the project site. After the sitework and foundations were completed by the Japanese crew, American carpenters were sent to the site for two weeks to supervise the assembly of the panels and other imported building components, working alongside a Japanese crane operator and other construction workers. The project served as a valuable training and demonstration opportunity for Mr. Shinohara and his workers, and on subsequent related projects they did the construction and panel assembly processes on their own.

This project expanded into a number of other opportunities to work with construction companies throughout Japan, including a series of projects in Shizuoka prefecture. The earliest of these, the Zebra office and Obata Showroom, were panelized and prefabricated in the United States, while the more recently completed Nagao Residence was designed specifically to take advantage of newly established factories in Japan focused on doing the panelization nearer the project sites.

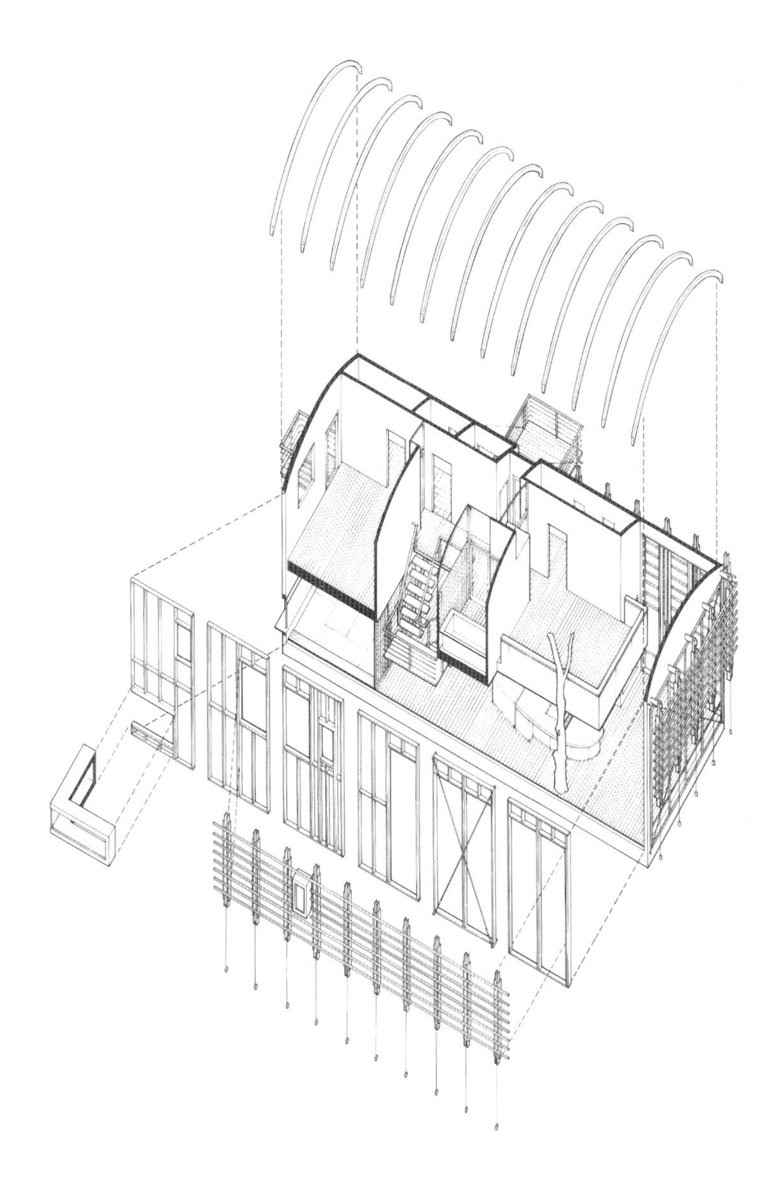

Left: Exploded axonometric drawing of Amerikaya prototype base module showing primary structural components and optional screen and window assemblies

OPPOSITE

Plans, exterior elevations, and building sections through Amerikaya prototype base module showing placement with greenhouse and garage structures on typical lot

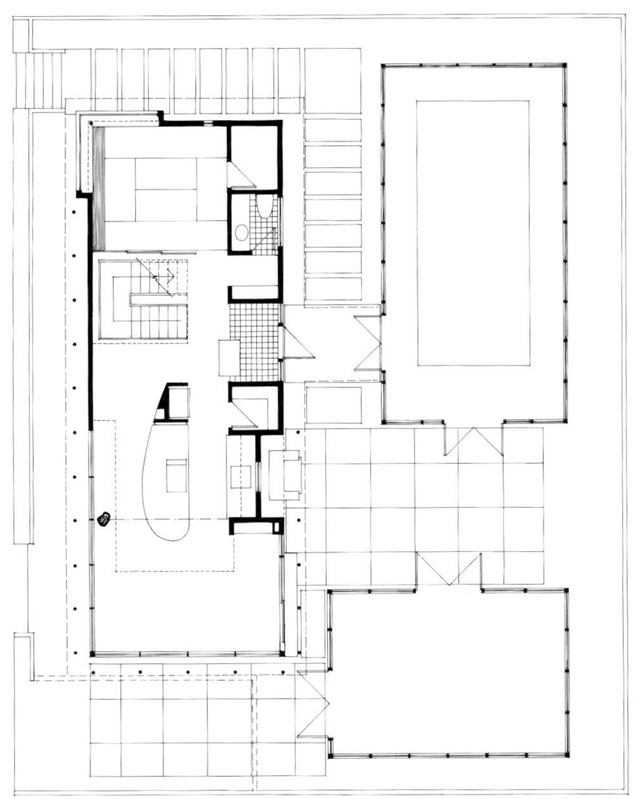

Ground floor

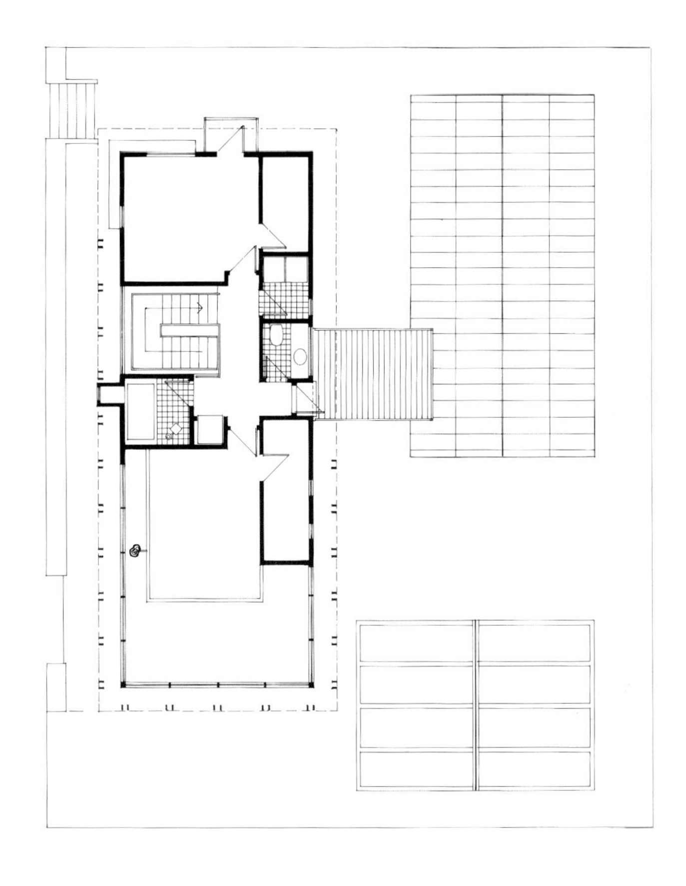

Upper floor

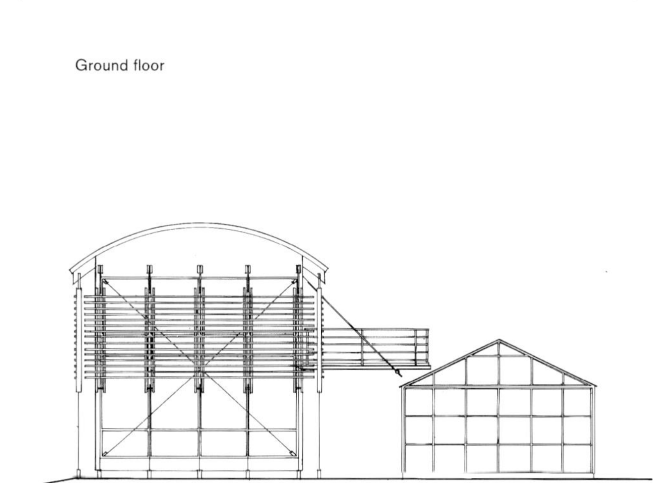

South elevation

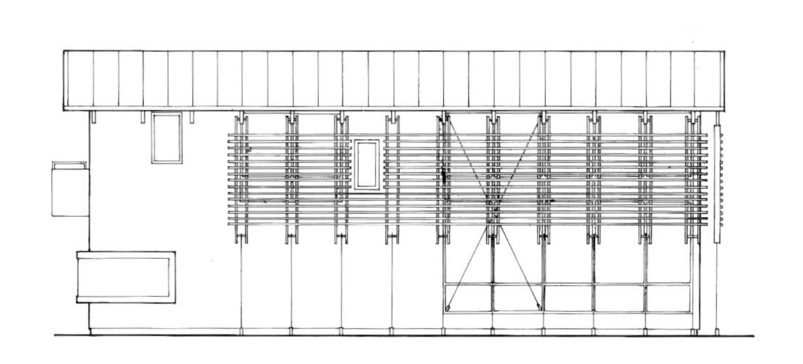

West elevation

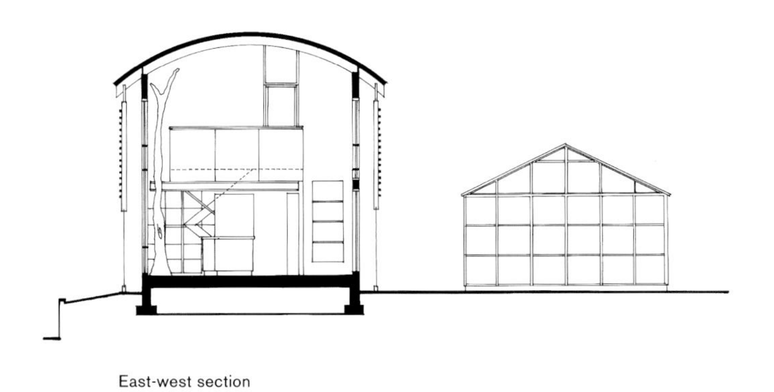

East-west section

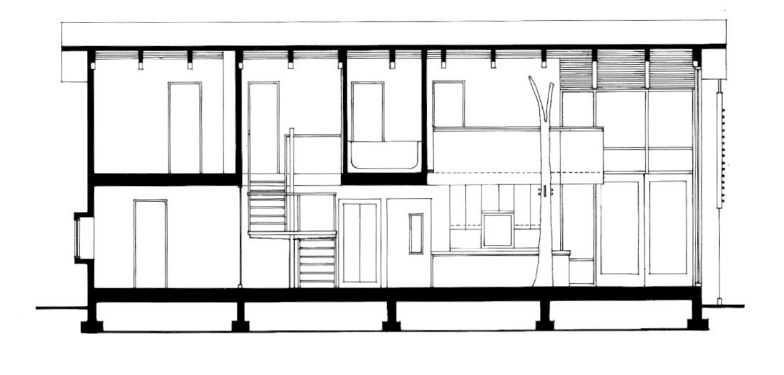

North-south section

Amerikaya & Garden Pacific Prototypes

Kosai Houses 1 and 2 / Japan

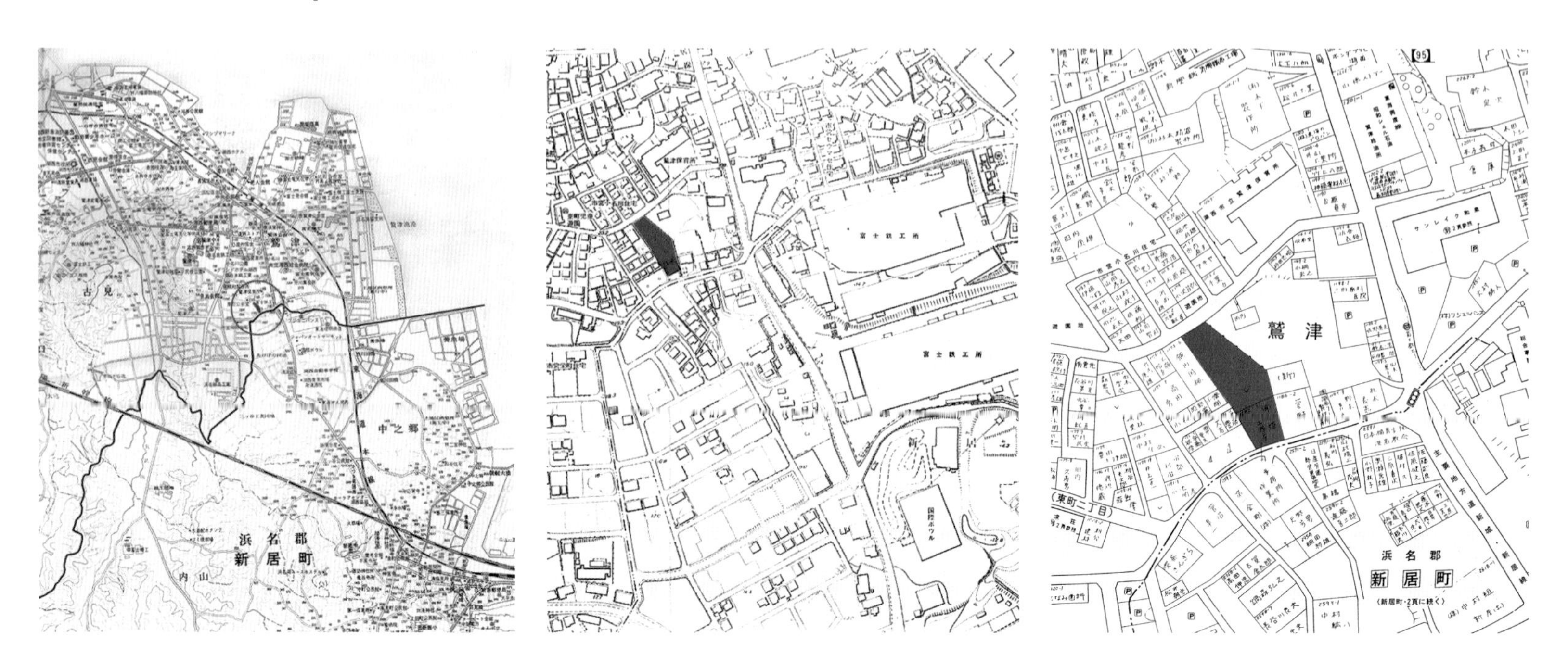

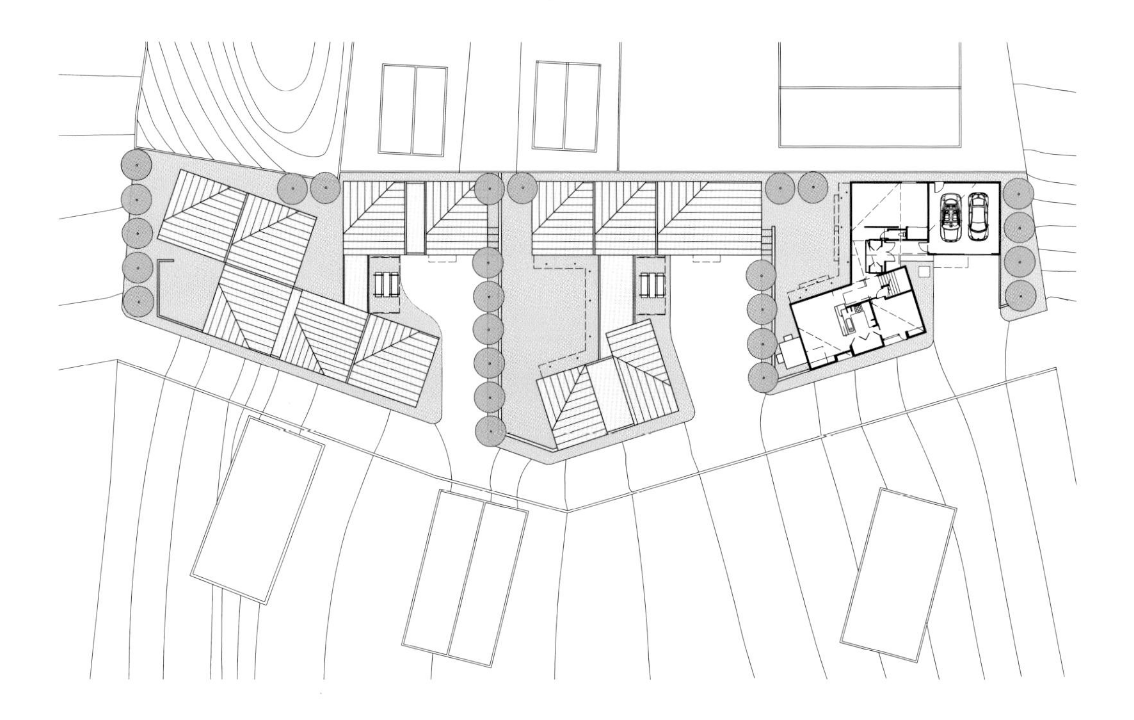

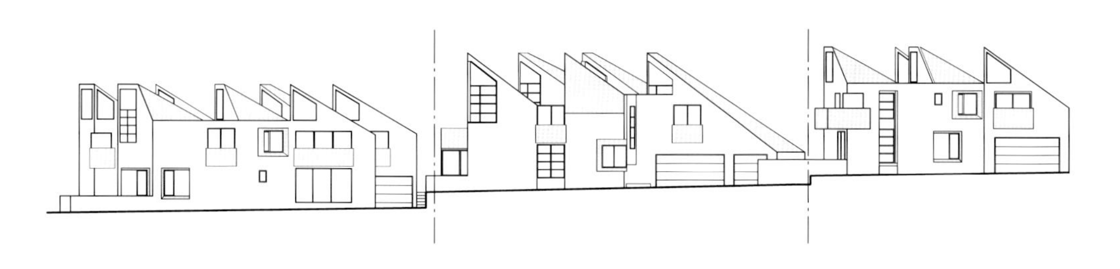

Top: Vicinity maps showing infill site in dense urban neighborhood
Middle: Site development plan
Bottom: Site elevation for series of Garden Pacific prototype houses showing varied configuration of panelized repetitive modules for each of three building sites.

OPPOSITE
Top: Ground- and upper-floor plans
Bottom: Construction sequence photos for northernmost house. Unlike the Amerikaya projects, these panels were fabricated in a Japanese factory using lumber imported from the U.S.

Panelized 2x4

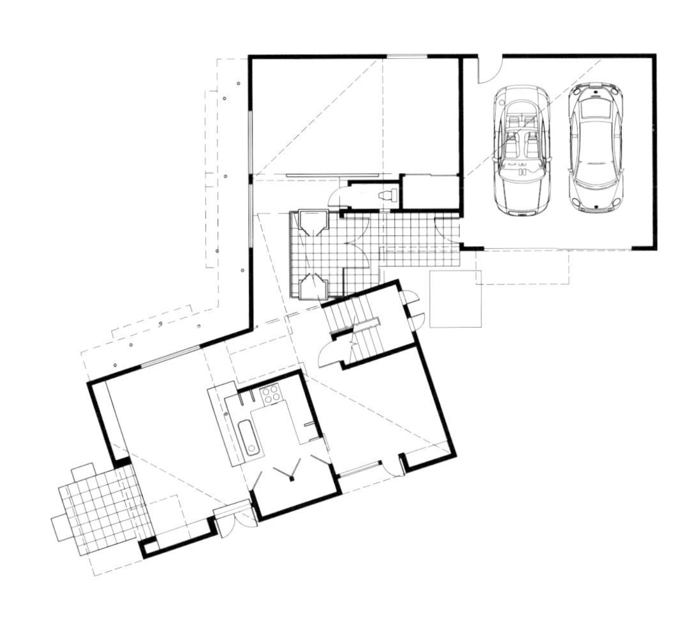

Ground floor

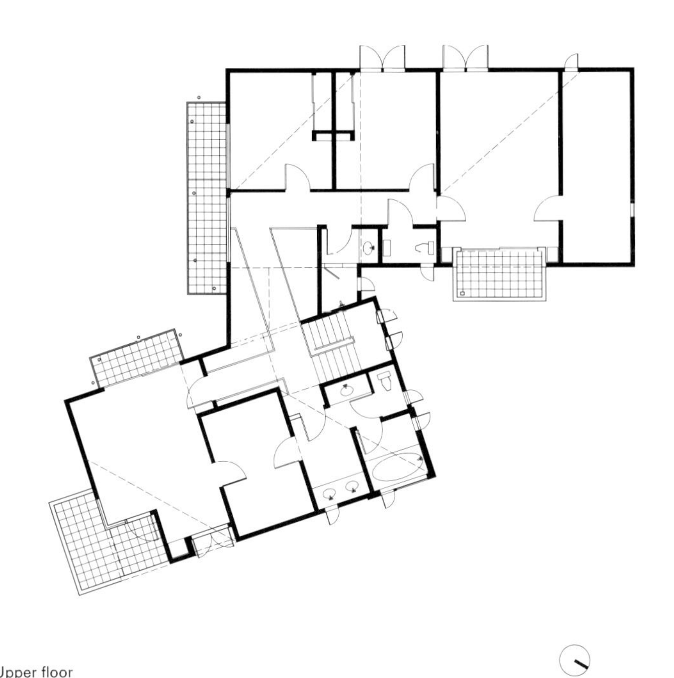

Upper floor

Chiba Multi-Family Accessible Housing

Chiba, Japan / 1997

This mixed-use building in metropolitan Tokyo employs the prefabricated framing-and-finish system developed in the Fox Island prototype, as well as in the Anderson Island cabin near Seattle, the Zebra showroom in Shizuoka, and the emergency community center prototype designed for temporary housing projects in Kobe following the Great Hanshin Earthquake. All of these projects were designed for cost-effective prefabrication, shipping efficiency, and rapid construction using our own balloon-framing-based panelizing methods and inexpensive stock materials; they are different from the more typical panel systems used in the Amerikaya projects, and more typically used in North America.

The Chiba project was designed as a prototype for a small construction company developing a series of very low-cost apartment buildings to compete with the common, relatively unpleasant developer housing blocks in which so many Japanese families live. A change in Japanese tax law in 1995, intended to encourage new low-cost housing production, created a sudden and very large market for the construction of cheap apartment blocks. The problem with this macro-economic policy shift was that the incentive system was structured such that the primary financial gains resulted from the simple presence of the buildings rather than from rental incomes and competitive attractiveness to prospective tenants. Consequently, there was a sudden surge in the construction of poorly planned concrete structures that were both unpleasant to live in and not very efficient to build.

These coinciding factors in turn created a market for inexpensive construction that was also pleasant to live in, exploiting advantages of the prefabricated wood panel systems to the fullest extent possible and using the cost savings from construction efficiency to provide added space and comfort to the individual units. The specific construction rules of the Japanese Government Housing Loan Corporation, which financed these projects, combined with the already strict and complicated zoning constraints and fire codes in Japan and the specific rules of the tax incentive, formed a complex regulatory context shaping the design of this system.

This project maintains the demographic mix of the neighborhood by creating a variety of dwelling types and retail shop space on a very small site sandwiched between train tracks, a temple cemetery, and a dense area of traditional houses and rice farms. The building contains a small shop and accessible flats on the ground level with two-story townhouse apartments above, with the intention of accommodating older residents as well as young families in the same building. The materials and layout provide a flexible loftlike living environment with flow-through ventilation, bright daylight, and maximum opportunity for outdoor living on terraces, balconies, and trellis-wrapped rooftop decks.

Further developments of the three-story, two-unit modules of this project have been configured for other applications with ground-floor garages and recombined as townhouses and duplexes, including a six-unit townhome on a flat urban site in Anchorage, Alaska, a steep mountainside vacation house in Hawaii, and a large single-family residence near New York City.

Panelized 2x4

OPPOSITE
Ghosted rendered view showing relationship of balloon-framed panel structure to exterior building elements

Left: Site plan and rendered interiors
Right: (bottom to top) Ground floor, with accessible housing units, commercial space, and entry doors and stairs to upper townhouse units; second level, with bedrooms of townhouse units; top floor, with main living spaces; and rooftop terrace level

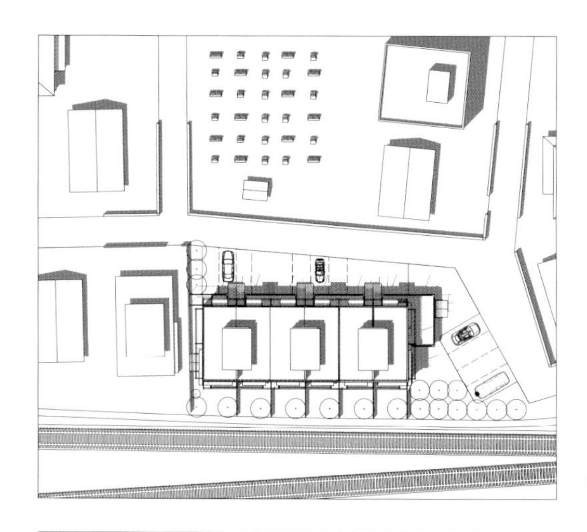

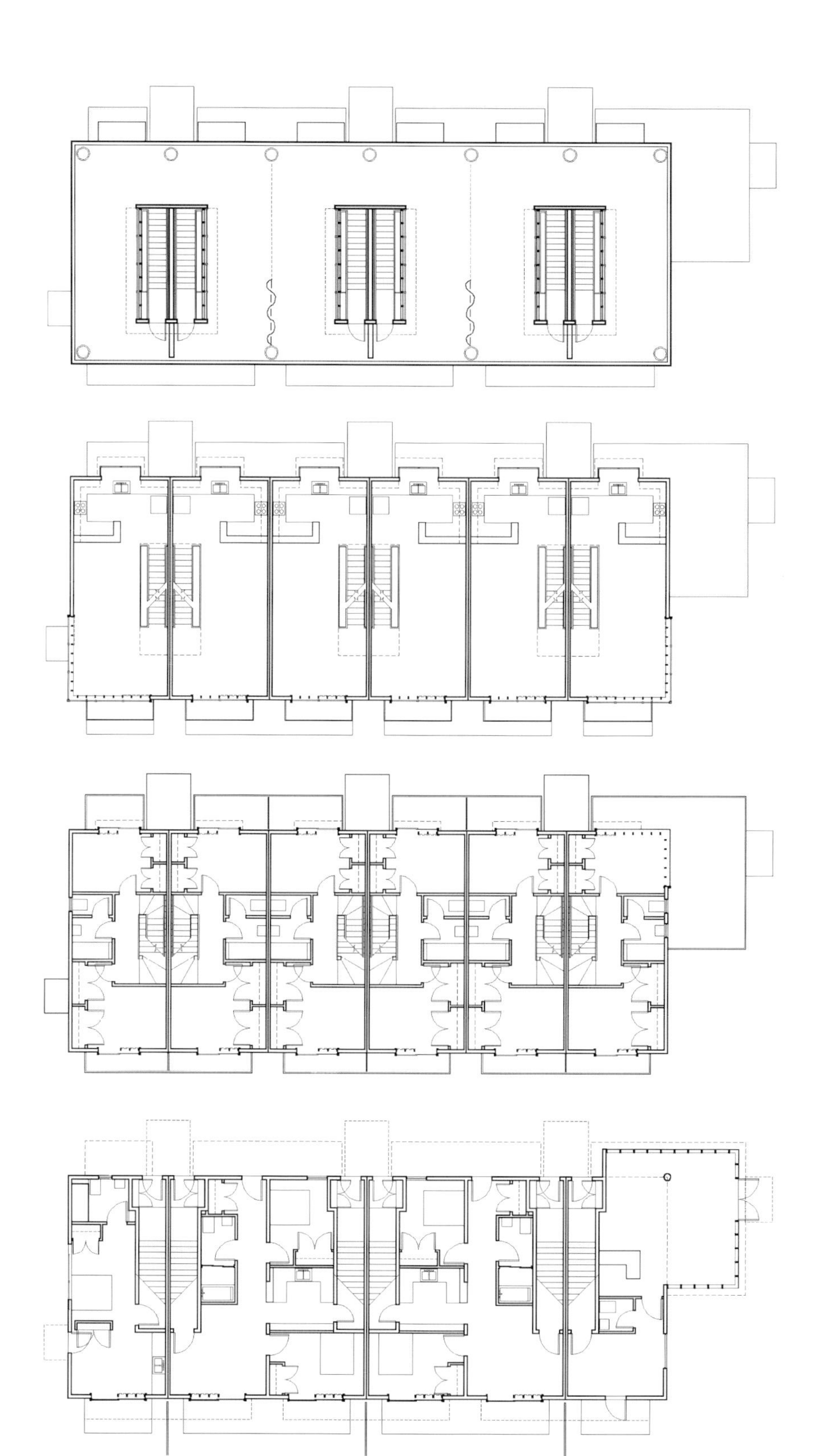

Chiba Multi-Family Accessible Housing

Parts List:
a. Stainless steel cable vine trellis
b. Steel frame for rooftop shade trellis
c. Private roof garden access stairs/ light wells
d. Aluminum windows
e. Diagonally braced 2x6 wooden stud framing
f. Panelized wall sections
g. Aluminum storefront
h. Panelized wood-joist floor with concrete topping slab and EPDM roofing
i. Double 2x6 panelized separation walls
j. Panelized 2x6 exterior wall framing
k. Striped EPDM roofing fabric screwed to sheathing
l. Prefabricated steel balconies
m. Prefabricated steel-and-glass entry canopies

Assembly Order:
1. Foundation/1st floor
2. 2nd floor/interior walls
3. 3rd floor/interior walls
4. Roof
5. Prefabricated enclosure wall system
6. Private garden access/ light wells
7. Entry canopies/balconies
8. Trellis system

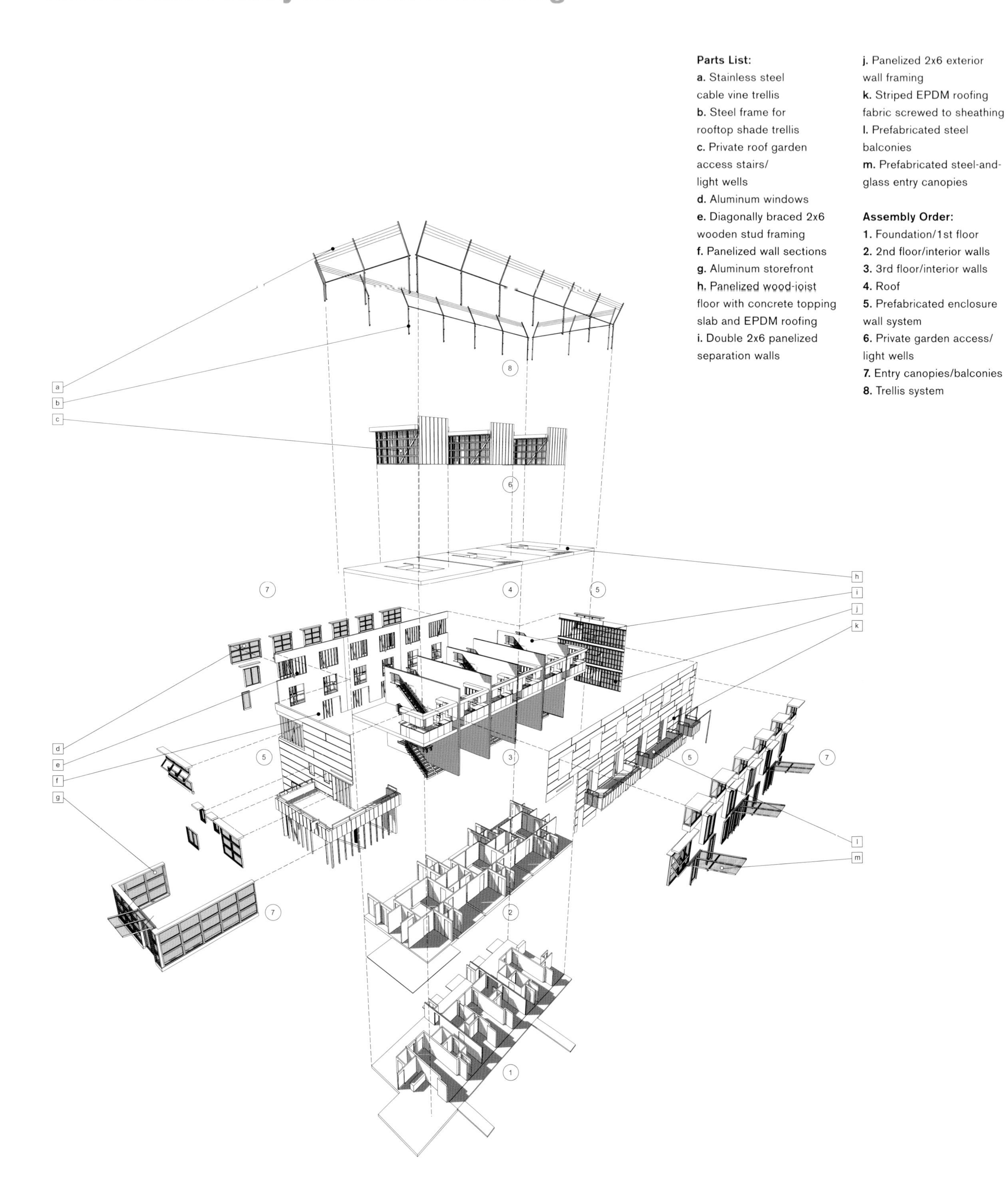

Panelized 2x4

Shaded perspective view with primary panel assemblies shown above

Parts List:

a. Panelized 2x6 wall studs continue through major window panels to minimize structural headers and to provide rigidity in shipping.

b. Structural OSB sheathing, finish siding, flashing, and windows are installed in factory before shipping. Flashing connections are given a random pattern on facade to allow field connection of panels without obvious "panel" joints.

c. After installation of wiring and plumbing from interior side of erected panels, interior surfaces are covered in drywall and finished on site.

d. Steel railings, trellises, cap flashings, and accessory components are all prefabricated and ready to bolt on in field.

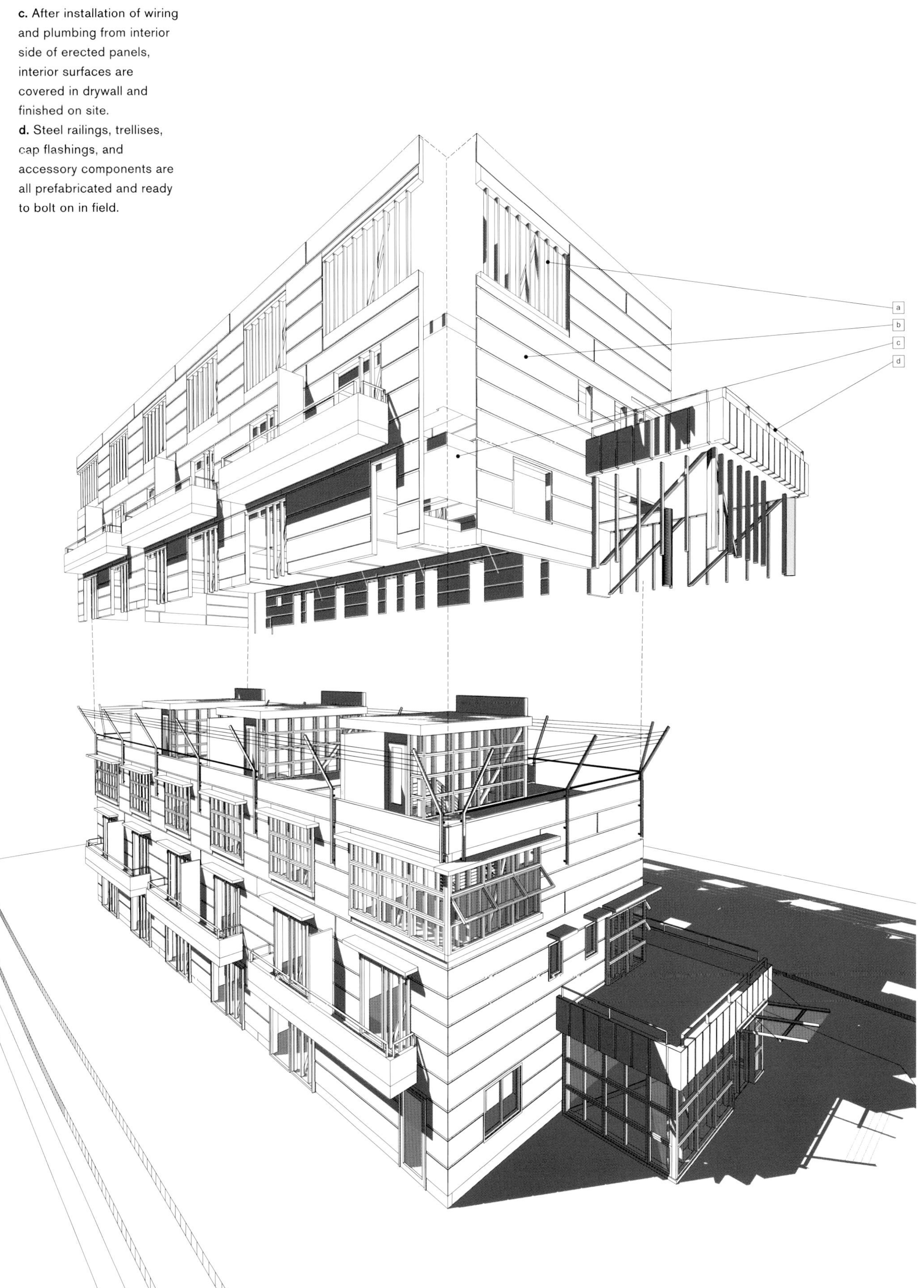

Chapter 2

CNC Timber Framing

Timber framing is a building system that has been used for thousands of years and is most often associated with the temple architecture of Japan, China, and Korea, as well as Northern European structures evolving from the building traditions of the Middle Ages. Traditional timber-framing techniques center around the intricate and often beautiful joinery work that connects the structural members, but these techniques are typically considered too labor-intensive and too weak for modern construction. With the advent of Computer Numeric Control (CNC) milling machines, however, much of the hand labor can be reduced, and exposed or concealed steel connectors can be used to make rigid connections. A resurgence of interest in timber-framing systems has contributed to a revival of historical forms, particularly in residential construction, but there has been relatively little use of this technology in modern design.

Traditional timber framing is most often used as a post-and-beam structural system, where loads are transferred through a building on linear paths through massive timber elements. With the wall enclosure systems independent of the structure, there are many opportunities to develop dramatically open spaces with large openings of windows and interior or exterior walls that remain separate from the structure. Timber framing is compatible with other prefabricated building systems, such as panelized stud walls or SIPs panel systems, which can be used to form the non-load-bearing portions of a building.

Log buildings are a particular kind of prefabricated timber-framed system, typically incorporating load-bearing solid wood wall sections with more purely post-and-beam systems for their roof structures. The log home industry has introduced significant technological advances in log construction in recent years, and the production capability of working with these processes has progressed more quickly than has any design evolution to take advantage of it. Although working with logs is still a niche area of the construction industry, there are many interesting opportunities for expanding their use into more building types and directions.

As more and more timber-frame manufacturers invest in this new generation of CAD/CAM machinery, there exist a growing number of sources for this technology in all regions, although the factories tend to be most often located in the timber-producing regions of the United States and Canada, and in Northern Europe. It is interesting to note that the majority of the CAD/CAM milling machinery used in the North American timber-frame industry is designed and built in Germany.

Harstine Island Timber House

Harstine Island, Washington / 1993

In many cases, the choice of building system is most strongly influenced by a project's location—its proximity to sources of materials and labor or ease of access by road, water, rail, or air. In other projects, it is the technical requirements related to the particular challenges of the site itself that might make one choose to build in steel, concrete, or wood, or with assemblies of panels, modules, or frames. In the case of the Harstine Island House, a third kind of site-specific factor drove the decision to select a rustic form of timber framing for the project—related less to the physical and practical requirements of the site and more to an appreciation of the place it occupied in the owners' experience.

The initial challenge in the design for this vacation home on a high island bluff in Puget Sound was to reconcile the idea of introducing a permanent structure onto a very narrow area, where the logical house site, a natural clearing in the forest at the edge of a bluff, had long been a cherished family campsite. Our strongest impression of the site on our first visit was that the patterns of use were already clearly established, both in terms of the physical development of the land and the emotional and spiritual role it played in the lives of its owners. While they were anxious to have a home on the site to extend its usefulness, we feared that the introduction of any new building here might endanger the very qualities of the place that made it most attractive.

In talking with the owners about how they used the campsite, it became clear that the journey to the site, and the ritual of arrival, were just as important to them as being there. During the two-hour drive from their home in Seattle, they would pass through decreasingly populated areas and increasingly beautiful natural surroundings, going in and out of dense forests before coming out of the trees to cross a bridge over the waterway surrounding the island, then re-entering the island forest. Arriving at their own land, they would lift the rough-hewn log gate and traverse a twisty, needle-strewn twin-track road until finally arriving at the clearing, high on the waterfront bluff with spectacular views to the east of the water, other islands, and Mt. Rainier in the distance.

We designed the building to frame this entry to the clearing and to define its edges but not occupy it. Although we were first told not to cut any trees, we argued that this restriction would result in filling the clearing with the house, and instead proposed to maintain the clearing as it was, positioning the house as a long, embracing wall around it, displacing the first rows of trees that had previously protected the clearing. In their place, the encircling structure of the house would be made of an open structure of standing logs, with a curtain wall of wood-framed glass and a one-room-deep U-shaped plan open toward the bluff and the views. The interior of the house, plainly visible from the clearing and glowing warmly with soft light at night, is completely paneled in pine boards on floors, walls, and ceiling, and the outer wall of the compound is very solid, with only a few small windows, completing the sense of a fortlike encampment, a kind of stockadelike protection of a precious open space at its center. With the extreme ends of the building extending almost to the edge of the bluff itself, the only access to the clearing is now through a covered porch framed by log timbers, which divides the structure into owners' quarters and a guest house and provides the final threshold for the experience of arrival at the clearing.

The peeled log structural elements are combined with a heavily insulated balloon-framed shell as a direct response to the owners' desire to have a rustic cabin that at the same time corresponds to their wish to make careful use of material and energy resources. Avoiding the usual log home problems of insufficient insulation, excessive air infiltration due to cross-grain shrinkage, and inefficient consumption of large-dimension trees, we chose to use peeled logs only where full use could be made of their structural qualities and high visual impact, mainly as vertical support columns and as roof rafters. Heavily insulated balloon framing was chosen for the exterior perimeter wall, clad on the outside with board and batten cedar and on the inside with pine paneling. This hybrid system enabled the majority of the wood consumption in the building to be from environmentally friendly small-dimension lumber products from plantation-grown small-diameter trees, while still maintaining a total environment of enveloping wood materials.

In addition to the requirement to minimize the impact of the building process on the site and to preserve the native vegetation, we worked together with the owners to develop a program that placed significant importance on the thoughtful use of natural resources, including not only the consumption of materials and energy in the original construction of the house but also in the long-term impact on the ecosystem over the course of the house's life. Natural, unprocessed materials were chosen for their low embodied energy requirements and their abilities to weather and age gracefully with a minimum of resource-wasteful maintenance. The choice of a locally sourced prefabricated log timber structural system was a central strategy in meeting these goals.

Reducing or eliminating requirements for energy utilization in the heating, cooling, and lighting of the house were also essential factors in the home's overall design. The placement of the house on the site, the layout of the rooms, and the placement of the windows and overhangs are all a result of thorough analysis of sun angles and solar orientation, planning for maximum solar gain in the cool months, natural convective cooling in warmer weather, and minimal requirement for supplemental electric lighting. The zoned heating system, decidedly more high-tech than the rest of the house, provides remote thermostat adjustment via telephone to allow the owners to manage heating requirements on an as-needed basis, promoting the energy-saving ability to leave the house minimally heated while they are away, yet still allowing them to arrive to a warm home on cool winter weekends.

Left: Site plan
Top: Upper-floor plans
Middle: Ground-floor plans
Bottom left: Log prefabrication and assembly process
Bottom right: View from covered entry to deck and living room

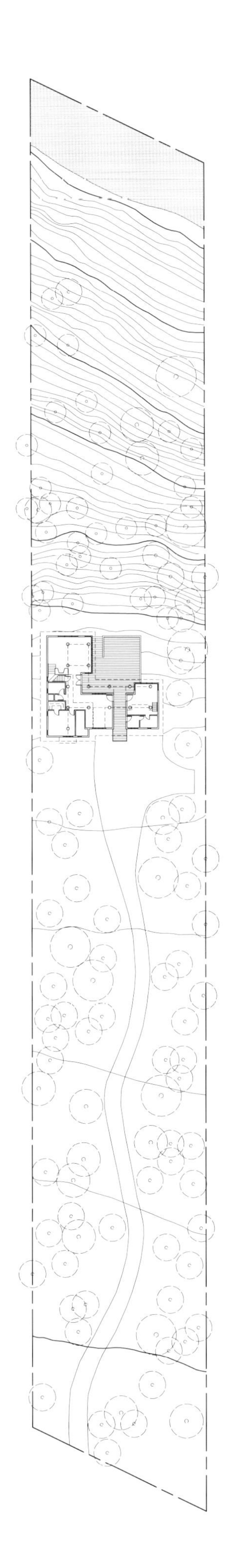

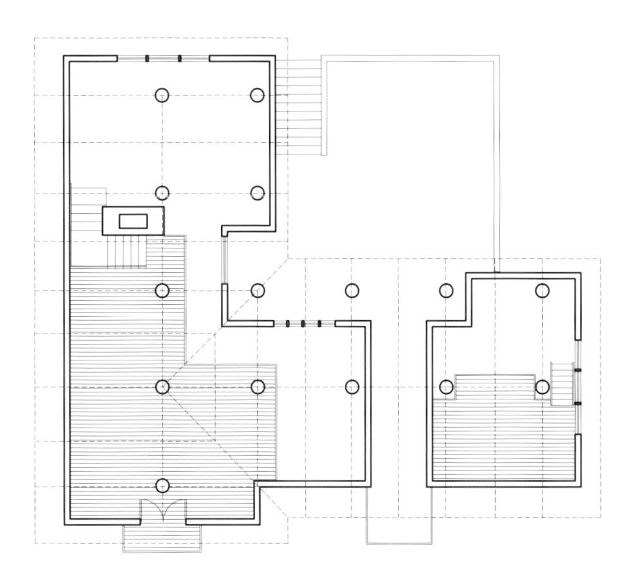

Upper floor

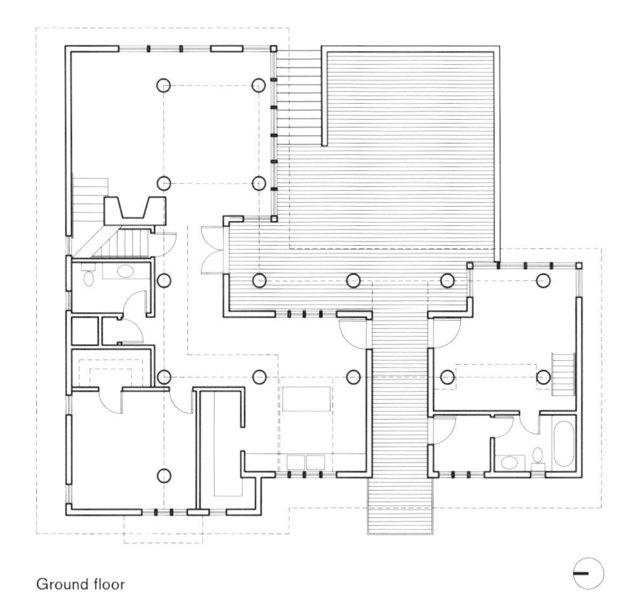

Ground floor

Mount Whitney Trailhead Prototype

Inyo National Forest, California / 2004

The National Forest Service requested a prototype design for a series of prefabricated trailhead structures that would integrate the increasingly chaotic program elements necessary at points of departure for some of the nation's most traveled and environmentally sensitive wilderness trails. The first of these prototypes was to be constructed at the Mount Whitney Portal, the eastern access point to the rugged trails ascending the continental United States' highest mountain peak. The August-to-May time frame for design and construction was short; the structure was required to be complete in time for the trail's centennial celebration, scheduled for the season opening just after the spring thaw. Aside from this necessity for offsite fabrication during the winter, while the site was buried under deep snow and ice, the larger requirement for prefabricated modular structures involved both the minimizing of construction activity on sensitive wilderness sites as well as the Forest Service's desire for repetitive, recognizably similar structures at various sites in the forest that would clearly announce the location of a trailhead with minimal need for additional signage.

As wilderness trails have seen more and more use over the years, the Forest Service has been challenged in both maintaining the trails physically and in educating the public in safe and environmentally sensitive hiking and mountain climbing. Increasing trail use has required a good deal of new infrastructure at the trailheads, including pit toilets, garbage cans, signs warning of particular dangers—such as bear attacks, avalanches, hypothermia, and dehydration—and educational signage introducing important flora, fauna, and geological conditions. All of this has been added haphazardly over the years, causing a substantial clutter undermining the natural wilderness experience. At mountain climbing trailheads such as the Whitney Portal, the situation is further complicated by the fact that most serious hikers who will attempt climbing all the way to the summit must depart in the wee hours of the night, in order to avoid the harsh heat of the valleys and the glacier's deadly crevasses that open in the afternoon sun. Some form of lighting was therefore required to illuminate the warning signs, the trail sign-in log, and the general facilities. All of this added up to the potential for a distinctly non-wilderness let-down at what should be an uplifting and celebratory point of departure. Greatly constricting the potential design options, the Forest Service had a rigid set of well-intentioned design guidelines that encourage a kind of "mountain style" appearance while actually making it difficult to achieve the rugged simplicity of the Forest Service's traditional buildings. It quickly became evident as well that there was a vocal wilderness hiking community that was generally opposed to any sort of structure or signage marking the trail. For all constituencies, the primary design objective was to create a simple, rustic, modular structure that would screen lighting and conceal signage and infrastructure within an envelope of enclosure blending harmoniously into the landscape while nevertheless clearly marking the spot, even at night.

The project was a collaboration between the Forest Service and the University of California, Berkeley, with 50 percent federal funding and 50 percent provided by private donations of money, materials, and services. We structured the project as a broader research opportunity to be developed as a series of public projects organized through the University's Center for Environmental Design Research. Graduate students Jeffrey Jordan and Christopher Nogoy led the project as principal designers, fabricators, and supervisors of the onsite construction. A larger group of graduate and undergraduate students participated in the fabrication and construction process, working in collaboration with rangers and trail builders from the Forest Service. The building is constructed of ruggedly milled pine logs—from fallen urban trees salvaged by The East Bay Conservation Corps in Oakland—together with hot-dip galvanized steel structures, fabricated in San Francisco, and a cross-bracing lattice of stainless steel wire rope fastened together with mountain climbing carabiners.

Evening views of completed trailhead structure

CNC Timber Framing

Top row: Offsite-fabricated steel and timber components ready for assembly
Middle row: Connection details
Bottom row: Completed structure

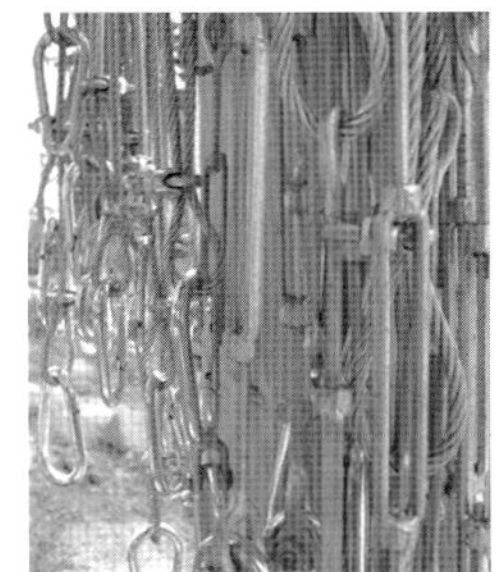

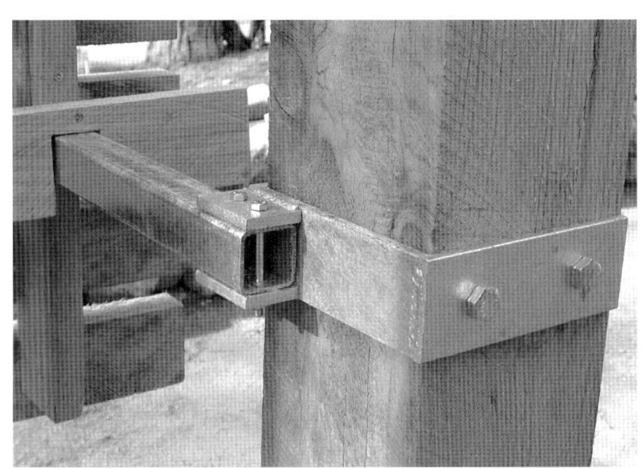

Marrowstone Island House

Marrowstone Island, Washington / 2003

Developed in collaboration with a fabricator using new-generation CAD/CAM timber-milling machines, this structural approach is a new application of ancient timber-framing techniques allowing for extensive use in offsite construction. The house is on an island in Puget Sound, making it advantageous for all major components to be fabricated in a nearby factory and brought to the site for rapid assembly. As a further test of the concept, we concurrently specified the same fabrication system and fabrication team for a project at an even more remote location—working with a Japanese builder on a structure on the tiny island of Iki, in the Korea Strait between Fukuoka, Japan, and Pusan, Korea. The large number of small, modular components used in these approaches to timber framing have proven to be a cost-effective and efficiently transportable prefabricated structural system when applied to particularly difficult or inaccessible locations, but they do not themselves provide a complete building solution as they require other systems of enclosure. Traditionally, the timber frame is infilled with mud, brick, or stone, while modern timber-frame construction techniques often use SIPs panels as non-structural prefabricated enclosure systems between the timber supports.

With the Marrowstone Island House, we began to explore timber-frame buildings in which the structure is dissociated from the building enclosure so as to fully articulate the function and beauty of the massive timber elements. These structures support umbrella roofs covering interchangeable prefabricated living spaces, resulting in a variety of options for modular approaches to the building enclosures. The system provides for highly flexible arrangements of various deck and modular box units adaptable to various sites and program requirements, with the indoor-outdoor living spaces particularly well-suited to seasonal occupation or moderate climates. The umbrella roof is framed with tripods of glue-laminated timbers that can be individually cut and milled by computer-operated machinery to accommodate a variety of sloping site considerations. Each tripod set has splayed legs providing shear resistance in all directions and eliminating the need for shear walls or other cross-bracing. The tripod timber frame is roofed with a clear polycarbonate plastic to provide a brightly lit, rainproof outdoor living space below.

Primary living, dining, and cooking activities take place on the giant wooden deck cantilevered out toward the water and sheltered from wind and neighbors by more solid living boxes sitting on the landward edge of the planking. One modular box contains a small indoor kitchen and family space with a sleeping deck above, protected by the high polycarbonate umbrella. The other box contains a small master bedroom, bath, and laundry on the deck level and two small bedrooms and bath on the second level. These boxes are intended to be cozy, enclosed retreats for bad-weather living. Access between the various box rooms is by outdoor passage on the main decks and by light steel stairs and catwalks above. The main entry and the stairs to the boat storage rooms and to the beach below passes through the gap between the two metal-clad living boxes.

Below: Site plan

OPPOSITE
Top: Wood scale model
Bottom: Model views of primary structural framing

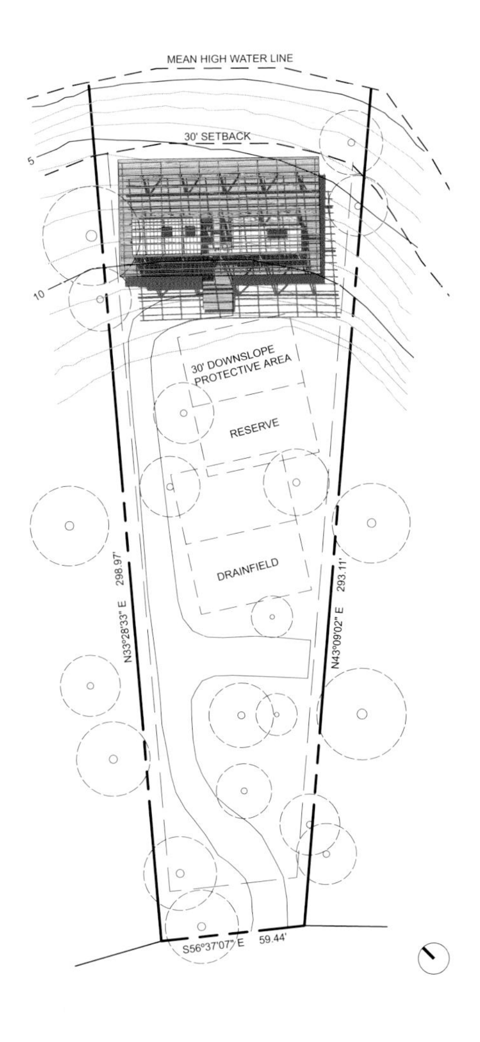

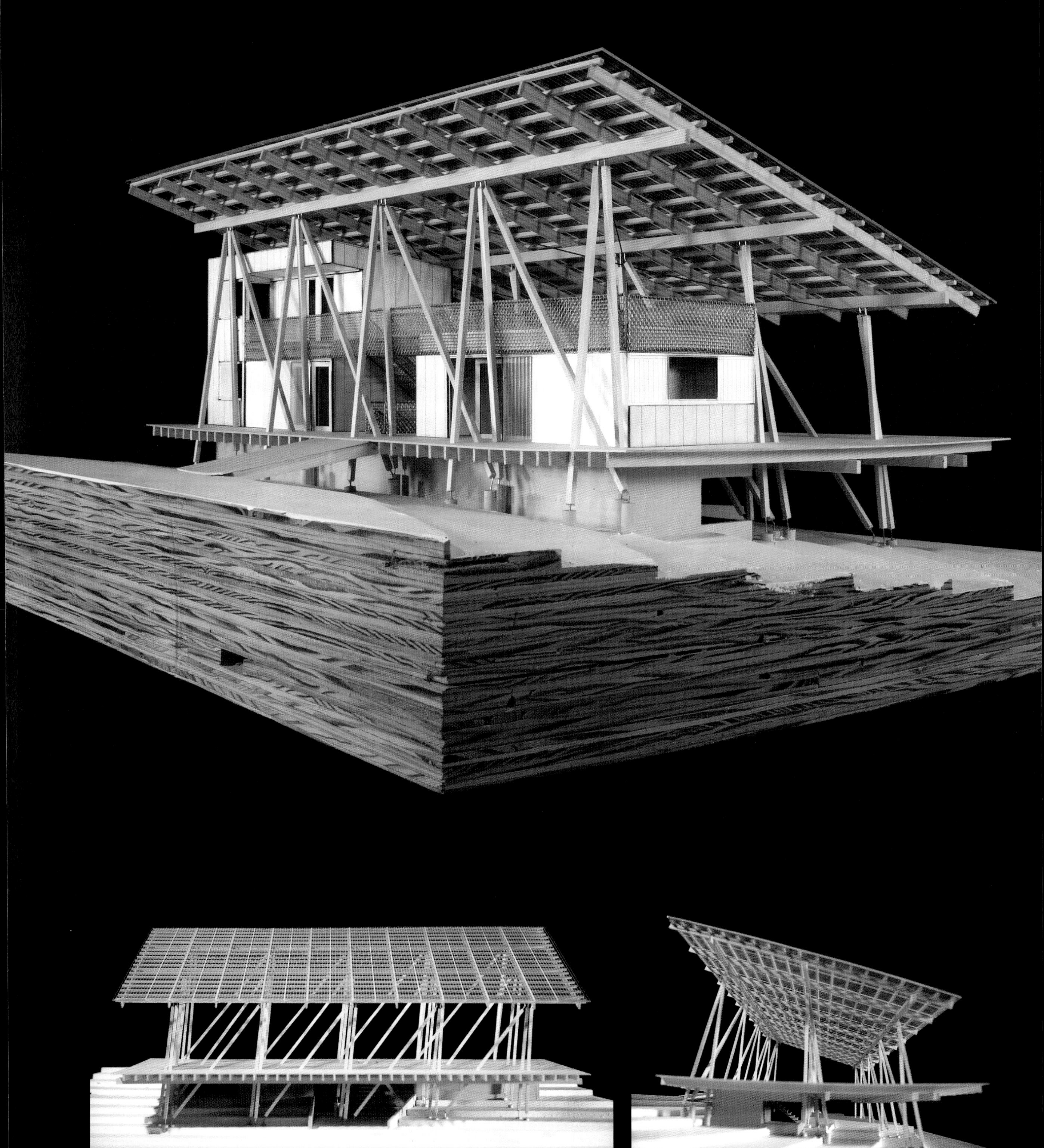

Marrowstone Island House

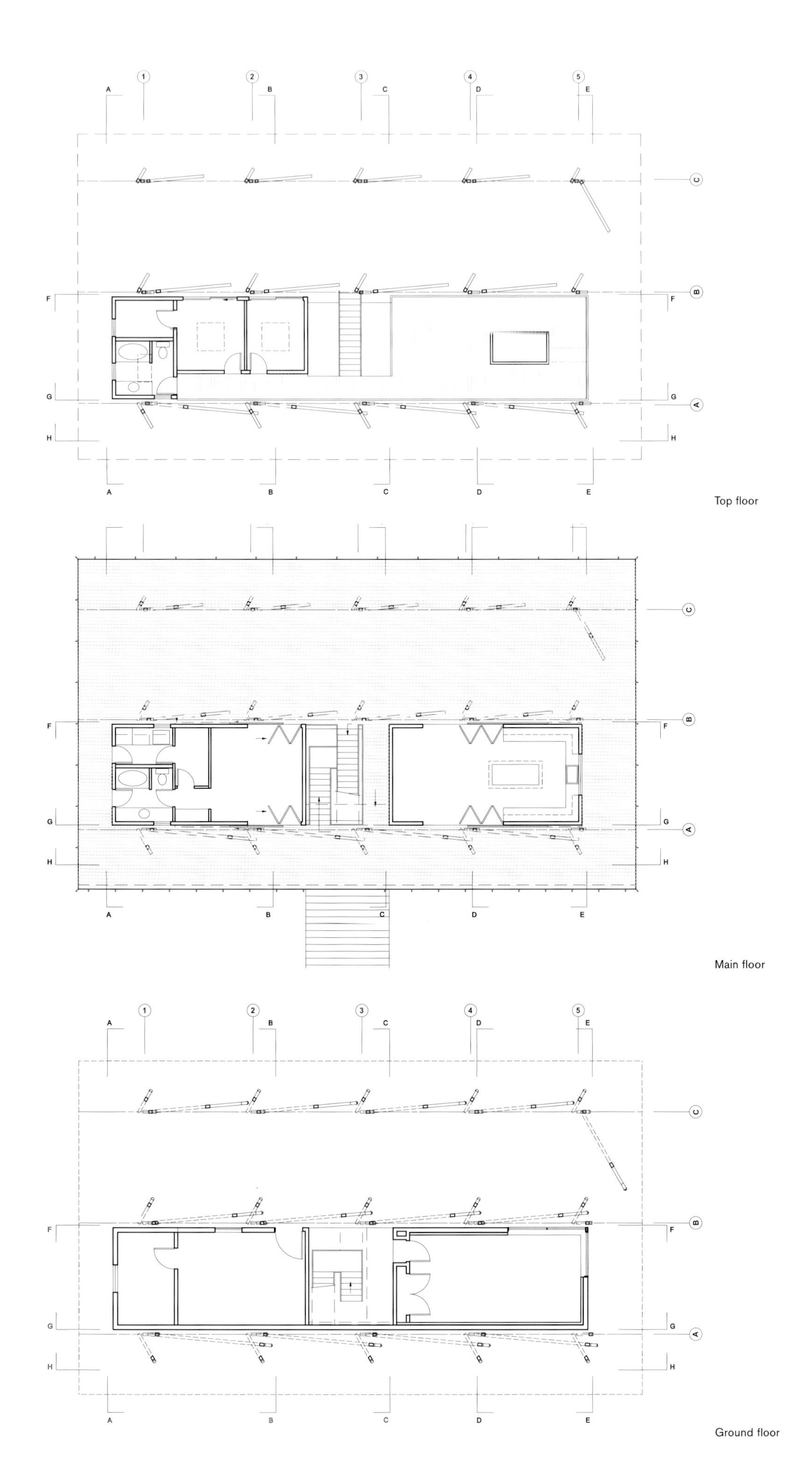

Top floor

Main floor

Ground floor

Left: (bottom to top) Ground floor, showing boat storage, mechanical space, office, and access stair to beach; main floor, with large covered wood deck area surrounding small living and sleeping space boxes; upper floor, showing two small bedrooms, bath, and sleeping loft roof deck above living space box and below covered roof

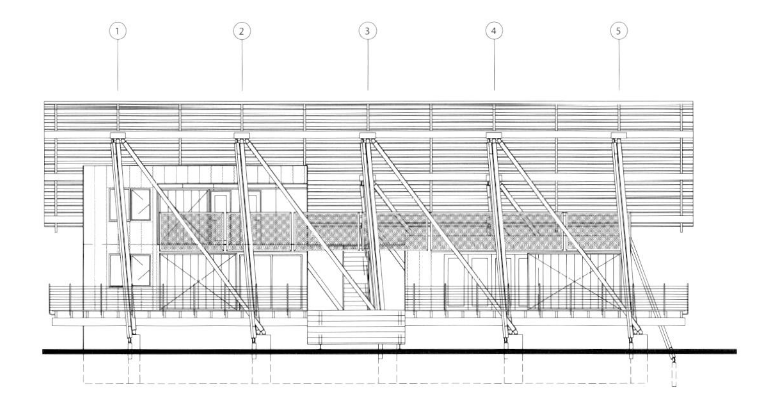

South elevation

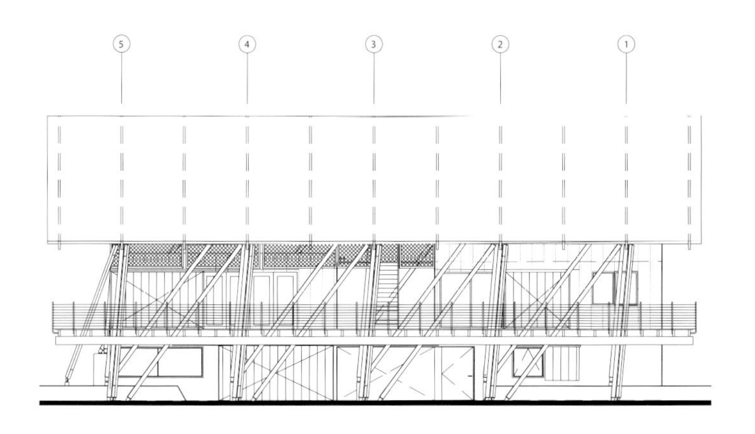

North elevation

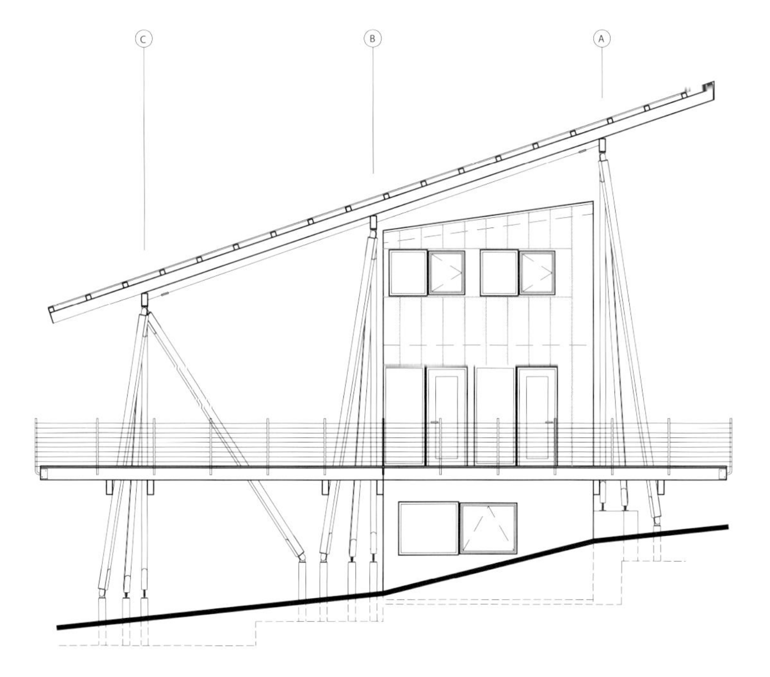

West elevation

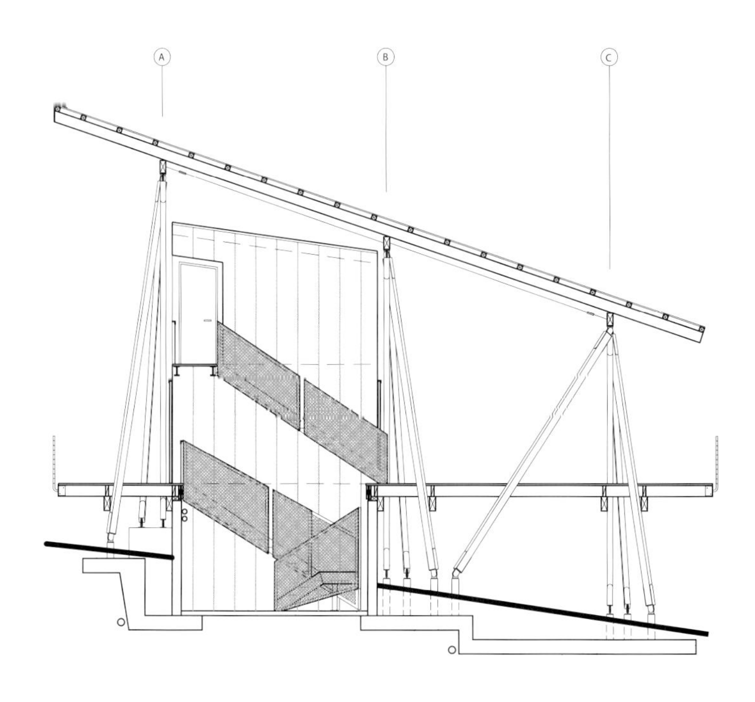

Section

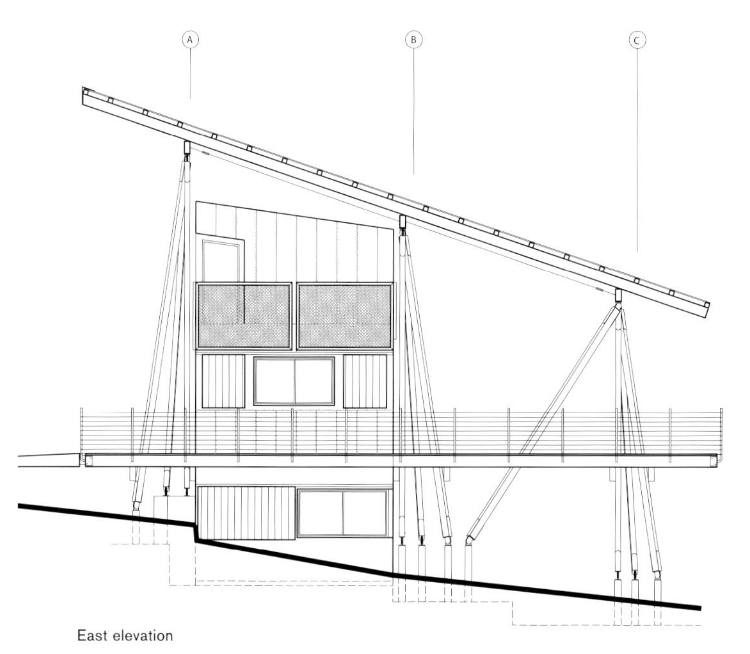

East elevation

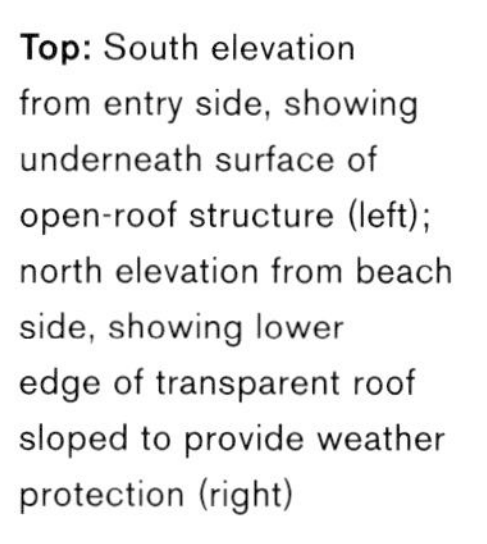

Top: South elevation from entry side, showing underneath surface of open-roof structure (left); north elevation from beach side, showing lower edge of transparent roof sloped to provide weather protection (right)

Middle: West elevation showing how tripod column length adjusts to sloping terrain (left); building section through exterior stair area showing access from beach level to deck level to bedroom/ loft level (right)

Bottom: East elevation

Marrowstone Island House

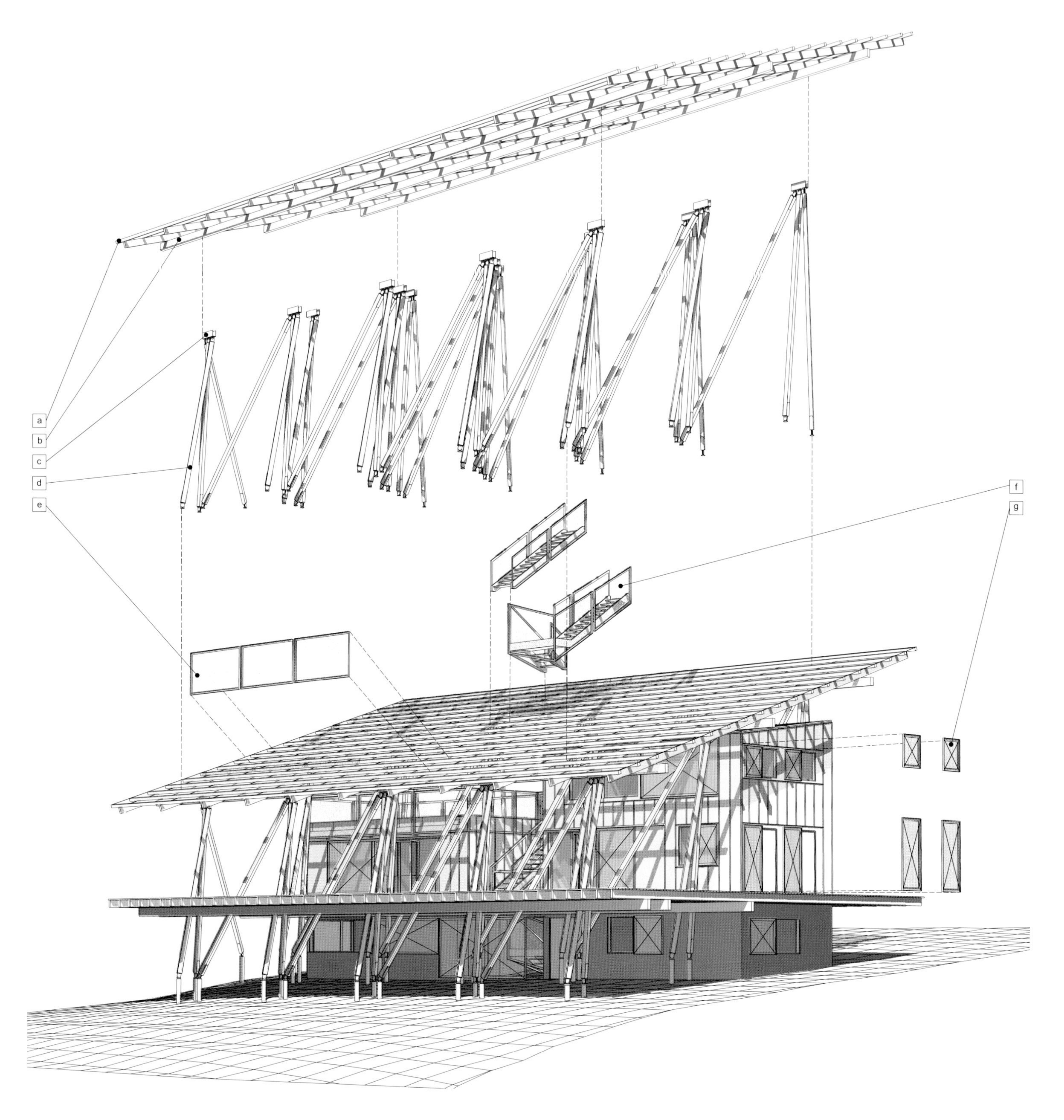

Rendered image of complete structure, with primary building components pulled out for clarity

Parts List:

a. Transparent polycarbonate roof panels
b. CAD/CAM milled glue-laminated beams
c. Steel glue-laminated saddle/tripod connector
d. CAD/CAM milled glue-laminated tripod
e. Steel/chain link railing system
f. Steel/chain link/bar grate stairs
g. Steel/polycarbonate storm shutters

CNC Timber Framing

Top row: Milling machine used for fabrication of timbers
Middle row: Steel joint assembly details
Bottom row: Completed tripod frames prior to installing roof purlins

Marrowstone Island House

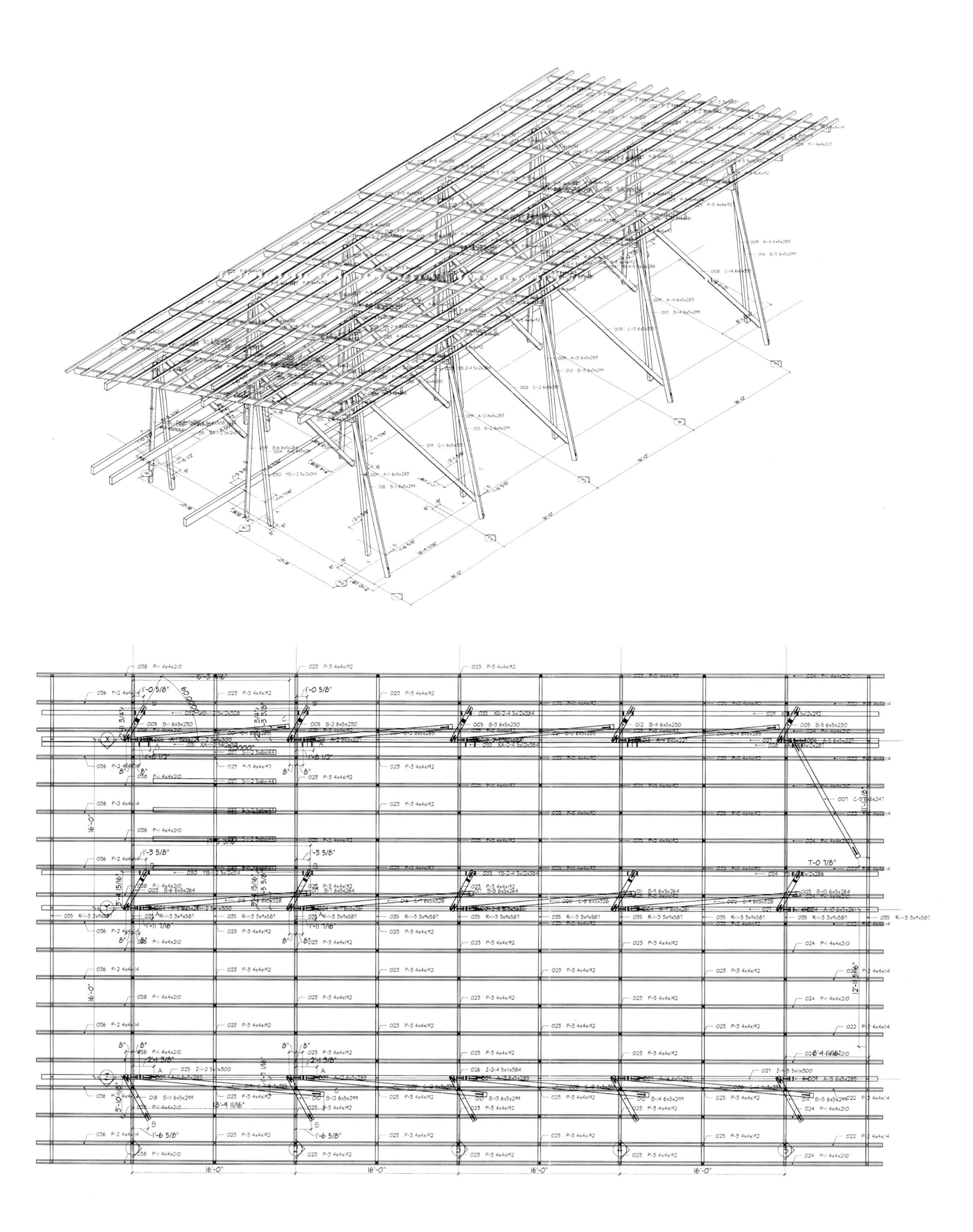

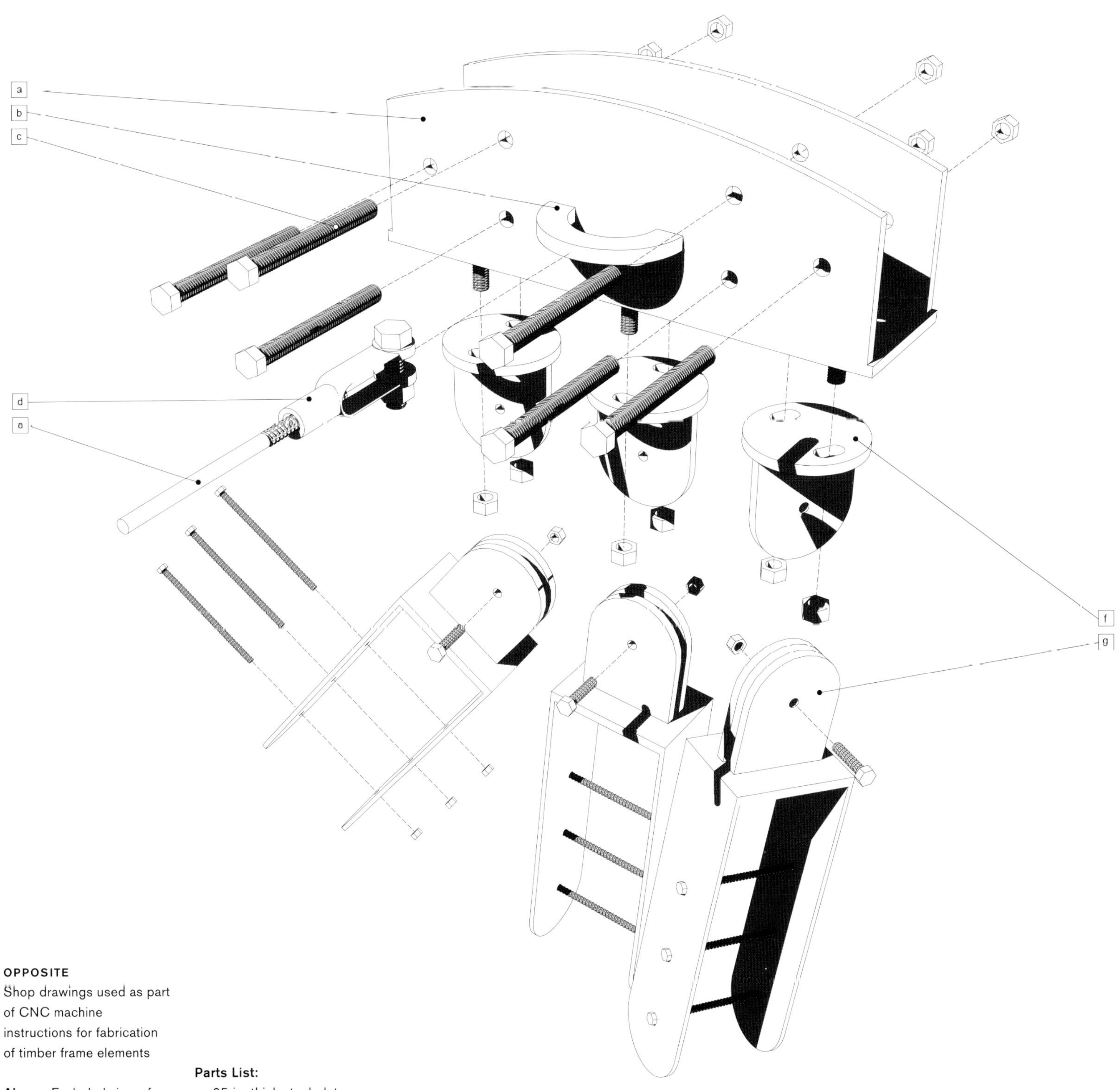

OPPOSITE
Shop drawings used as part of CNC machine instructions for fabrication of timber frame elements

Above: Exploded view of fabricated structural steel connector assembly, showing adjustable-angle pin connection system that allows repetitive use of identical connections for timbers of different lengths and angles; collar yokes extend well down sides of timbers to allow sufficient grip for those members used in tension.

Parts List:
a. .25 in. thick steel plate glue-laminated beam cradle/ tripod axis
b. Turnbuckle connection at cradle
c. Cradle/glue-laminated through bolts
d. Turnbuckle
e. .5 in. diameter steel rod cross-bracing
f. Swiveling tripod leg cradle connection
g. Swiveling tripod leg cap

Marrowstone Island House

Completed timber structure and enclosed living space boxes framed and sheathed

OPPOSITE
Top: Underside of completed roof structure
Bottom: Steel connections

Magnolia Gallery

Seattle, Washington / 2001

This project is a large addition to an existing waterfront home to accommodate the owners' growing collection of contemporary art. Not wishing to modify the 100-year-old Craftsman-style home that sits at the core of the estate, the new additions are conceived as a series of lanternlike pavilions that are distributed among the landscaped cliffside gardens. The large increase in building area is concealed by the integration of multiple building elements within the dramatic topography and mature landscape of the site, creating a range of different indoor and outdoor exhibition spaces for the varied collection. To increase the flexibility of the galleries for art display and for large-scale entertaining, a modular system of multi-layered sheer curtains are hung on tracks that allow definition of spaces within the open plan.

Situated on a steep and rocky waterfront site, the only practical means of access for settling this building into the mature garden was to do as much fabrication as possible off site, then barge all components into the site at high tide, erect a temporary construction crane, and assemble the prefabricated timber components on the shoreline. Using the same timber-frame system that we developed for the Marrowstone Island House, we designed a forest of splayed sticks that would angle inward and outward from the rectangular building enclosure to anchor into the jumbled rock ledges below. Minimally interfering with the landscaped garden slope, the judiciously planted sticks loft the gallery high into the air and display the primarily sculpture and ceramics collection within a dense field of timbers.

At one early design meeting, the owner of the house showed us a sketch hc had made the night before after waking from an intensely vivid dream, which he had titled "The House That is Not There." From that sketch, we developed an idea for a largely transparent, shadow-scattered structure tucked into and emerging from the garden and the fog. Wishing to avoid introducing a new object into the garden, we focused on designing an armature to accommodate specifically imagined living events, connected by a system of passages through the garden with multiple pockets of enclosure and little distinction between interior and exterior experiences. The existing spaces of the old house are linked to the new system as pockets of defined activity attached to the transparent garden-viewing passages. The newly introduced pockets, which loosely define program elements of bedrooms, baths, and entertaining and dining areas, are developed as additional pockets of space, usually draped in warm, heavy curtains.

One of the important functions of the house from the outside is to provide events that engage with the garden, including the opportunity to create outdoor lighting. As one views the house from the waterfront, and while descending the stairs from the road down into the garden, the internally lit pavilions provide lanternlike points of light pooling outward into the garden.

The new portions of the house are primarily planned to be built from a CNC-milled glue-laminated wood post-and-beam structural system, which is supported by screw anchor foundations drilled deep into the steep, unstable hillside. Conventional foundation systems would have required too much disruption to the delicate site, while the prefabricated steel screw anchor elements allowed for attachment points to be distributed across the face of the slope and located more or less as dictated by rock outcroppings and avoidance of existing garden features. The posts attached to the foundation anchors are thus not vertical, aligned with the structures they support, but slightly askew, like sticks and trees in nature. The combined effect of the many angled columns also provides much of the triangulated structural shear support needed in the house. The experience of moving through the screen of sticks is intended to be similar to the differential expansion and movement of space encountered in walking through a forest of small trees. Both the shadows cast into the house and those cast outward into the garden will intensify the layers of linework defining the walls of the house. Although much of the new space of the house is transparent in order to merge with the space of the garden, the layers of wood structural members will provide a warmly protective screen around the inhabitants.

To provide dense enclosures within the screens of glass and wooden sticks, heavy silk curtains of saturated color can be drawn around defined areas of intimate space. These curtained enclosures hang from modular fabricated steel tracks integrated into the ceilings, with two or more layers of cloth on separate tracks. The outer layer is heavy and opaque, while inner layers, though of the same color, are increasingly sheer and diaphanous. The layers can be drawn in various configurations to provide specific views, light qualities, privacies, protections, and warmth.

Top: Series of sketch studies exploring relationship between timber-frame system, slope, and existing structures on site
Bottom: Sketch models of new gallery structure as seen from beach below

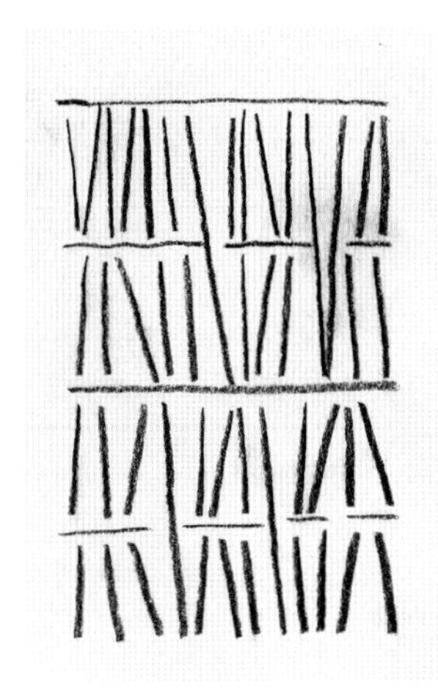

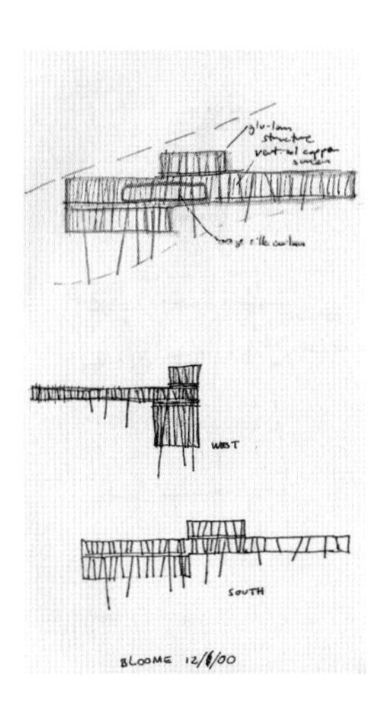
WEST
SOUTH
BLOOME 12/8/00

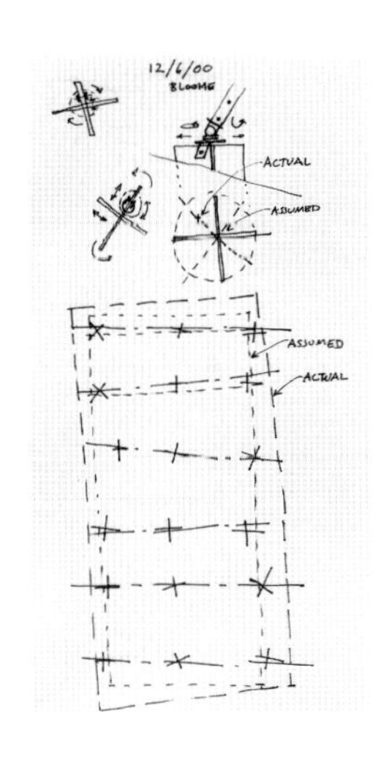
12/6/00
BLOOME
ACTUAL
ASSUMED
ASSUMED
ACTUAL

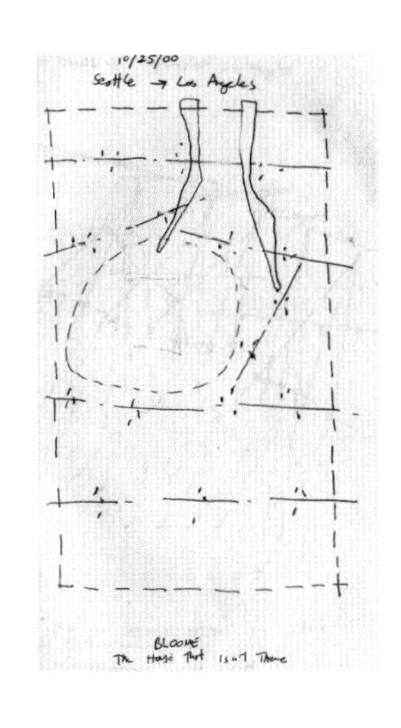
10/25/00
Seattle → Los Angeles
BLOOME

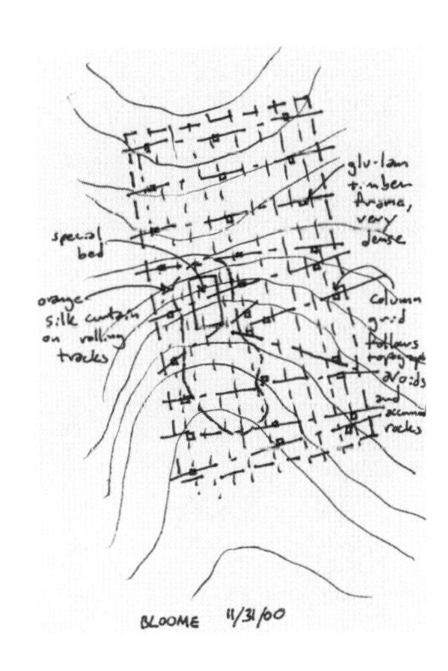
glu-lam
timber
frame,
very
dense
special
bed
orange
silk curtain
on rolling
tracks
BLOOME 11/31/00

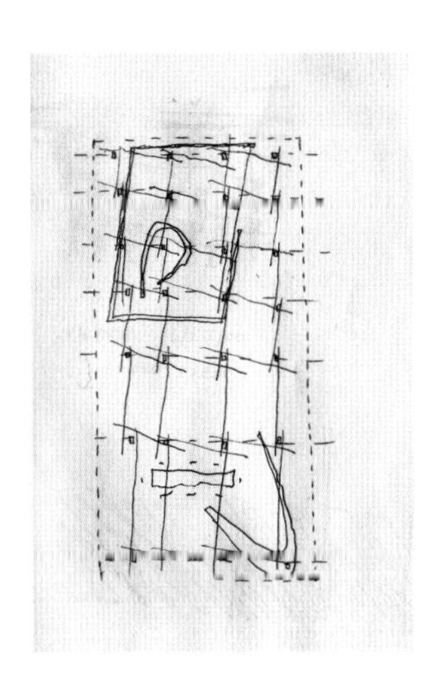

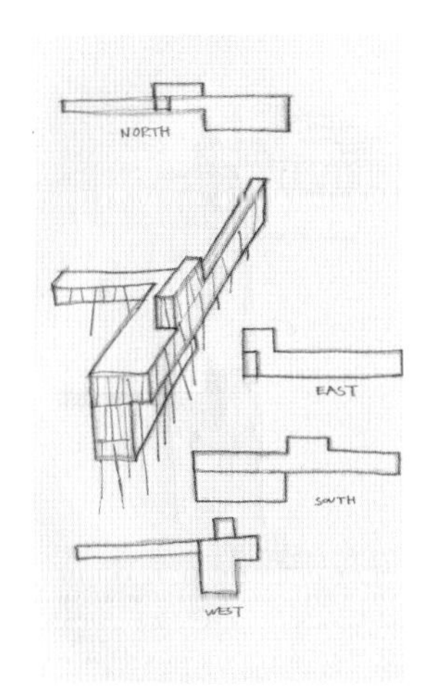
NORTH
EAST
SOUTH
WEST

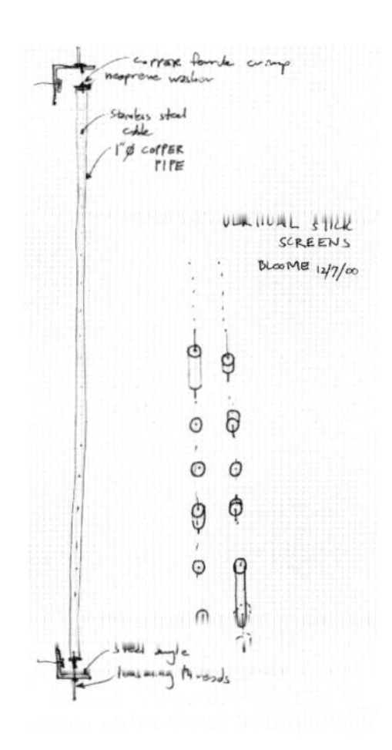
SCREENS
BLOOME 12/7/00

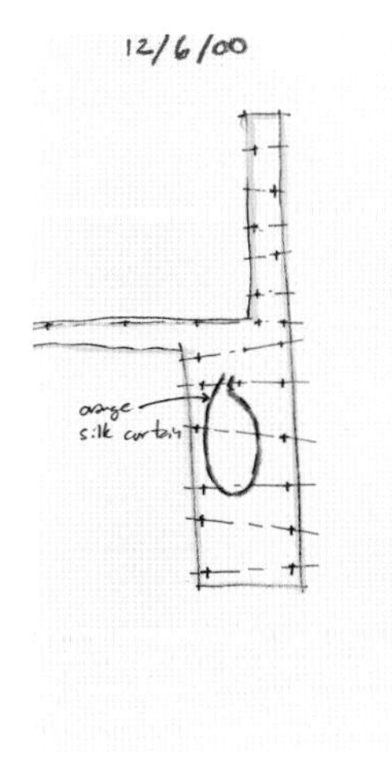
12/6/00
orange
silk curtain

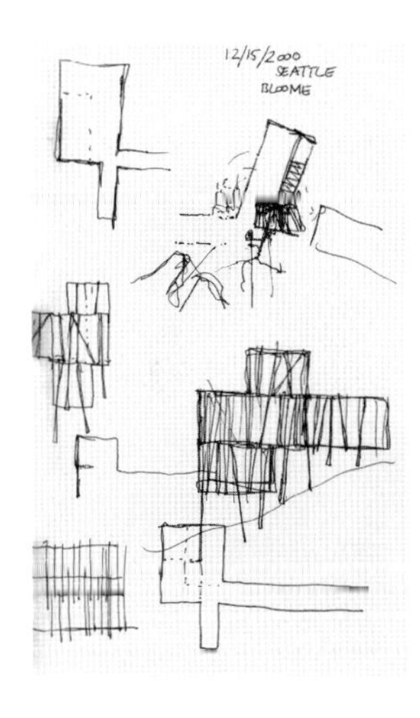
12/15/2000
SEATTLE
BLOOME

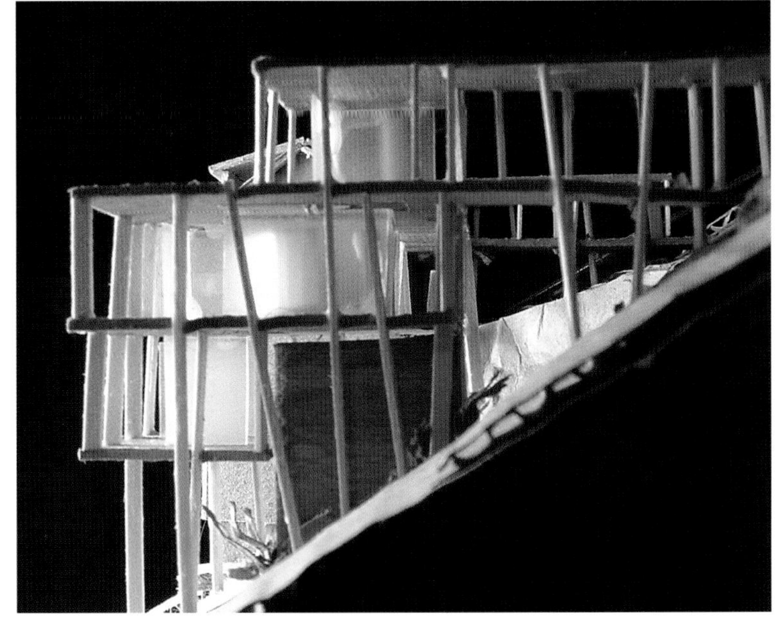

Magnolia Gallery

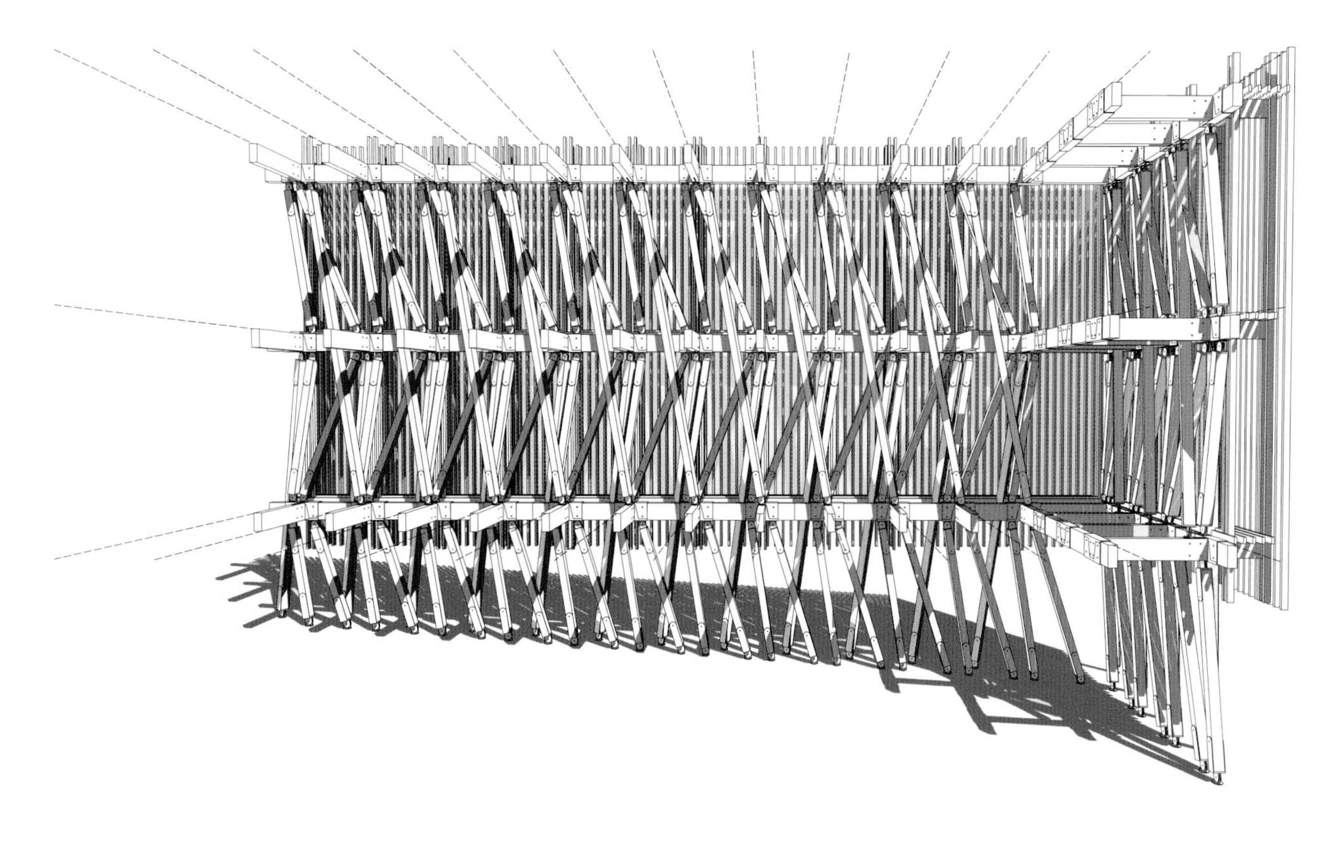

Top: Site plan showing new program additions linking existing structures
Bottom: Section view of structural frame, enclosure windows, and cedar slat shading system as viewed from building interior

CNC Timber Framing

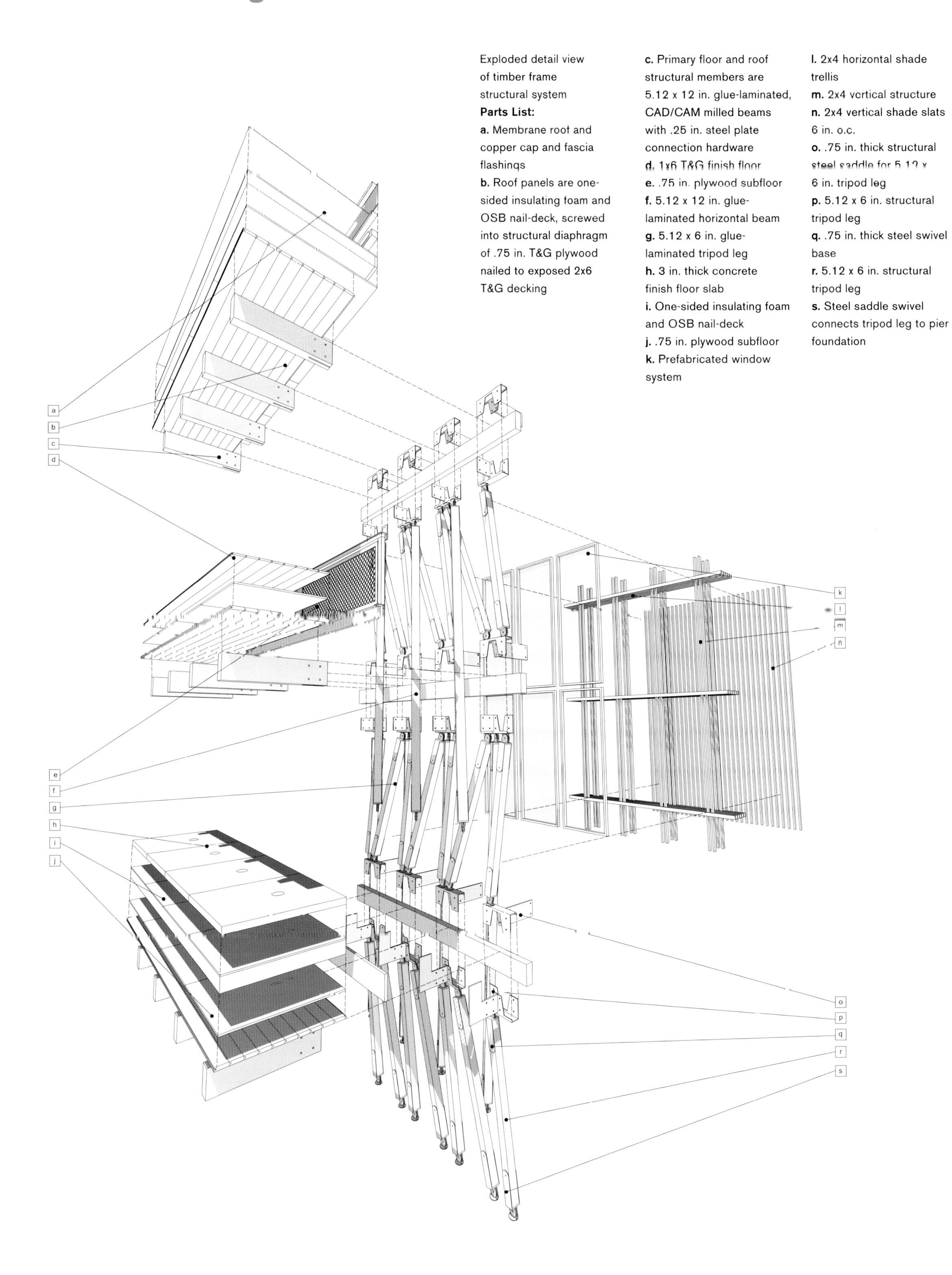

Exploded detail view of timber frame structural system

Parts List:

a. Membrane roof and copper cap and fascia flashings

b. Roof panels are one-sided insulating foam and OSB nail-deck, screwed into structural diaphragm of .75 in. T&G plywood nailed to exposed 2x6 T&G decking

c. Primary floor and roof structural members are 5.12 x 12 in. glue-laminated, CAD/CAM milled beams with .25 in. steel plate connection hardware

d. 1x6 T&G finish floor

e. .75 in. plywood subfloor

f. 5.12 x 12 in. glue-laminated horizontal beam

g. 5.12 x 6 in. glue-laminated tripod leg

h. 3 in. thick concrete finish floor slab

i. One-sided insulating foam and OSB nail-deck

j. .75 in. plywood subfloor

k. Prefabricated window system

l. 2x4 horizontal shade trellis

m. 2x4 vertical structure

n. 2x4 vertical shade slats 6 in. o.c.

o. .75 in. thick structural steel saddle for 5.12 x 6 in. tripod leg

p. 5.12 x 6 in. structural tripod leg

q. .75 in. thick steel swivel base

r. 5.12 x 6 in. structural tripod leg

s. Steel saddle swivel connects tripod leg to pier foundation

Arboretum of the Cascades

Preston, Washington / 2002

Working with Charles Anderson Landscape Architecture in Seattle, we designed four primary structures as interpretive centers and visitor facilities at various strategic points within the master plan for a new arboretum of native Northwest forestland. The nearby presence of a major highway, factory, and warehouse buildings caused us to suggest a revised theme to the arboretum's educational mission and development plans. Although initially asked to create a space as natural and removed from human intervention as possible, we proposed to instead focus our design concept on the relationship between the built and natural environments, taking advantage of the positive aspect of the site's easy access and visibility from the most highly traveled highway leading in and out of Seattle from the east. Rather than routing the arboretum entrance away from existing development and further encroaching on the forest, we repurposed the parking and other infrastructure of a defunct adjoining factory as a launching point at the edge of the forest from which to enter into an exploration of more appropriate examples of the interface between buildings and the natural environment. Wishing to demonstrate multiple strategies for buildings to relate sensitively to their sites, we designed all of the new structures, to be introduced into the forest itself, around a CNC-milled timber-frame system but deployed it in a different way for each interpretive center to create varying experiences: being underground with the roots of the trees, on the forest floor to focus on this habitat, and raised high on stilts up into the forest canopy itself. Instead of having a sharp contrast between the building and the surrounding landscape, the design concept is to provide a stepped progression of experience that is also a model and metaphor for the relationship between human intervention and the natural landscape.

The building is a completely foreign object within the natural landscape, but it is rendered in forms or materials taken from that landscape, successfully blending into a harmonious whole. This is juxtaposed with secondary "built" objects—trees planted in unnatural, buildinglike formations, showing human intervention in nature from another perspective.

Below: Drawing studies exploring relationships between introduced structure and existing forest density
Bottom: Images of site showing new growth in recently logged areas adjacent to stands of mature timber

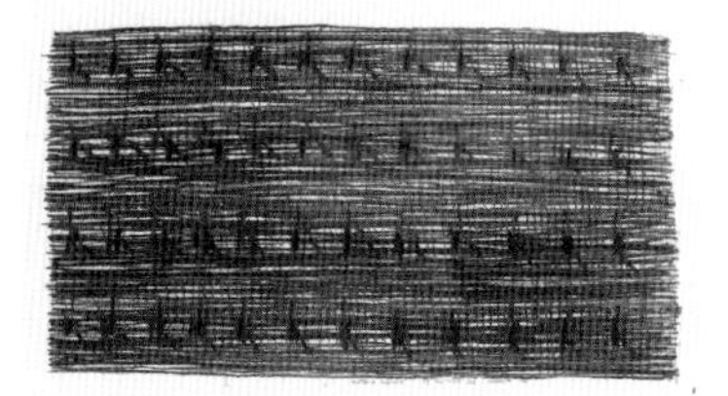

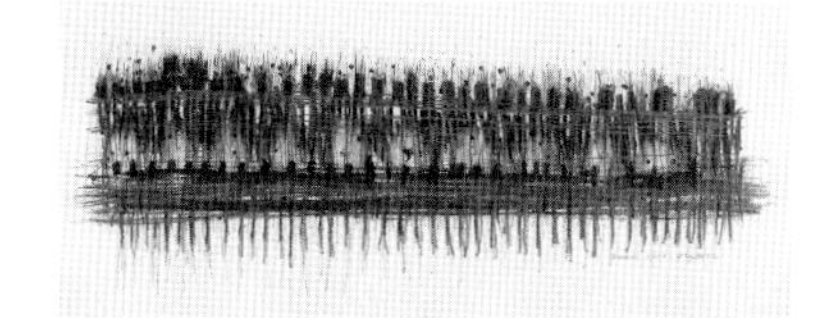

Site plan:
1. Visitors' Center
2. Salmon stream wetland remediation
3. Visitor Parking
4. Day Cabin visitor retreats
5. Skid road
6. Roots Walk
7. Forest Floor Pavilion
8. Canopy Walk
9. Moss Pit Amphitheater
10. Giants' Picnic
11. Giant Cedars Walk
12. Donkey Creek
13. Rock face
14. Interstate 90
15. Industrial park neighbors
16. Upland meadows
17. Secret location
18. Lowland meadows
19. Iron bridge
20. Nurse log bridge
21. Cedar suspension bridge
22. Dead Man bridge
23. Alder swamp

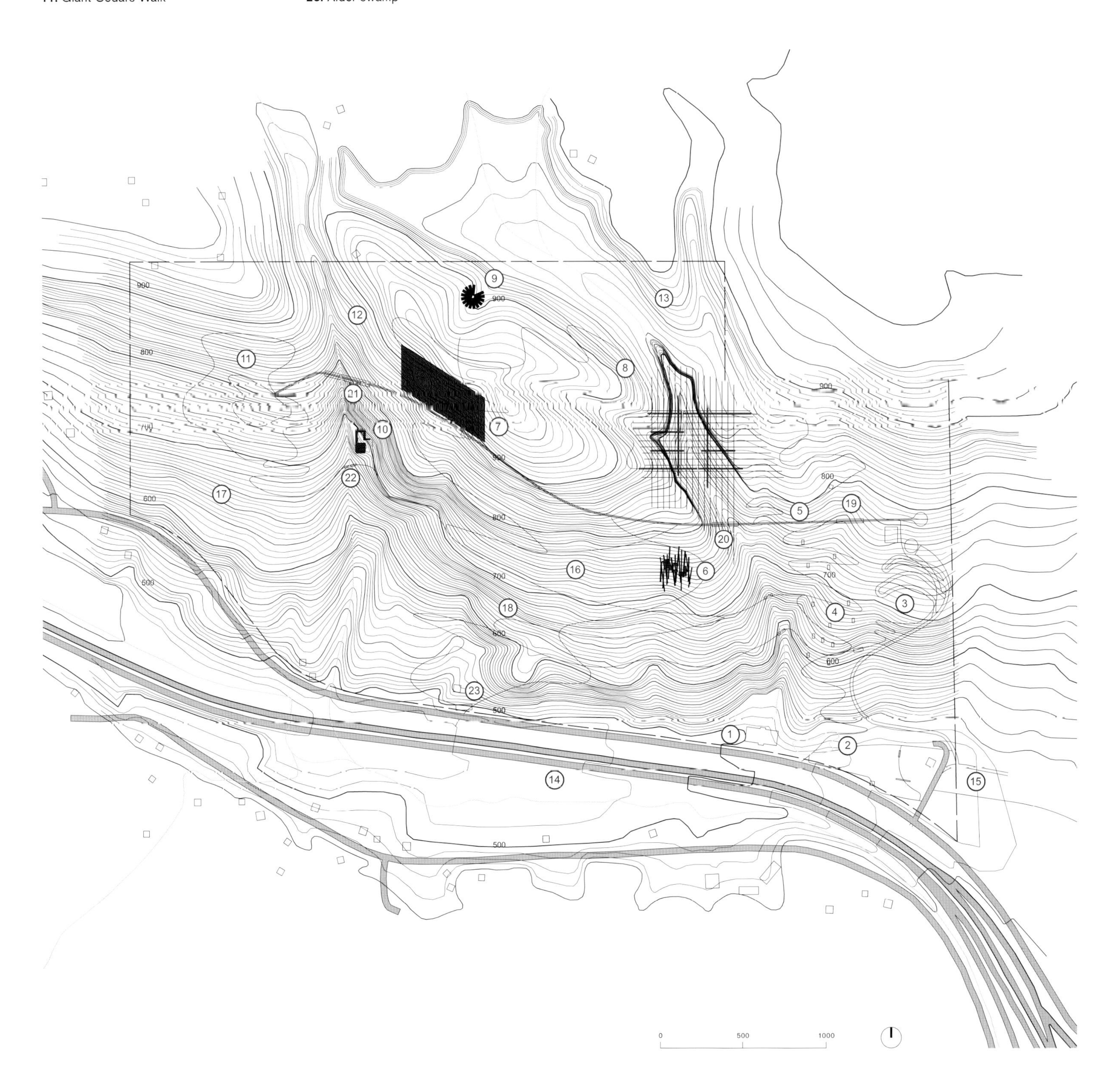

Arboretum of the Cascades

Pavilion of the Giants

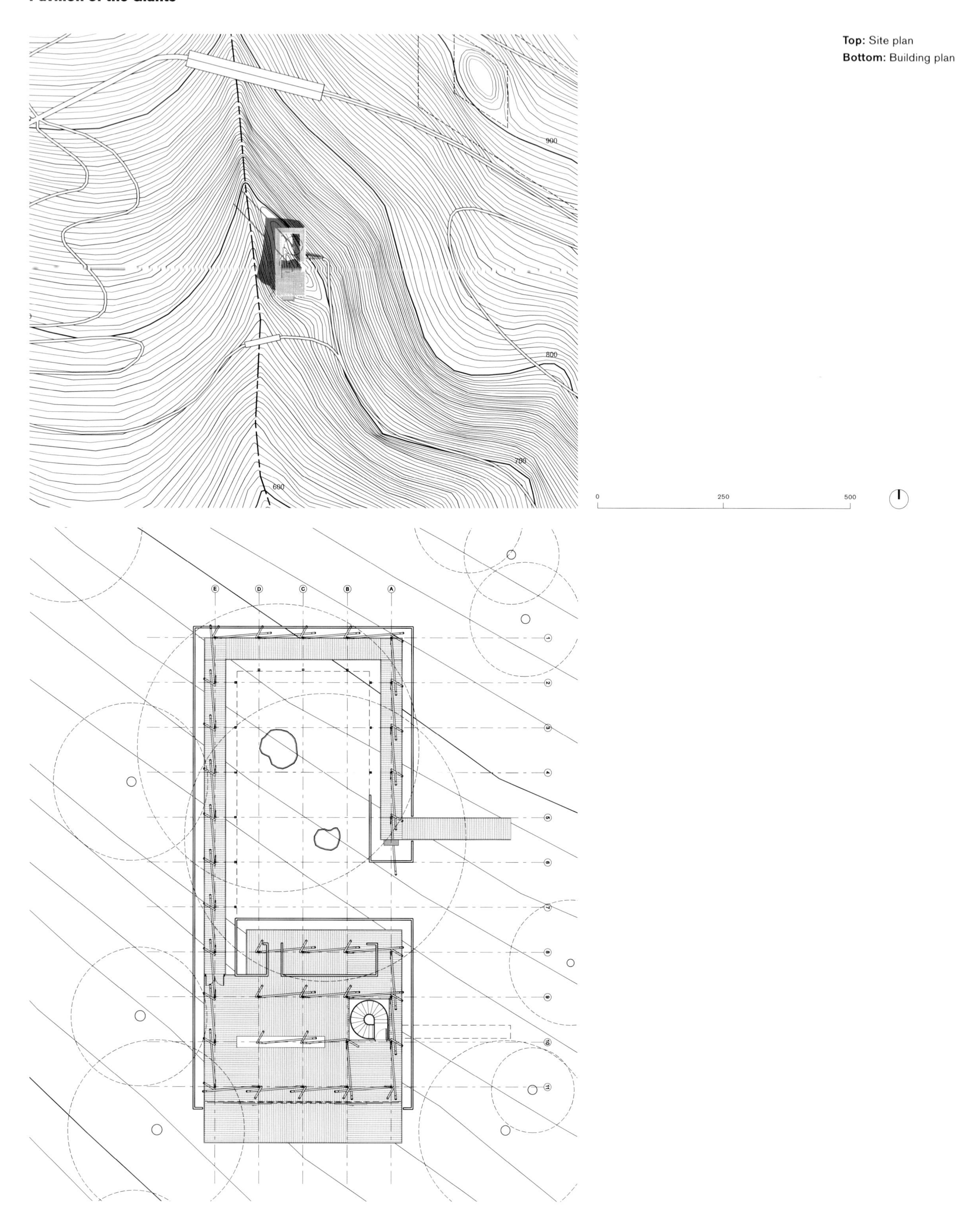

Top: Site plan
Bottom: Building plan

CNC Timber Framing

Rendered view of pavilion looking north from trail below

Arboretum of the Cascades

Pavilion of the Giants

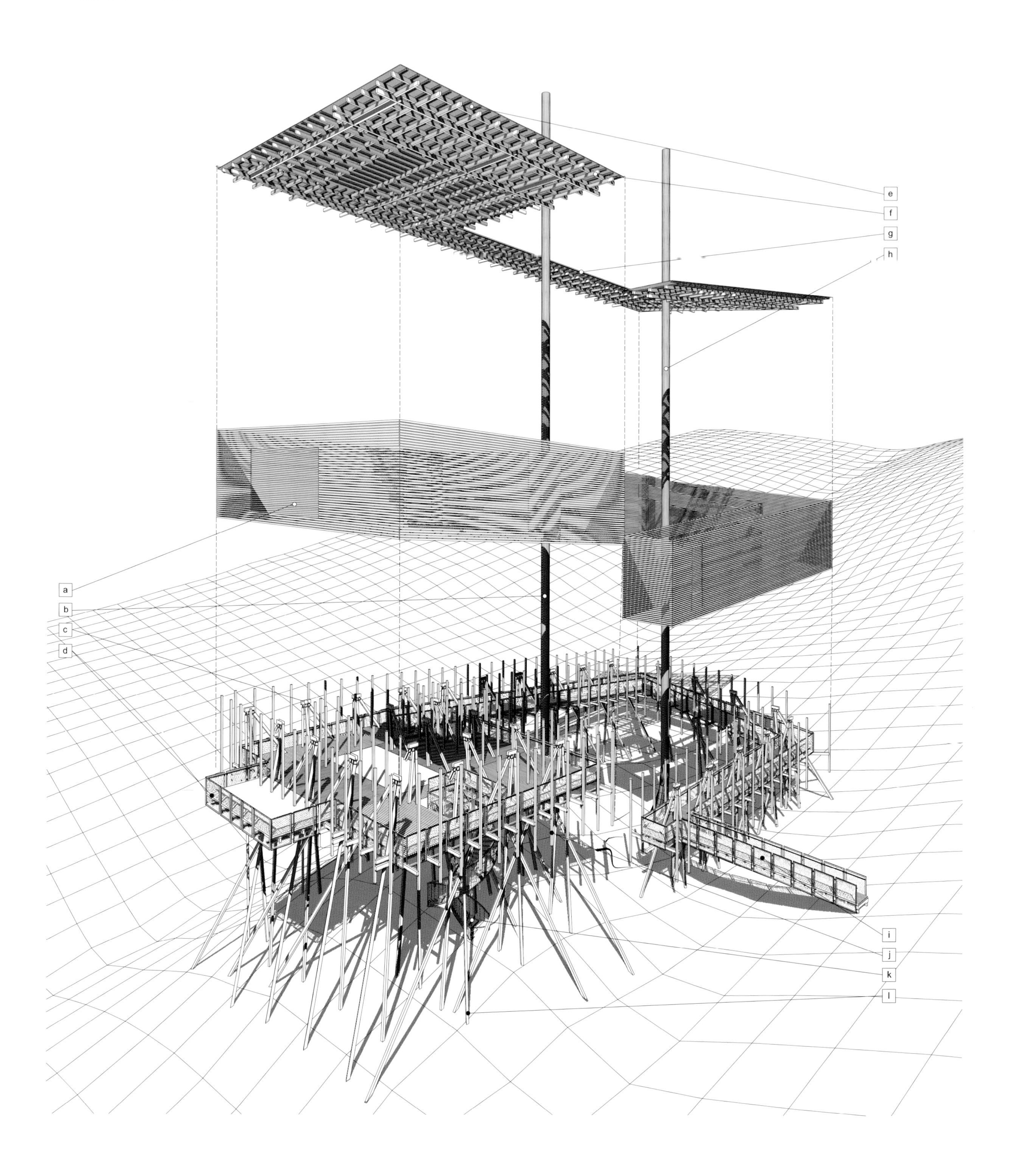

CNC Timber Framing

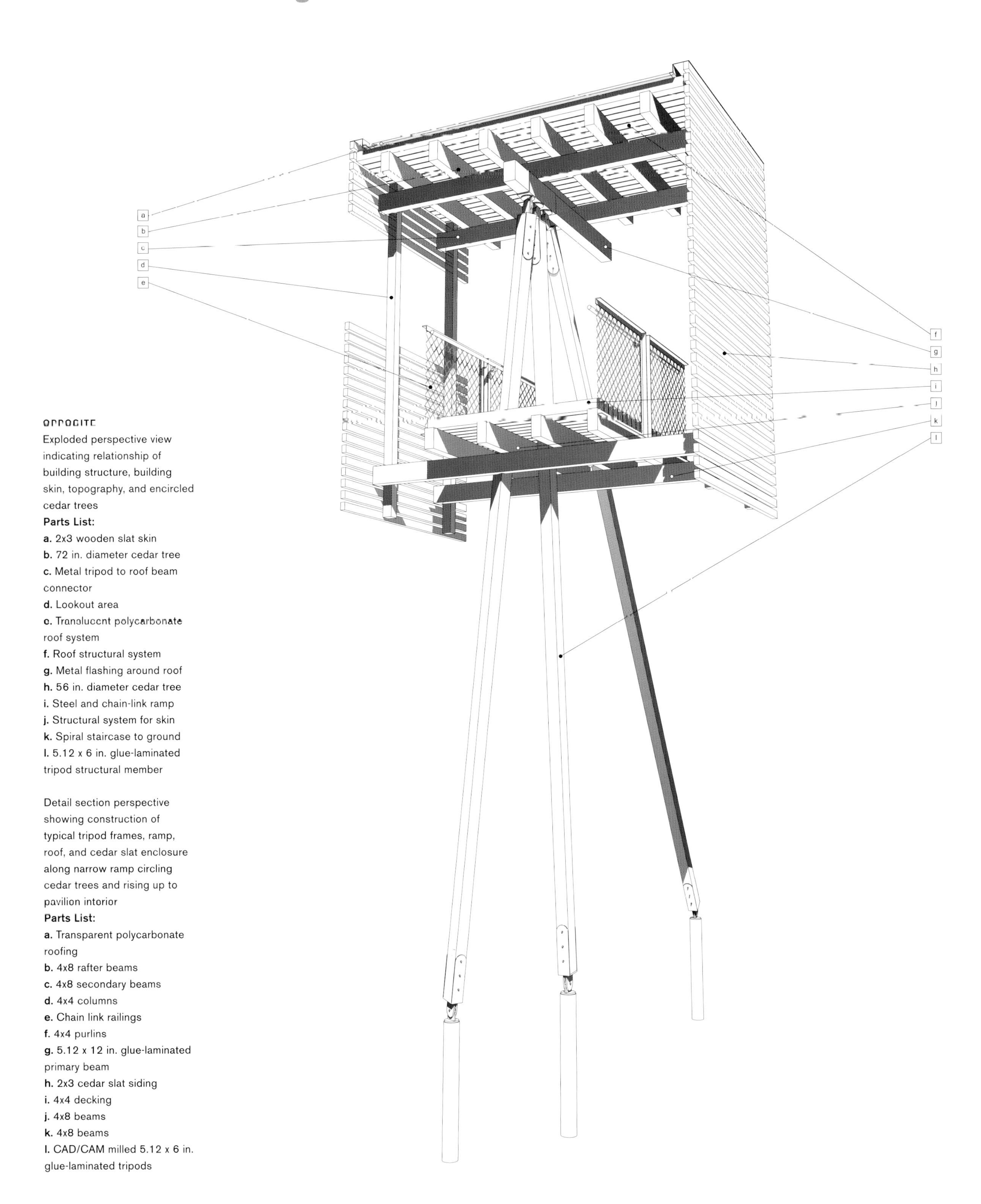

OPPOSITE

Exploded perspective view indicating relationship of building structure, building skin, topography, and encircled cedar trees

Parts List:

a. 2x3 wooden slat skin
b. 72 in. diameter cedar tree
c. Metal tripod to roof beam connector
d. Lookout area
e. Translucent polycarbonate roof system
f. Roof structural system
g. Metal flashing around roof
h. 56 in. diameter cedar tree
i. Steel and chain-link ramp
j. Structural system for skin
k. Spiral staircase to ground
l. 5.12 x 6 in. glue-laminated tripod structural member

Detail section perspective showing construction of typical tripod frames, ramp, roof, and cedar slat enclosure along narrow ramp circling cedar trees and rising up to pavilion interior

Parts List:

a. Transparent polycarbonate roofing
b. 4x8 rafter beams
c. 4x8 secondary beams
d. 4x4 columns
e. Chain link railings
f. 4x4 purlins
g. 5.12 x 12 in. glue-laminated primary beam
h. 2x3 cedar slat siding
i. 4x4 decking
j. 4x8 beams
k. 4x8 beams
l. CAD/CAM milled 5.12 x 6 in. glue-laminated tripods

Arboretum of the Cascades

Roots

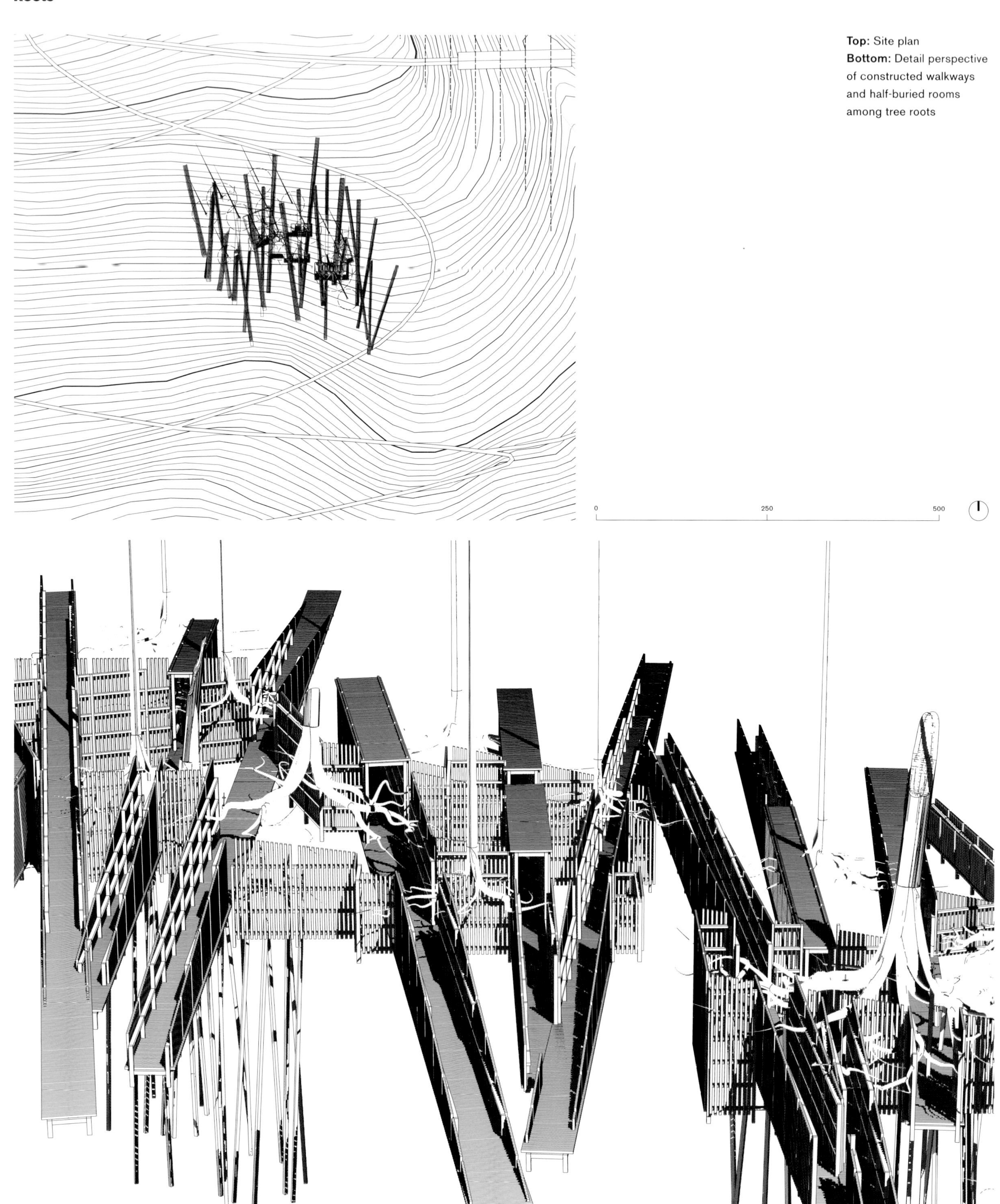

Top: Site plan
Bottom: Detail perspective of constructed walkways and half-buried rooms among tree roots

Perspective overview of excavated Roots exploration structure

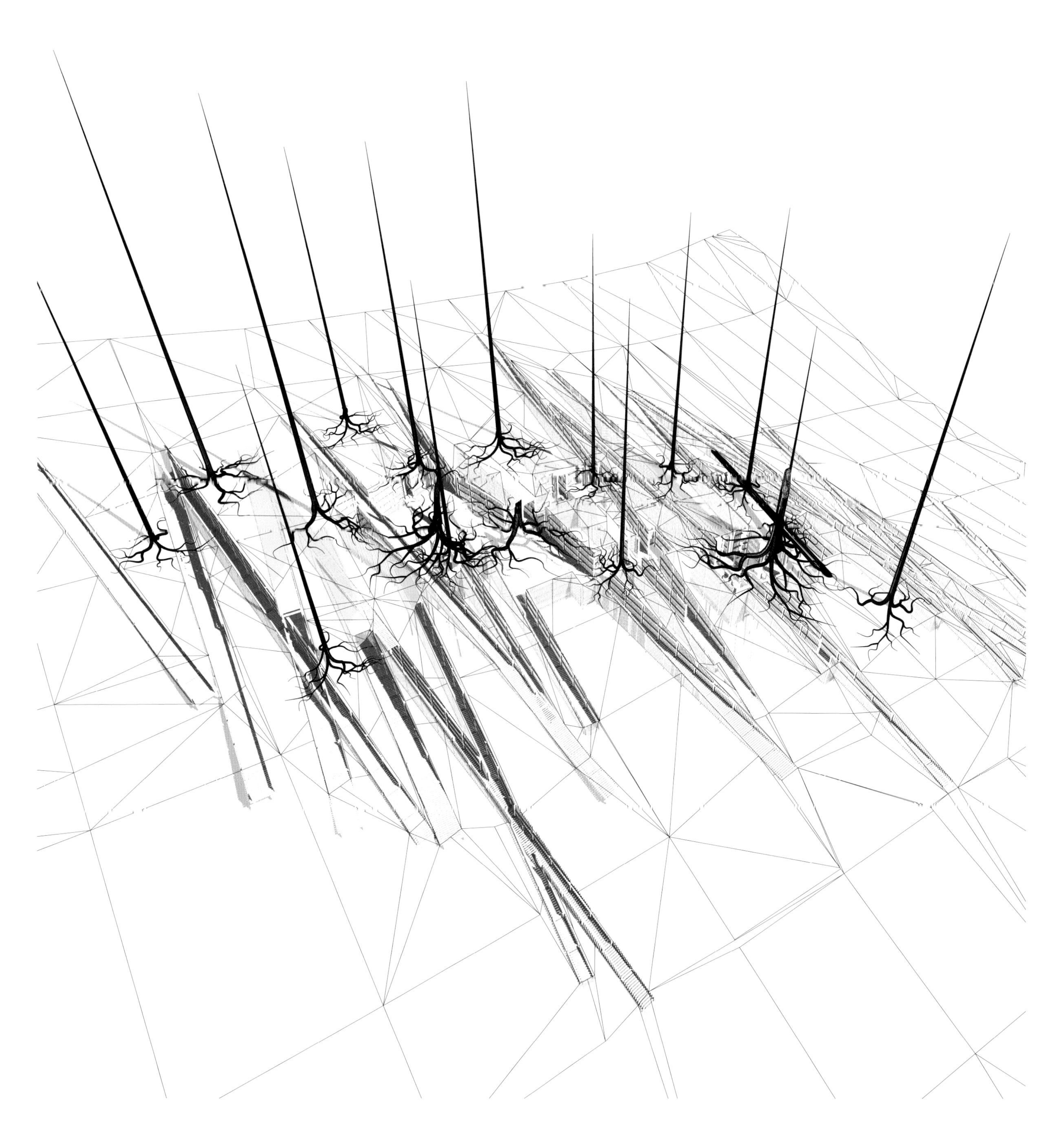

Arboretum of the Cascades

Roots

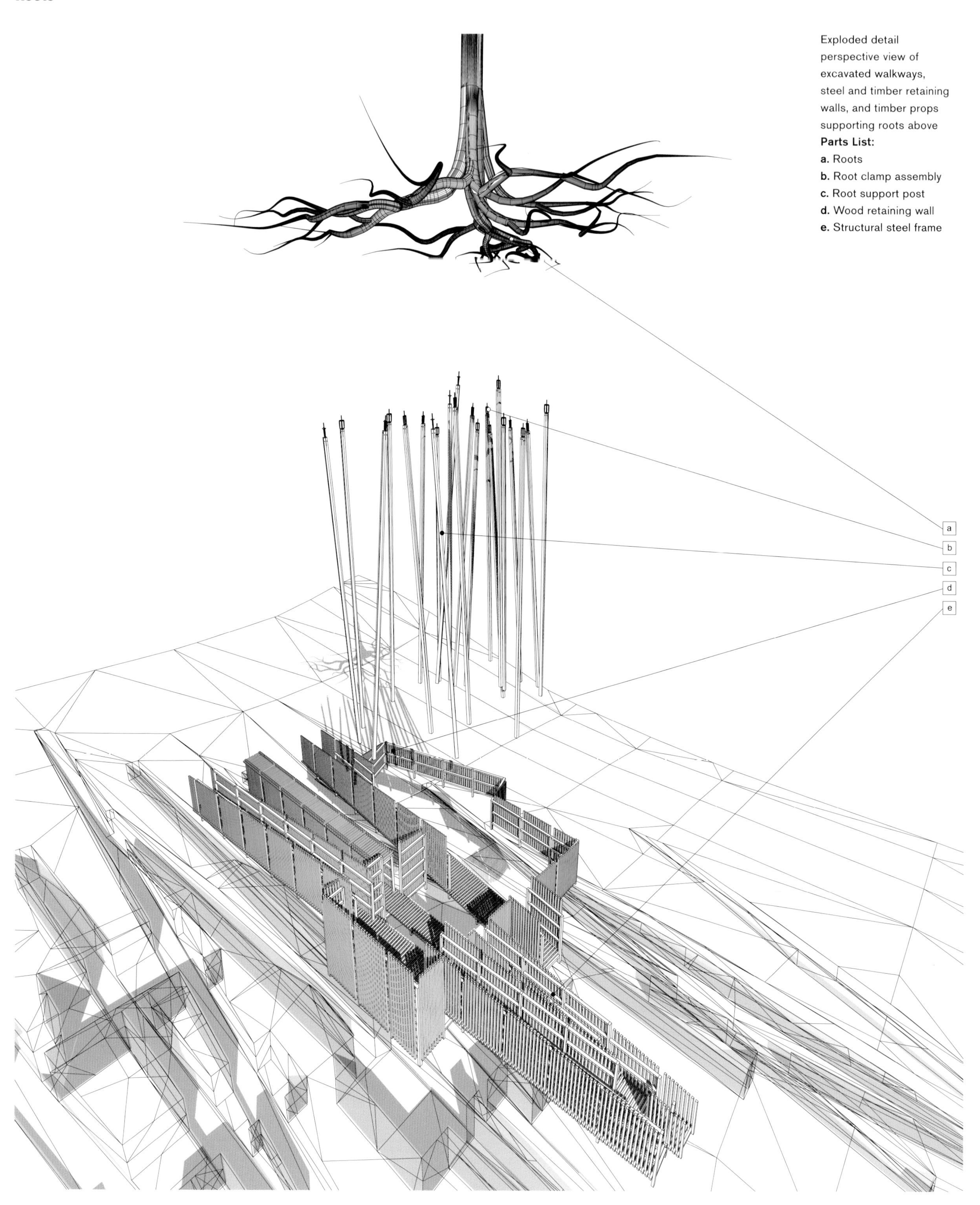

Exploded detail perspective view of excavated walkways, steel and timber retaining walls, and timber props supporting roots above

Parts List:

a. Roots
b. Root clamp assembly
c. Root support post
d. Wood retaining wall
e. Structural steel frame

Top: Detail perspective describing the attachment of timber support props to tree roots

Parts List:

a. Tree root

b. 5.12 x 6 in. glue-laminated timber support props with root clamp hardware at top

Bottom:

Parts List:

a. 1.5 in. steel pipe staple to secure root

b. 1 in. steel rod driven into earth as underground locator

c. Tree root

d. 1 in. steel pipe saddle

e. 1.5 in. steel pipe to accept locator rod

f. .5 in. steel plate attachment

g. 5.12 x 6 in. glue-laminated timber support prop inserted as earth is removed

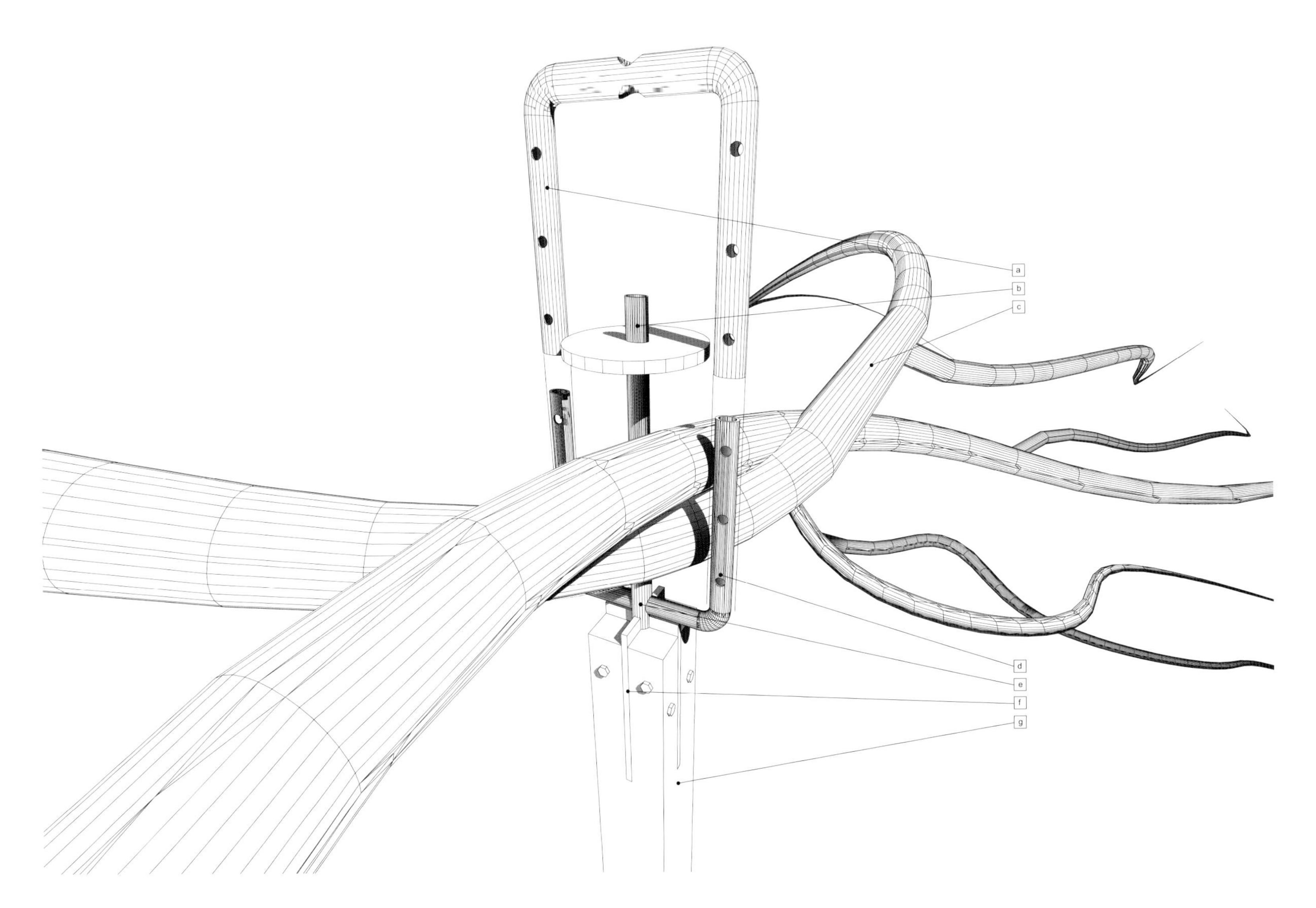

Arboretum of the Cascades

Forest Canopy

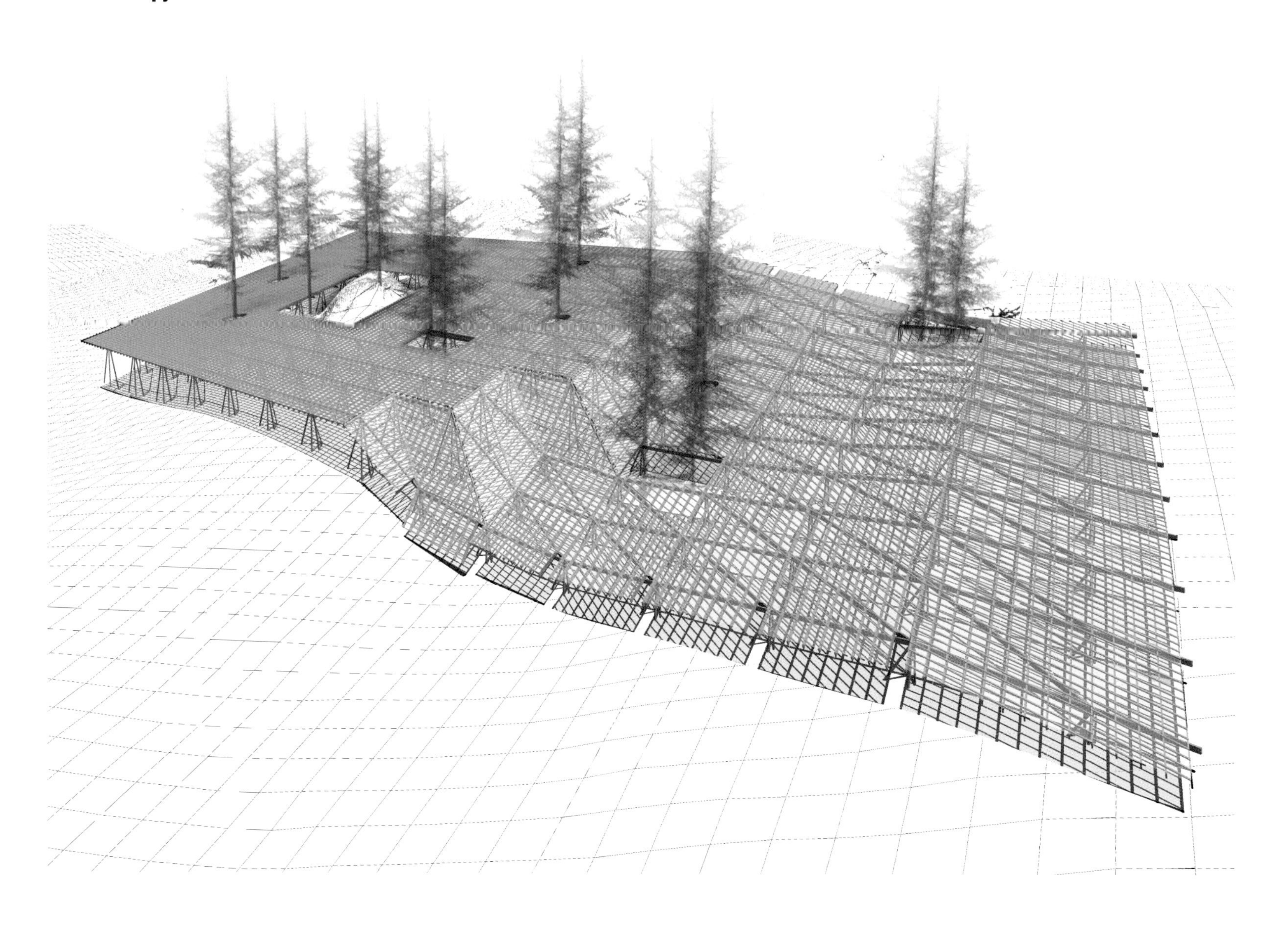

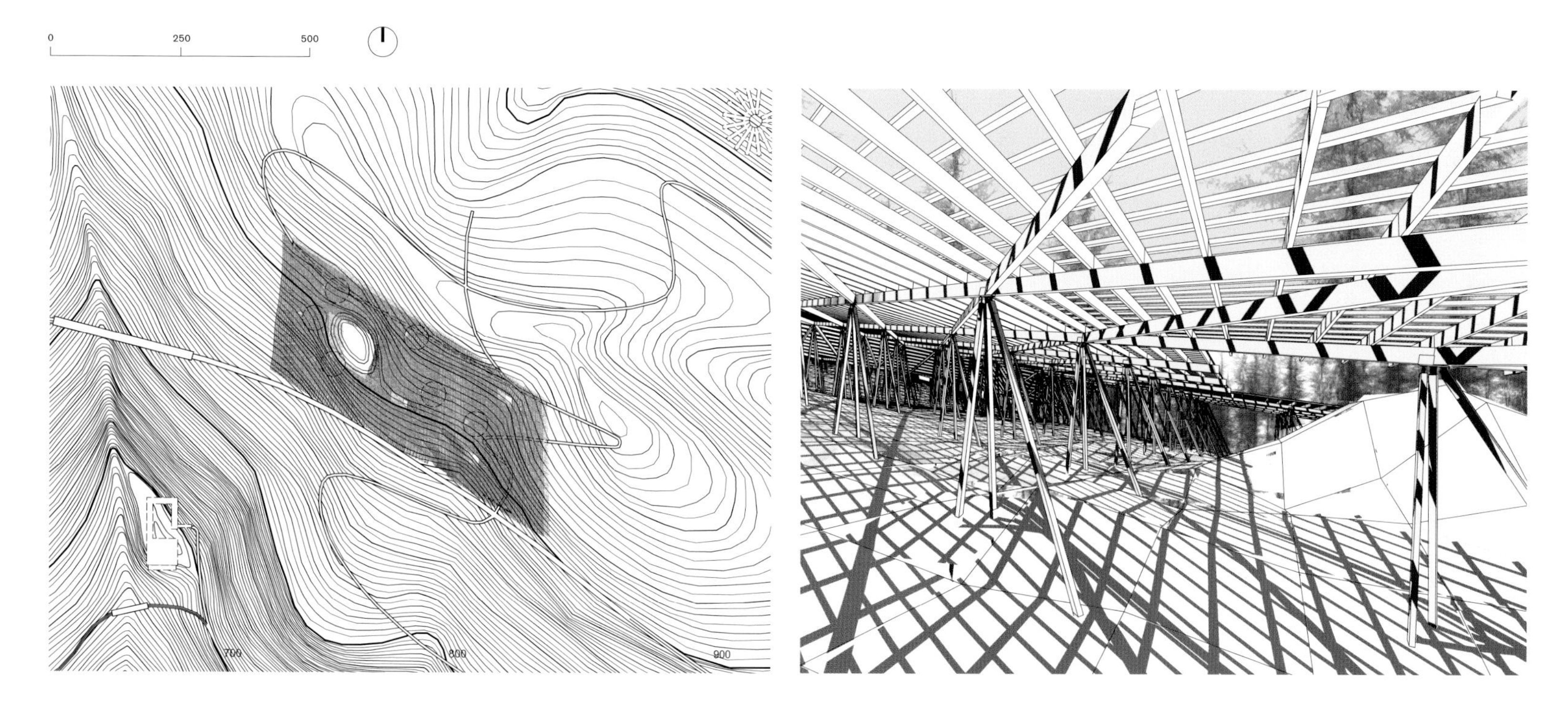

CNC Timber Framing

OPPOSITE
Top: Overview perspective; left side of structure remains in plane with ground rolling beneath. Right side of structure undulates directly with topography.
Bottom left: Site plan, Forest Floor interpretive structure
Bottom right: Interior perspective

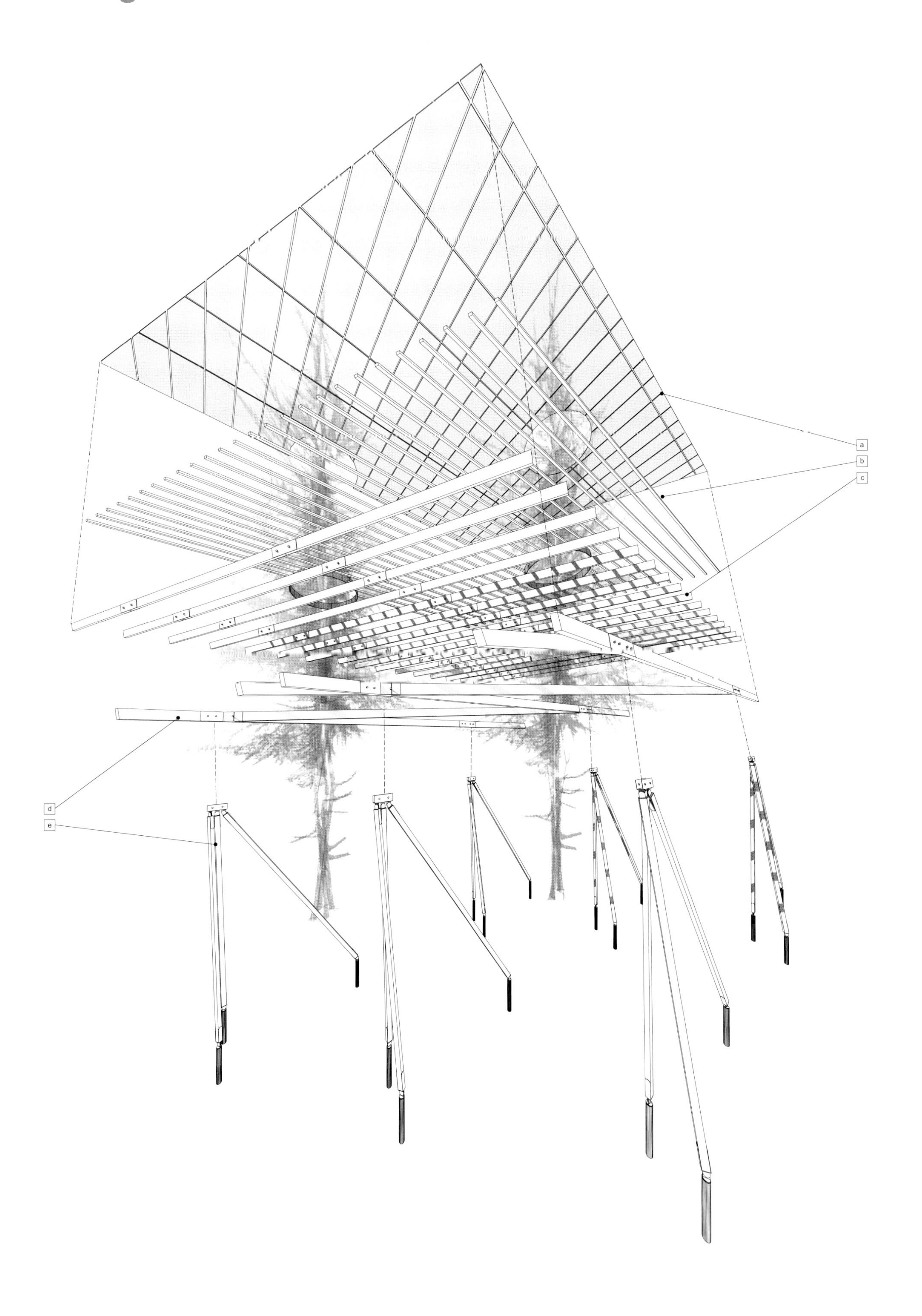

Exploded detail perspective view
Parts List:
a. Clear polycarbonate roof panels
b. 4x4 cedar purlins @ 24 in. o.c.
c. Secondary beam structure, 3.12 x 9.5 in. glue-laminated timbers @ 48 in. o.c.
d. Primary beam structure, 5.12 x 13.5 in. glue-laminated timbers
e. Timber tripod structures adjusted in length to meet rolling topography. 23 ft. o.c. north to south, 44 ft. o.c. east to west. 5.12 x 6 in. glue-laminated verticals with hot-dip galvanized steel saddle attachments

Arboretum of the Cascades

Forest Canopy

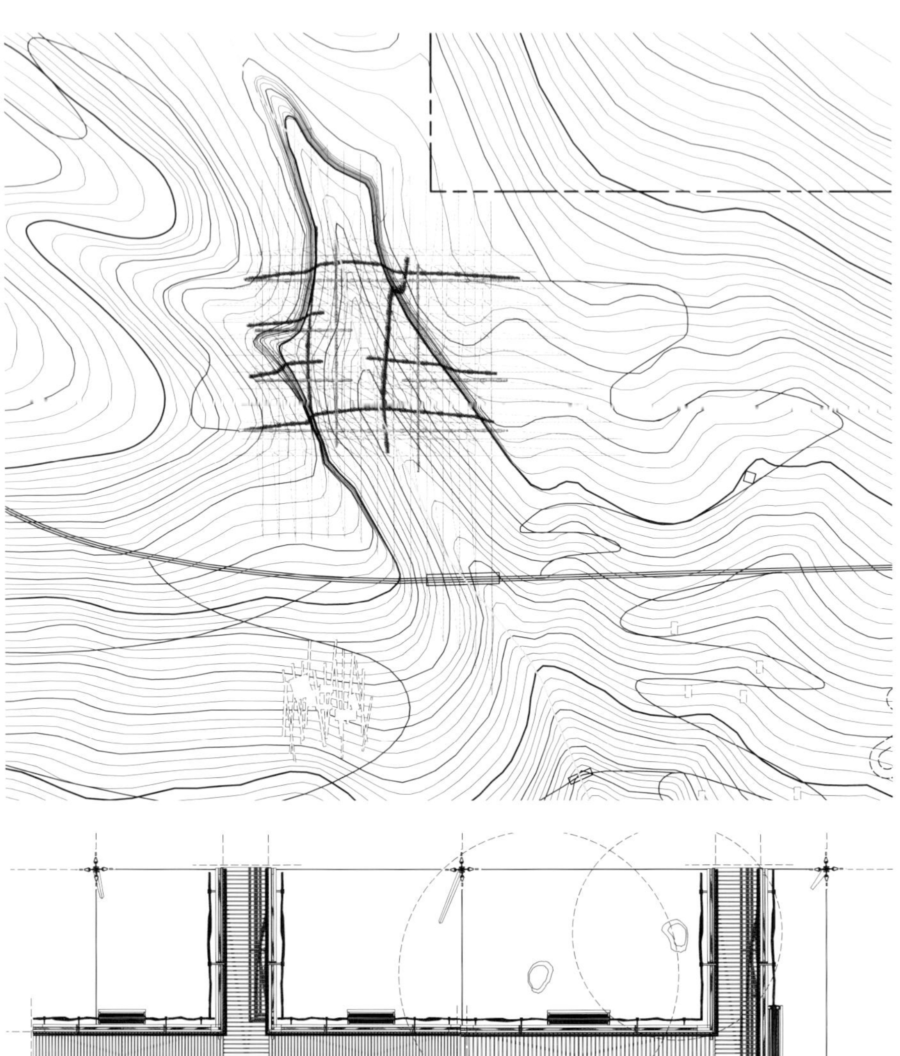

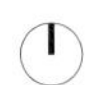

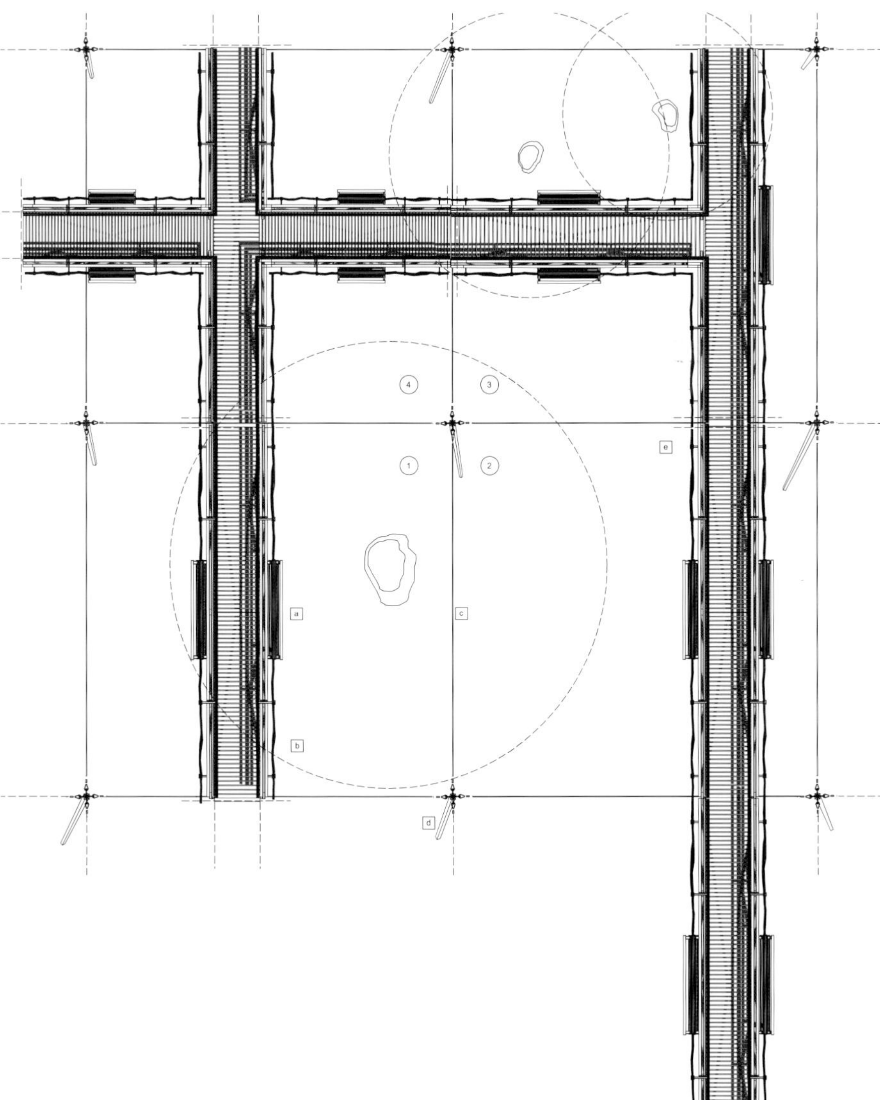

Top: Site plan
Bottom: Detail plan of forest canopy structure lofting hiking trails up into the treetops on steel suspension cables
Parts List:
1. Typical prefabricated bridge
2. Prefabricated cantilevered lookout bridge
3. Prefabricated 3-way connection bridge
4. Prefabricated 4-way connection bridge
a. Hot water manifold for micro-climate control
b. Hydroponic water and nutrient hose
c. 1 in. steel cable
d. Glue-laminated timber column
e. Bridge coupling hardware and cable attachment

OPPOSITE
Top: South elevation view of Forest Canopy trail indicating timber props and cable suspension crossing canyon below
Bottom: Aerial view of Forest Canopy interpretive trail system looking from northeast down slope toward southwest

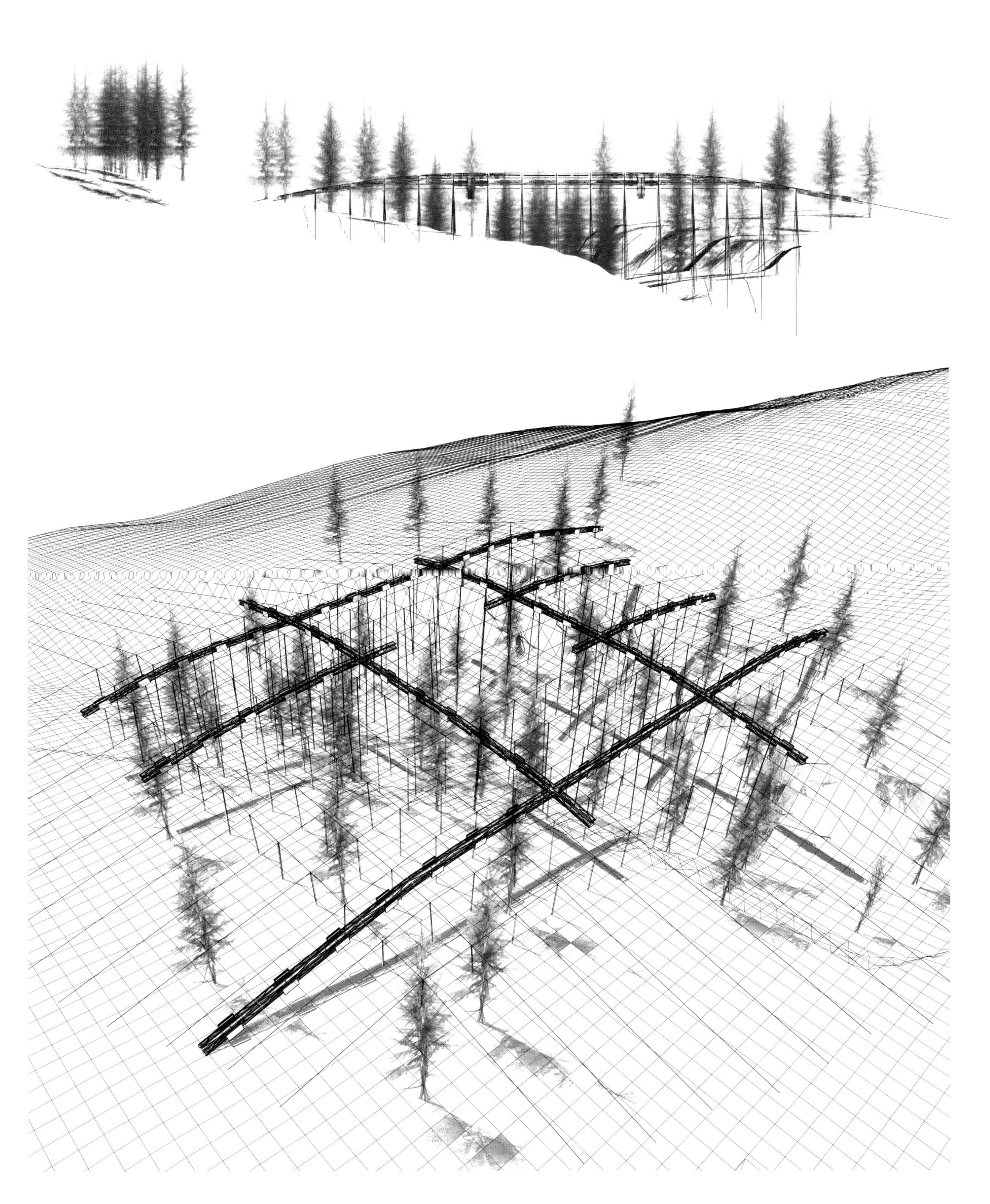

Arboretum of the Cascades

Forest Canopy

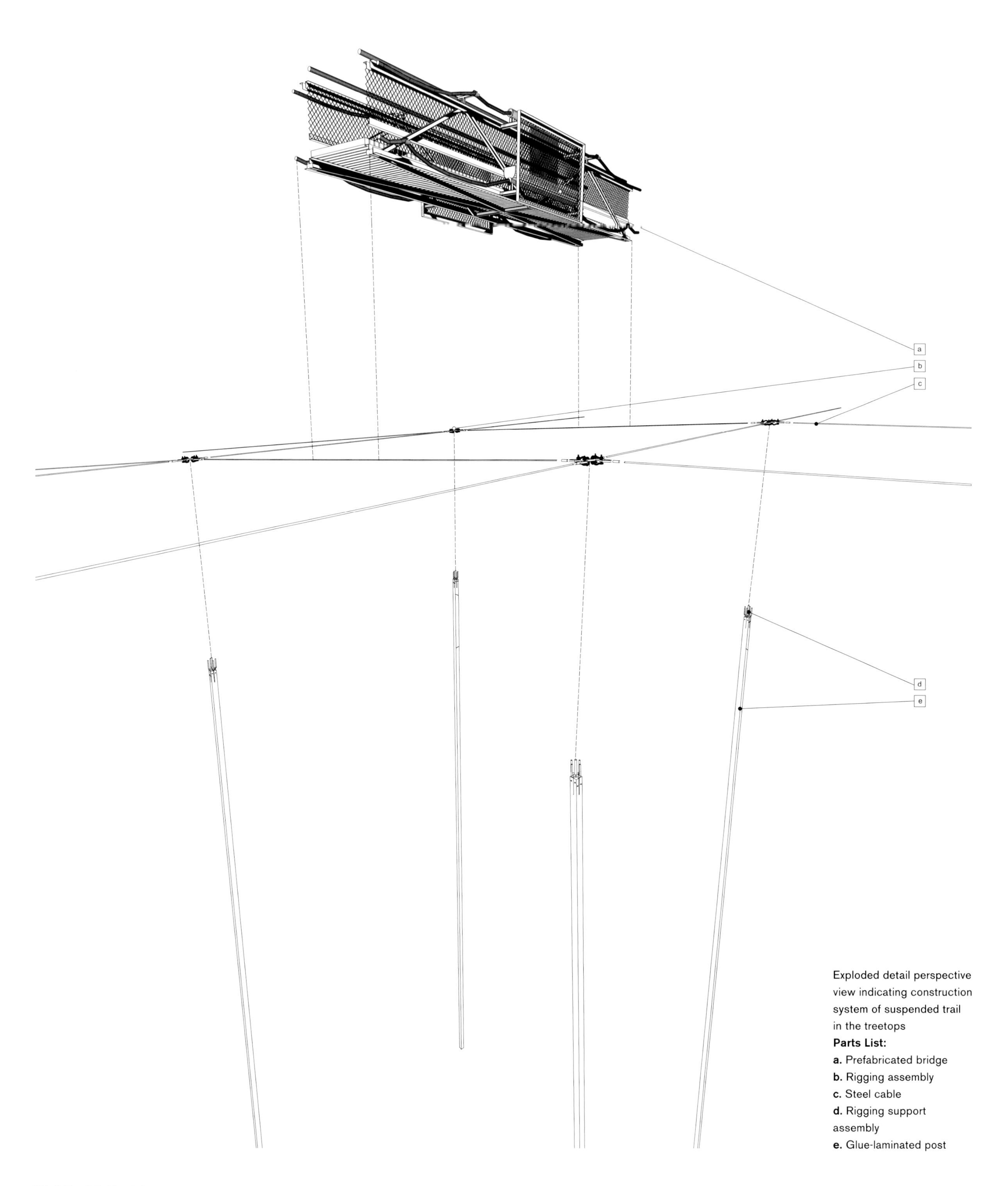

Exploded detail perspective view indicating construction system of suspended trail in the treetops

Parts List:

a. Prefabricated bridge
b. Rigging assembly
c. Steel cable
d. Rigging support assembly
e. Glue-laminated post

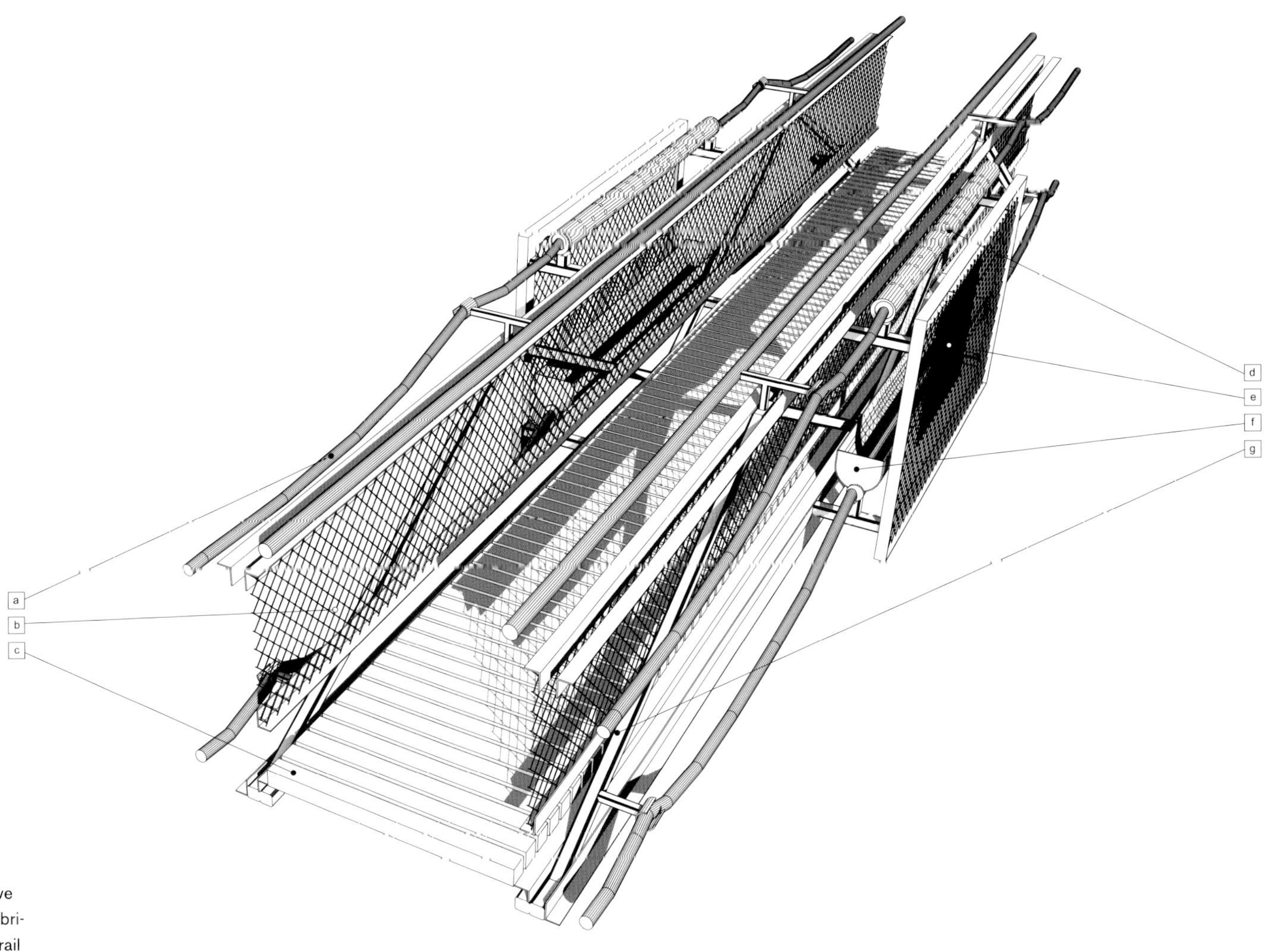

Top: Detail perspective view of modular prefabricated forest canopy trail structure, indicating structural system and environmental control systems, growing chambers and trellis plates for interpretive displays, and propagation of canopy epiphytes and wildlife

Parts List:

a. Hydronic water and nutrient hose

b. Chain-link aluminum fence

c. 4x4 cedar decking

d. Hot water manifold for micro-climate control

e. Aluminum growing trellis

f. Peat moss soil medium

g. Welded aluminum truss

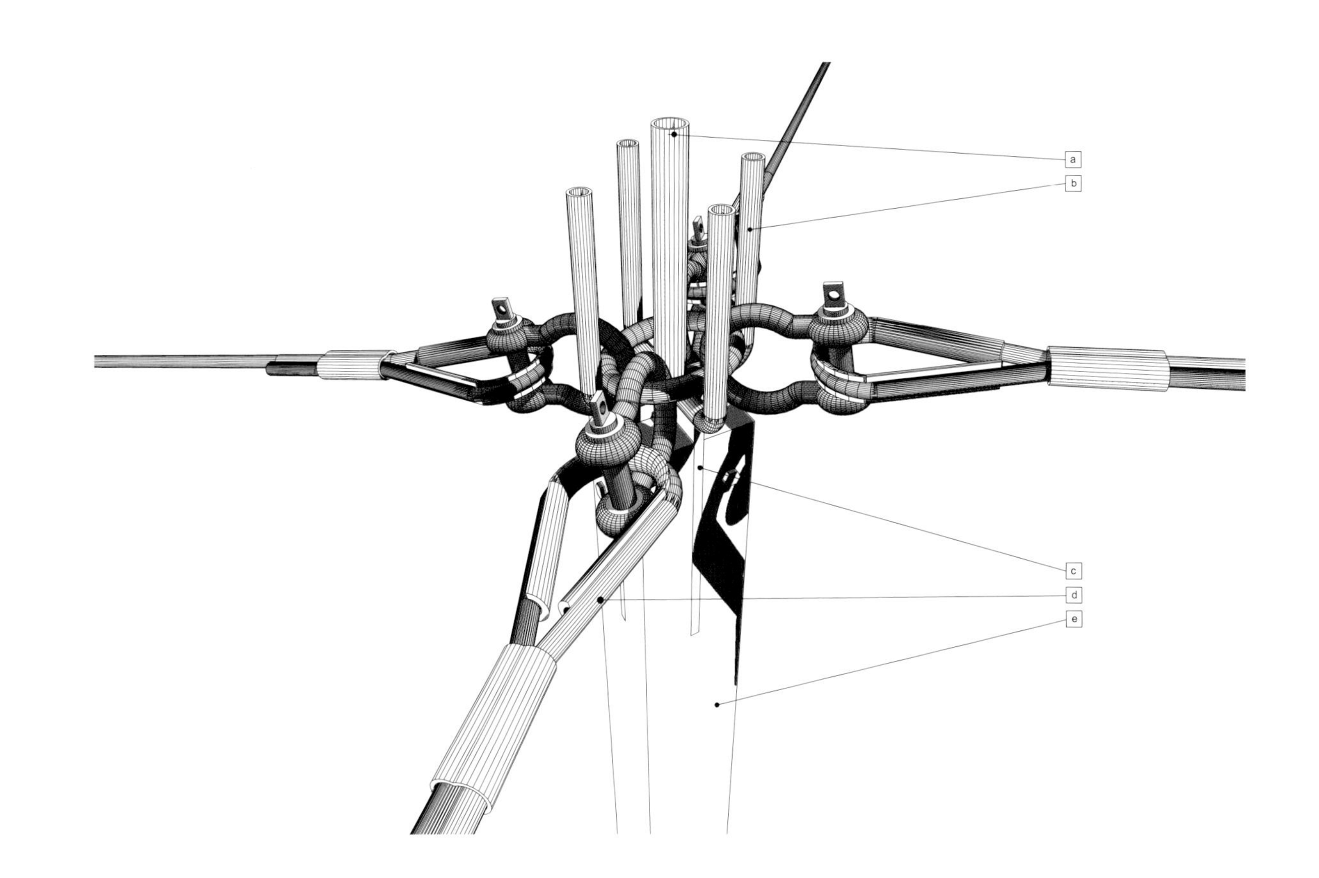

Bottom: Detail perspective view of prefabricated suspension cable grid and timber support columns

Parts List:

a. 2 in. steel pipe

b. 1.25 in. steel pipe saddle

c. .5 in. steel plate

d. 1 in. diameter stainless steel cable fittings

e. 7.12 x 7.5 in. glue-laminated timber column

15

Chapter 3

Concrete Systems

Building in concrete has a very long history, but considering concrete as a viable prefabricated material relies on several relatively recent innovations. Concrete block, or Concrete Masonry Units (CMU) could be seen as the earliest application of offsite fabrication for this material. The system relies on a small number of unique shapes and forms that can be combined in endless ways to create endless variety of buildings. This is both a blessing and a curse in terms of its success as a prefabrication system, and makes it worth some attention as a model for understanding offsite fabrication as a whole. The most common CMU in North America has the nominal dimensions of 8 x 8 x 16 inches, and is a standard, non-proprietary design made by hundreds of manufacturers in many places. The same reliance on standardized, non-proprietary masonry units can be found throughout much of the world, although the standards might apply to much smaller regional areas. The adoption of the metric system as the de facto international standard of measurement has had the greatest effect on materials that are shipped across borders and over long distances, while the low technology requirements for producing and building with masonry units, and the relatively high cost of transporting them, has allowed a greater regional variation than, for instance, in steel products. The modularity of the design and interchangeability between manufacturers makes it easy to design for and build with. The small scale of the modular units makes them relatively adaptable to a wide variety of applications but also contributes to higher labor costs for their assembly, since so many units must be placed and mortared on site to make up a large wall.

When compared with the even smaller unit sizes of fired-clay bricks, which could be considered the world's most ancient engineered and manufactured building component, a CMU structure can be assembled much faster and provides significant structural advantages, which is why CMU has virtually replaced brick as a structural building material in North America and other industrialized parts of the world since its widespread introduction in the early 1900s. A similar material, used more commonly outside of North America, is the hollow clay tile, which provides many of the same advantages as CMU yet is made from indigenous clay and other materials more commonly available in other regions.

A somewhat more recent improvement on the standard CMU is what is known as Autoclaved Aerated Concrete (AAC), sometimes called "foamed" concrete. Aluminum powder is added to a mixture of cement, lime, water, and sand, causing the material to expand as it hardens in molds of any desired shape, which is then cured in a

1. CAD/CAM milled polystyrene foam insulation panels shaped to accomodate reinforcing steel in IHI composite wall panels.

2. Wall panel with welded steel frame, cross-brace rods, reinforcing mesh, and foam insulation ready for concrete placement.

pressurized steam chamber. It can be made using fly ash from coal-burning power plants, which can help solve the disposal problems associated with this byproduct. The process was discovered by Swedish scientists in 1914 but not commercialized for building components until mid-century, when it emerged throughout Europe as a logical extension of the masonry building tradition. Its use soon spread to Asia and the Middle East and, in the last twenty years, to Australia and South America. It was not until the mid-1990s, however, when European manufacturers opened the first factories in the United States, Canada, and Mexico, that AAC became more widely used in North America.

With a density of just 20% that of conventional concrete, AAC materials afford a number of advantages in construction, including superior thermal and sound insulation, the ability to be cut and drilled far more easily than conventional concrete, and the ease of transportation and assembly due to its lighter weight per unit of volume. It is possible to make much larger components off site that can be moved by hand or with light equipment, allowing for further reductions in assembly time and cost. The disadvantage of AAC's lower density is a significantly reduced compressive strength compared to standard concrete, which limits the use of AAC to low-rise structural applications or non-load-bearing uses in combination with other structural systems.

Concrete that is cast on site may seem an unlikely fit in a discussion of prefabricated building systems, but when the focus shifts to the formwork that is required to give shape to the liquid concrete, there are more obvious connections. The most common traditional forming systems use lumber and plywood or lumber form-board panels to shape the concrete that are then removed after a long day or more, when the concrete is sufficiently hardened to support itself. These forming systems are relatively inexpensive for smaller projects, but tend to be only partly reusable, with considerable waste after each use; transporting, building, then disassembling and moving these forms to the next job involves heavy, labor-intensive work. A new generation of lightweight removable forming systems has improved the process, substituting steel, aluminum, or plastic forms for the wood panels and temporary structures that hold them in place. These can yield more complex molds to make any shape desired and also allow working with larger-sized modules, opening new design opportunities and further decreasing labor and production costs.

The use of large blocks of foam polystyrene for milling into complex formwork provides a number of possibilities for cost-effective casting of complicated concrete shapes. Particularly when the formwork is carved into shape with multi-axis CAD/CAM milling machines, as has become standard practice now in many industries—for forming the interior core of giant fiberglass Mickey Mouse statues at Disneyland, for example—this type of directly computer-controlled form making has begun to make attainable complex shapes that were previously unaffordable in typical building construction.

One particularly interesting direction in concrete construction makes use of leave-in formwork, typically made from some form of lightweight foamed polystyrene, which may or may not be mixed with cement. Usually called Insulated Concrete Formwork (ICF), these foam building components first provide the support for the curing concrete, then remain in place to provide insulating properties and serve as a substrate for other finish surfaces on the interior and exterior of the wall. As opposed

3. Computer-controlled nozzles pass over steel framed panel to deliver precisely measured quantities of high-strength concrete.

4. Completed concrete wall panel after finishing with computer-controlled power trowels. Computer calibration of concrete placement affords minimal material waste.

to CMU, these ICF components are proprietary systems and come in many forms and configurations, usually composed as interlocking blocks, planks, or panels, in increasing order of module size. Although these component-based systems are not usually interchangeable or easily combinable from one manufacturer to the next, they all follow roughly the same principles. There is some debate over the environmental friendliness of using petroleum-based plastics as formwork for concrete, but since many of the ICF manufacturers use a high proportion of post-consumer recycled polystyrene in their products, turning the waste-disposal problem into a highly efficient thermal insulation benefit, it can well be argued that the net benefits make ICF construction a very good solution.

Pre-cast concrete has primarily been used for large buildings and infrastructure projects such as highway bridges, but is becoming increasingly available in a variety of new forms applicable to smaller scale construction and residential projects. Among the most important areas of technical research and new applications in pre-cast systems are those that involve fiber-reinforced concrete. This system replaces some or all of the typically cumbersome and labor intensive steel reinforcing bars of standard concrete construction with short strands of fiberglass, carbon, or steel fiber suspended in the concrete mixture itself. While standard concrete mixes supply only compressive strength and rely on the reinforcing bar cages for tensile strength, fiber reinforced concrete has both compressive and tensile strength within the matrix of the concrete itself. This advance in concrete technology offers the potential for much lighter, smaller section panels that can accommodate far more complex shapes and curves and will allow for more mobile and less costly offsite concrete construction. Other systems combine concrete with steel framing or other materials to create structural hybrid solutions, a highly automated approach to very high-quality, offsite fabricated, lightweight structural systems that we are currently using for a series of projects in Los Angeles and a waterfront home on a challenging cliff-side site near Seattle. All of these approaches take advantage of the many positive properties of concrete, which include low cost and ready availability of raw materials, great flexibility of configuration due to its liquid-to-solid conversion, fire resistance, thermal mass, sound insulating capabilities, and low maintenance requirements.

Concrete Tower Houses 1 and 2

San Francisco, California / Seattle, Washington / 2005

We have begun a new series of prefabrication projects using a highly innovative composite panel structure of rectangular hollow steel tube frames filled with polystyrene foam insulation and covered on both sides with a relatively thin slab of fiber-reinforced concrete. These composite panels are the patented products of the IHI Corporation in Vancouver, British Columbia, Canada, which has built a sophisticated factory to robotically weld the frames and place and finish the concrete slab surfaces. To this point the major product of this firm has been industrial structures, liquid storage tanks, and floating roadbeds for Canada's far north oil fields. For building construction, these concrete panels are able to serve as interior and exterior finish surfaces; have excellent thermal, sound, and fire insulation properties; and are designed for erection by steel workers with industry-standard skills. The panels have application even as structural systems for high-rise buildings and long span structures, but they have been particularly interesting for us in housing applications on steep hillside sites with complex foundation and retaining wall requirements. With their experience in producing underground tanks and industrial buildings in extreme environmental conditions, we are working with IHI to produce strong, watertight, underground foundation walls that can be prefabricated and immediately inserted into place as excavation proceeds, eliminating the need for temporary shoring; slow and dangerous site work on forms, rebar, and concrete placement; and cumbersome waterproofing applications, all allowing for immediate back fill and restabilization of the site without the typical weeks or months of risk and weather worry.

The tower houses shown here represent two quite common urban site conditions on infill lots in west coast cities, where dense neighborhoods have a good number of open lots that have long been considered too difficult to build on. With the disturbance of sometimes poorly constructed buildings on adjacent lots, these sites are often unstable, and a properly constructed building will actually increase the stability of the adjacent homes and hillsides, as well as beneficially increasing the density of urban neighborhoods where housing is in high demand. These types of sites are extremely difficult to build on, however, both as a result of the hillside complexities and because of the disruptions to a closely packed community if the construction period stretches over many months. These conditions require innovative design for the construction process itself, and prefabricated concrete panels that serve as foundations and overall structure are ideally suited to these types of sites.

The Seattle house is sited at the base of a potentially unstable waterfront bluff, requiring an uphill wall that will stabilize the slope and an access bridge from the top of the house that will span the slope above without adding any additional weight to the hillside. The solution is a garage and entry well uphill from the house, with an enclosed bridge that floats across the environmentally sensitive crest of the bluff and lands on top of the tower. The San Francisco house works almost in reverse, with the street access from the bottom of the slope. In this case, local height restrictions require that the house actually tunnel back into the slope, disappearing underground until emerging again at the top of the hill. In both cases, the access point is at the distant end from the most desirable public spaces of the home. In Seattle, the entry sequence must flow through the house down onto the beach level. In San Francisco, the entry sequence must climb up to the primary spaces that have spectacular views from the top of the hill. In both cases, there must be a hollow vertical core serving as the functional and spatial heart of the home. The particular strength and above- and below-ground continuity of the prefabricated concrete panels make it possible to safely and affordably build spatially rich, vertically integrated houses on these difficult hillside sites.

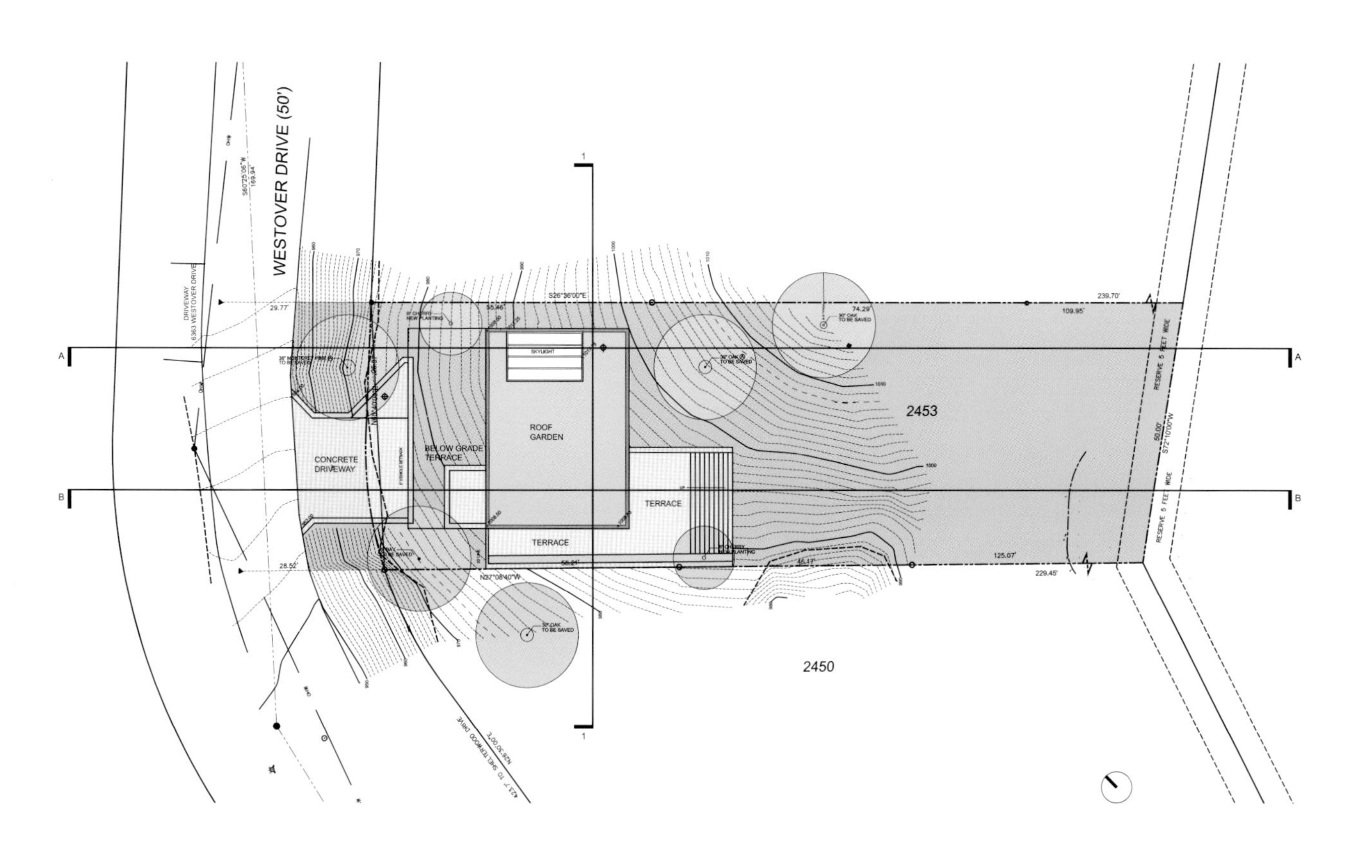

Site plan, Tower House 1

OPPOSITE
Early study models indicating relationship between precast panel structure and hillside excavation strategy

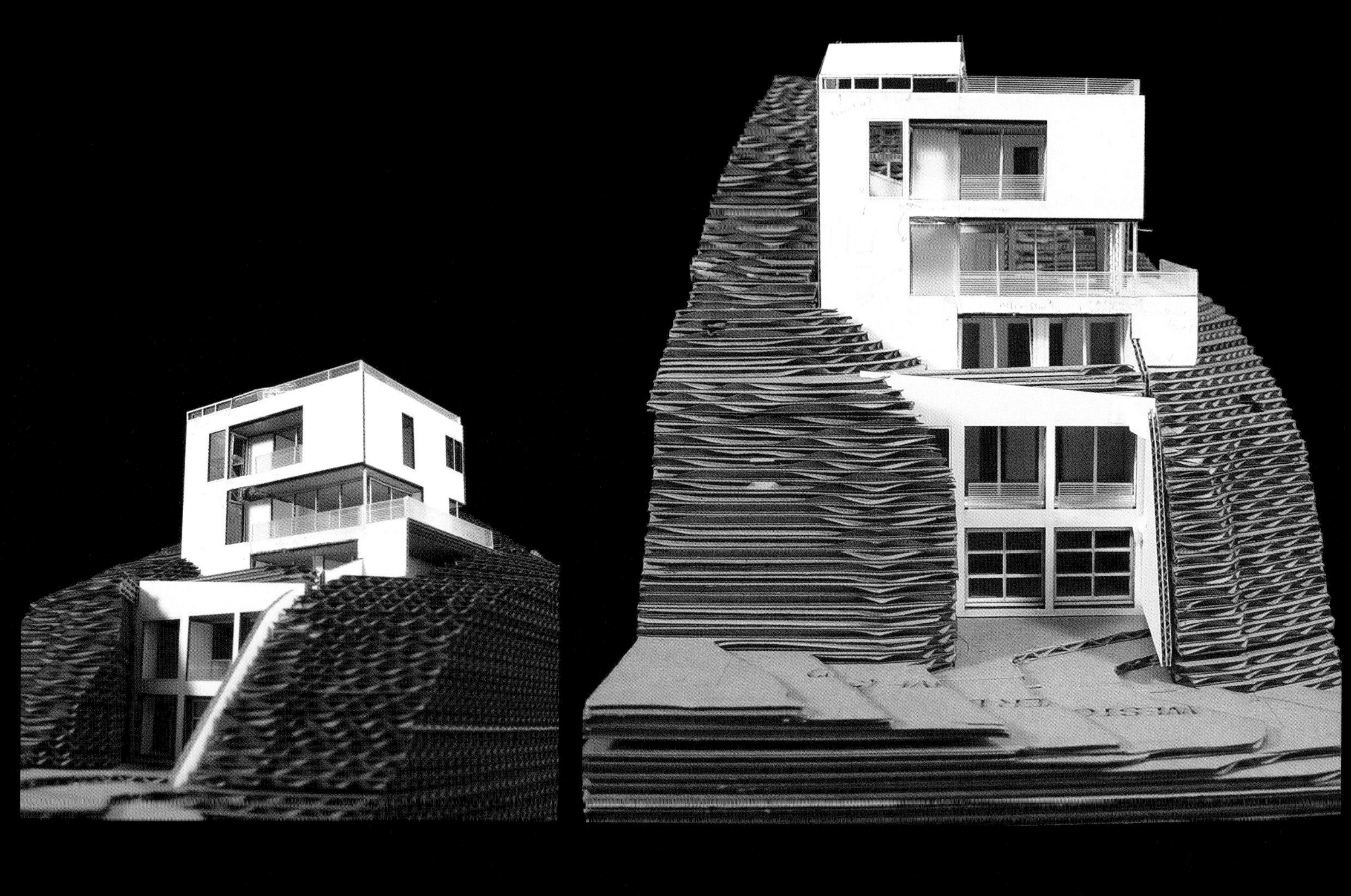

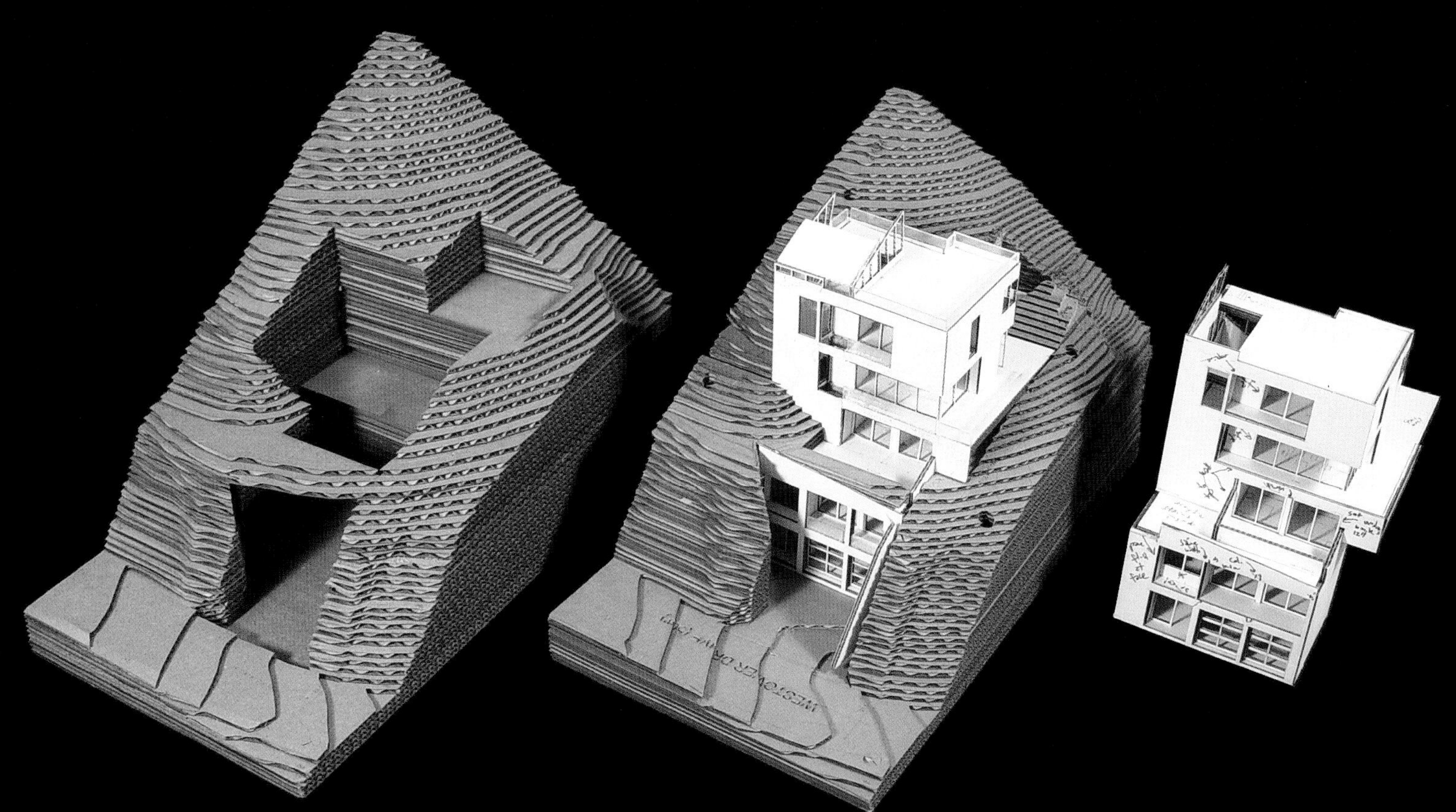

Concrete Tower Houses 1 and 2

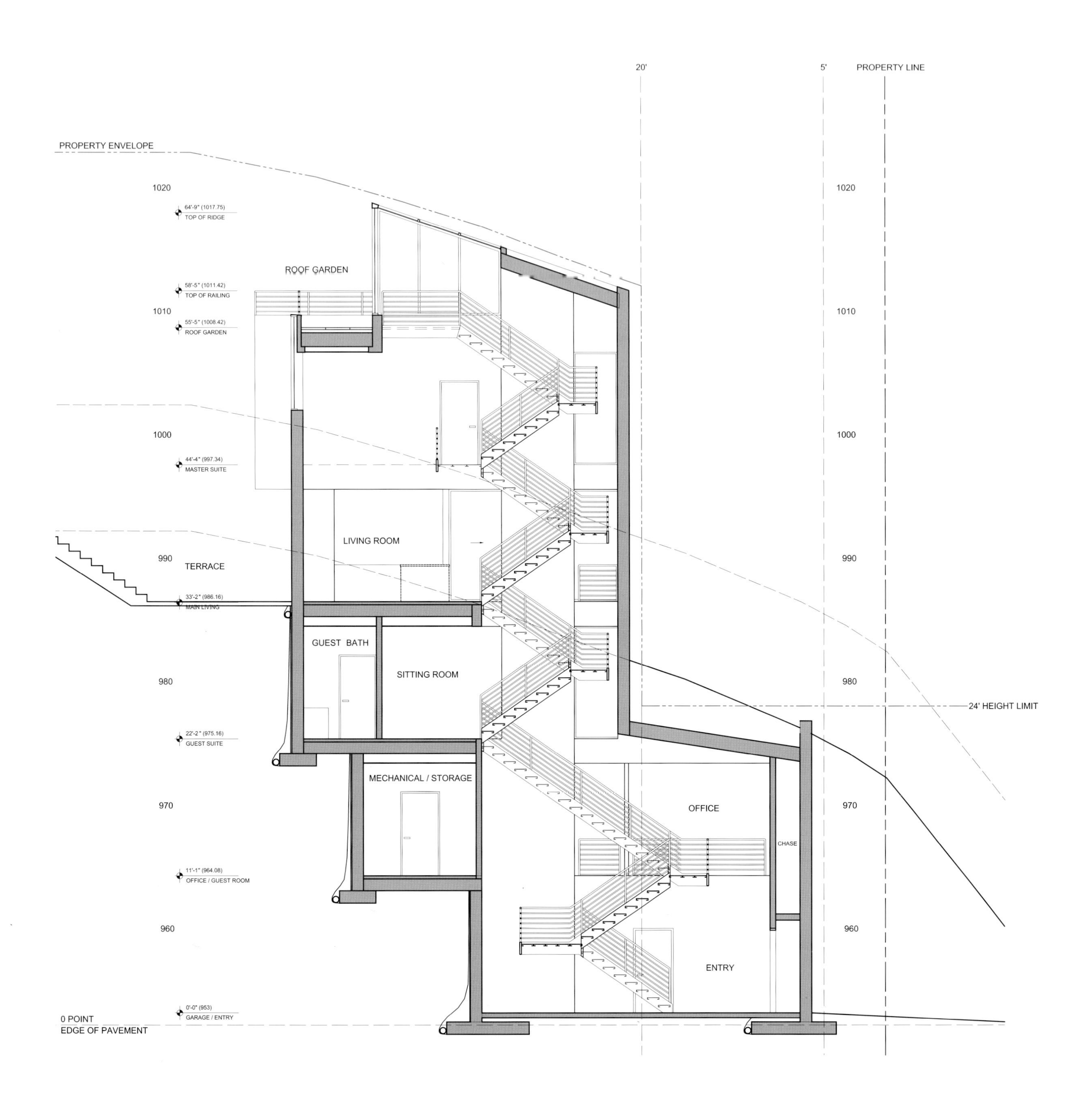

Cross-section,
Tower House 1

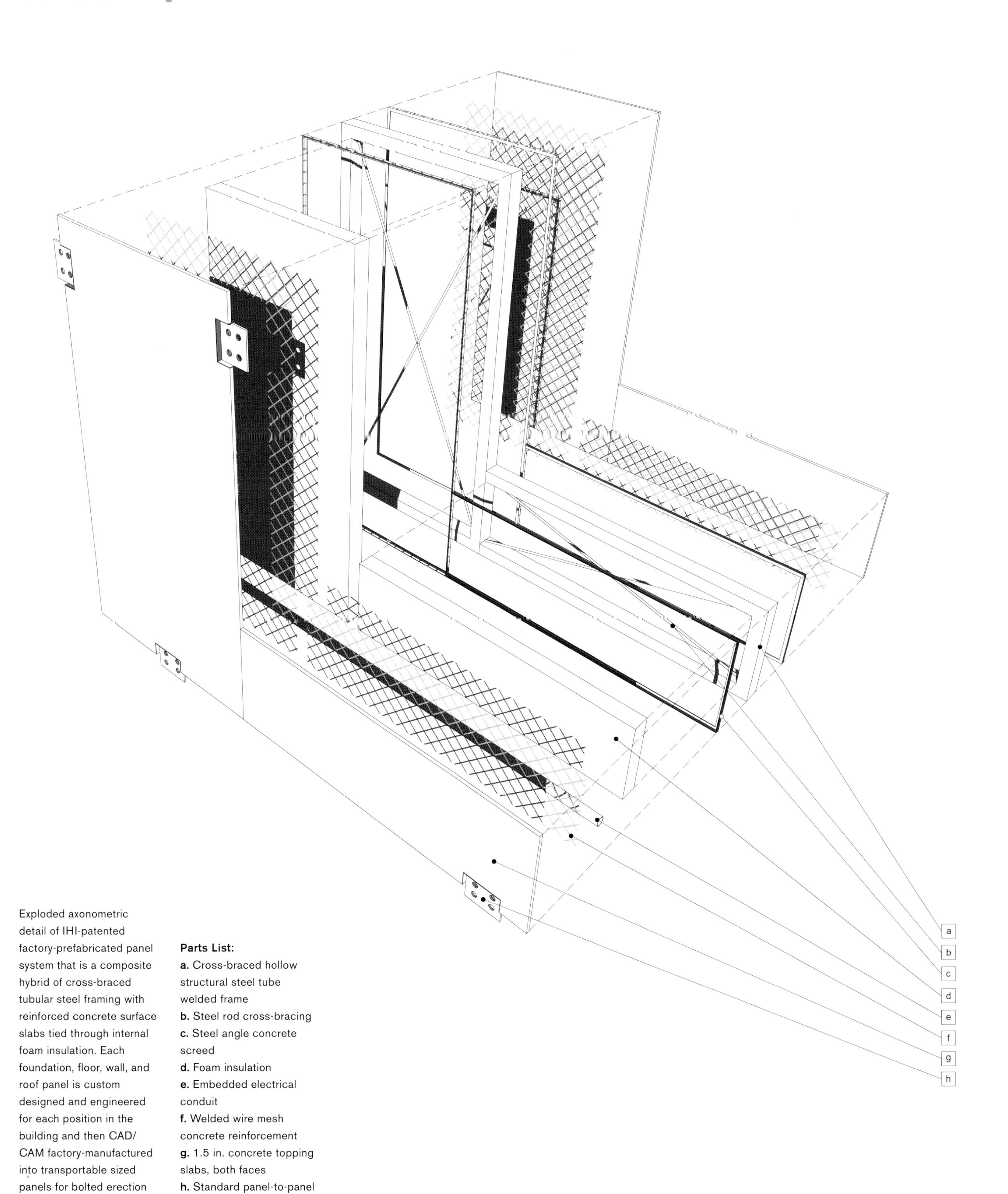

Exploded axonometric detail of IHI-patented factory-prefabricated panel system that is a composite hybrid of cross-braced tubular steel framing with reinforced concrete surface slabs tied through internal foam insulation. Each foundation, floor, wall, and roof panel is custom designed and engineered for each position in the building and then CAD/CAM factory-manufactured into transportable sized panels for bolted erection on site.

Parts List:
a. Cross-braced hollow structural steel tube welded frame
b. Steel rod cross-bracing
c. Steel angle concrete screed
d. Foam insulation
e. Embedded electrical conduit
f. Welded wire mesh concrete reinforcement
g. 1.5 in. concrete topping slabs, both faces
h. Standard panel-to-panel connection plates

Concrete Tower Houses 1 and 2

Top: Site plan, Tower House 2
Bottom: Floor plans, Tower House 2

OPPOSITE
Top: South elevation, Tower House 2
Bottom: East-west section, Tower House 2

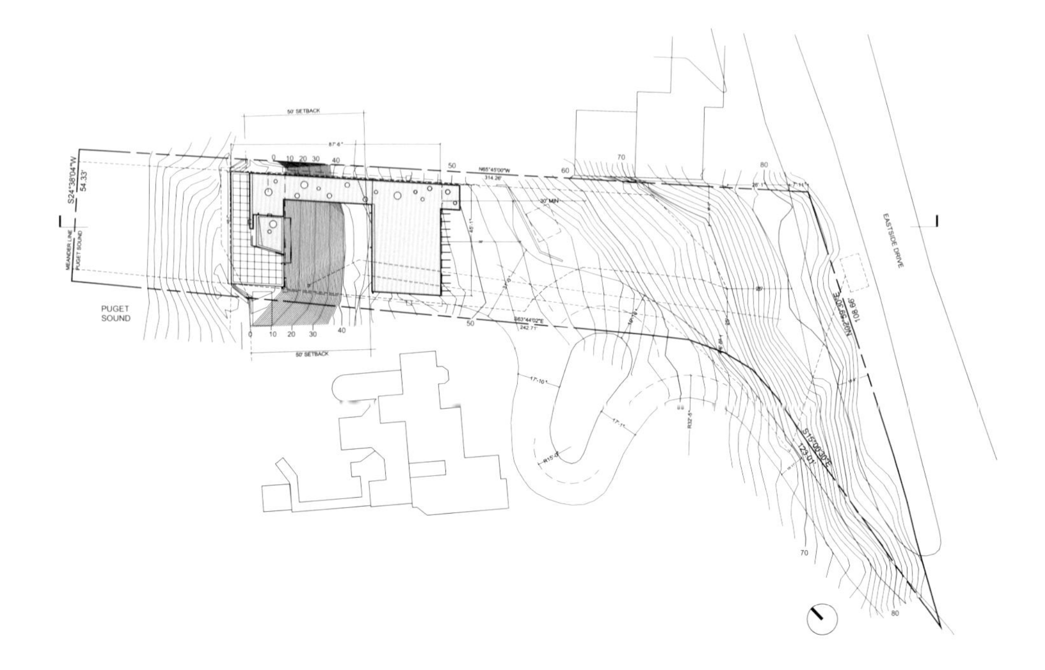

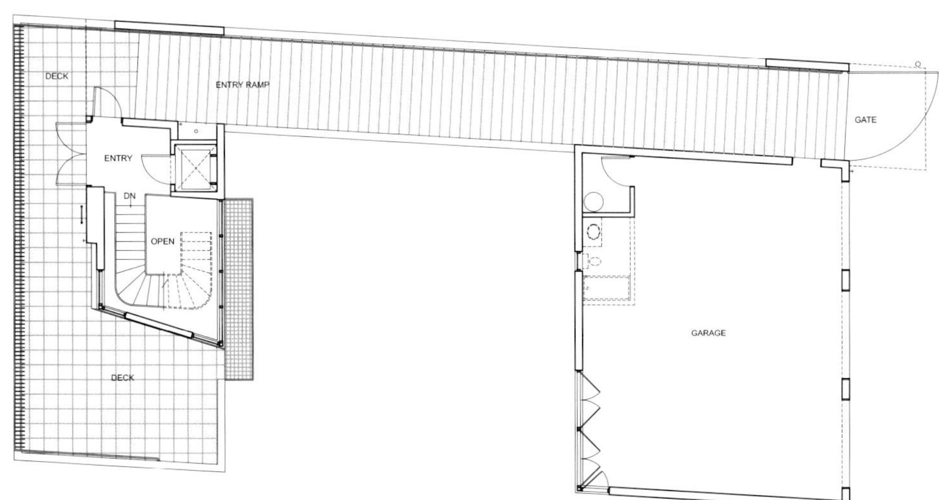

Level 4

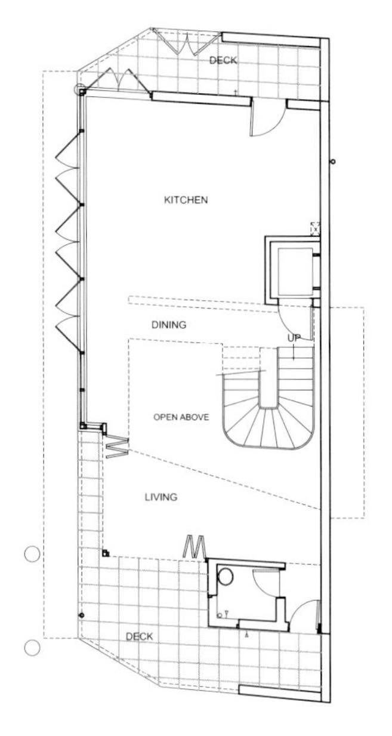

Level 1

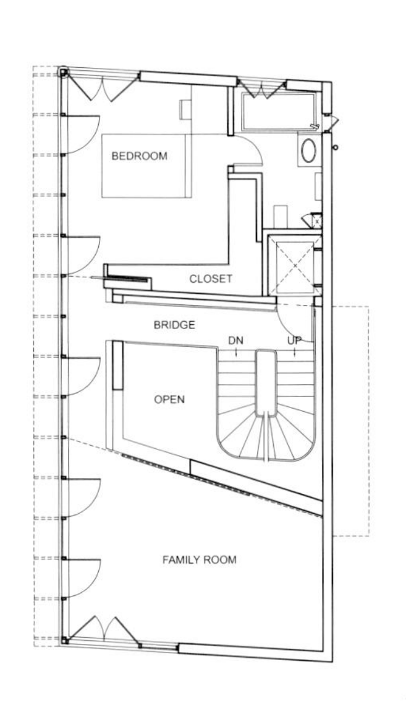

Level 2

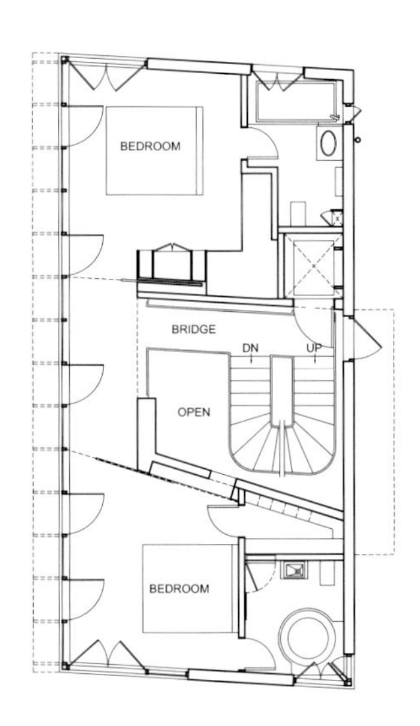

Level 3

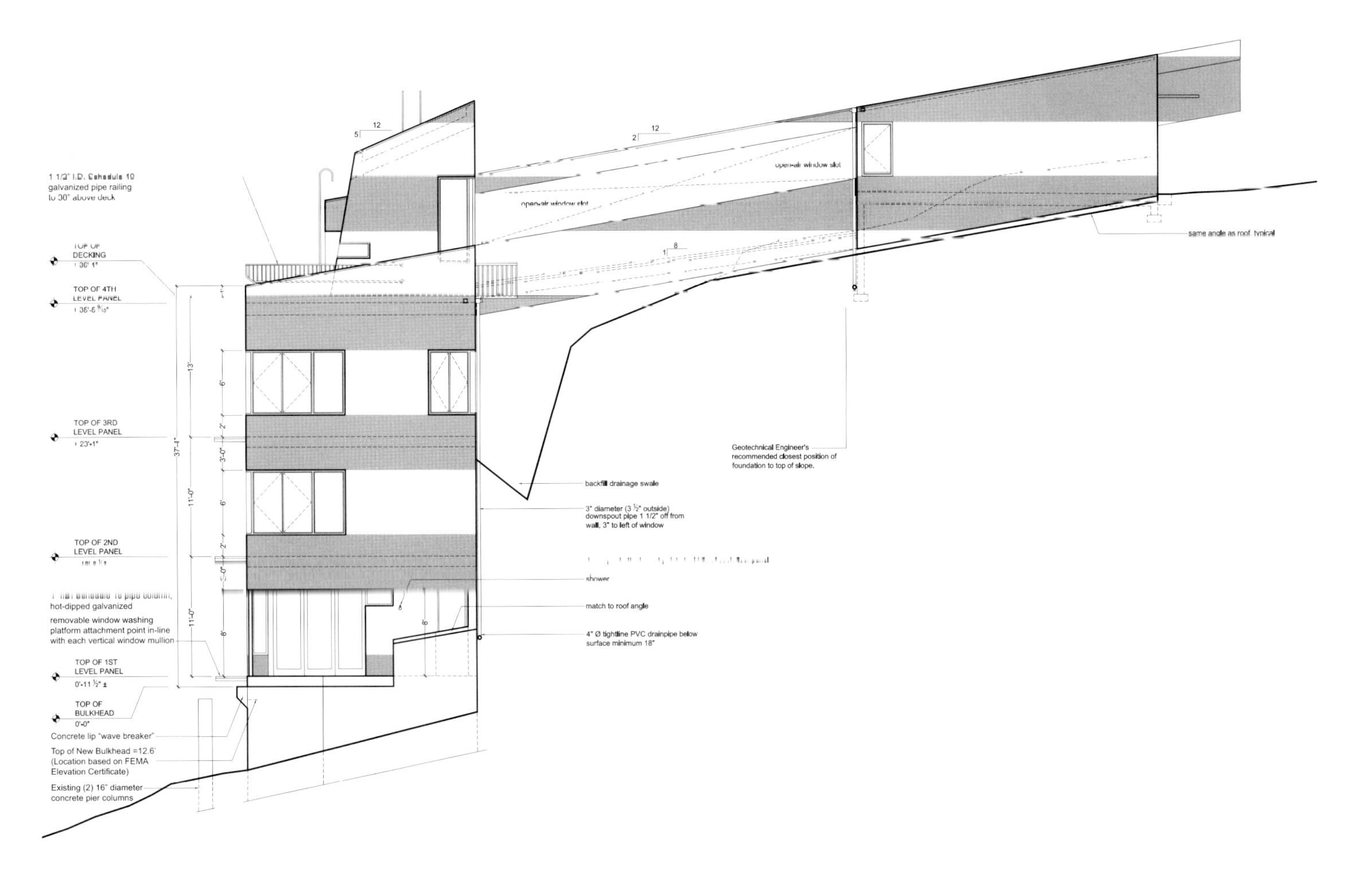
galvanized pipe railing
TOP OF 4TH LEVEL PANEL
TOP OF 3RD LEVEL PANEL
+ 23'-1"
TOP OF 2ND LEVEL PANEL
hot-dipped galvanized
removable window washing platform attachment point in-line with each vertical window mullion
TOP OF 1ST LEVEL PANEL
TOP OF BULKHEAD
0'-0"
Concrete lip "wave breaker"
Top of New Bulkhead =12.6'
(Location based on FEMA Elevation Certificate)
Existing (2) 16" diameter concrete pier columns
same angle as roof, typical
Geotechnical Engineer's recommended closest position of foundation to top of slope.
backfill drainage swale
3" diameter (3 ½" outside) downspout pipe 1 1/2" off from wall, 3" to left of window
shower
match to roof angle
4" Ø tightline PVC drainpipe below surface minimum 18"
37'-4"
13'
11'-0"
11'-0"

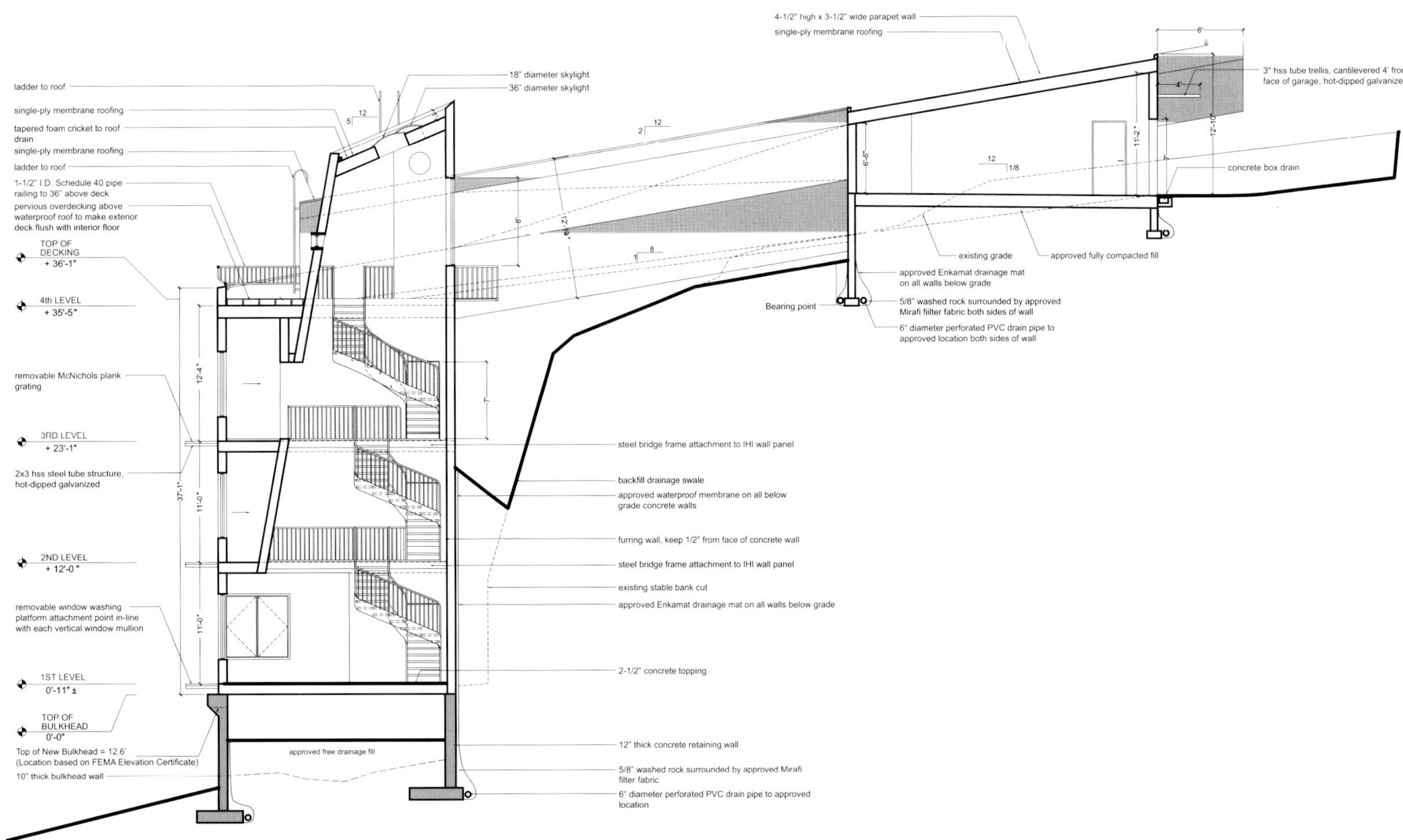
4-1/2" high x 3-1/2" wide parapet wall
single-ply membrane roofing
3" hss tube trellis, cantilevered 4' from face of garage, hot-dipped galvanized
18" diameter skylight
36" diameter skylight
ladder to roof
single-ply membrane roofing
tapered foam cricket to roof drain
single-ply membrane roofing
ladder to roof
1-1/2" I.D. Schedule 40 pipe railing to 36" above deck
pervious overdecking above waterproof roof to make exterior deck flush with interior floor
TOP OF DECKING
+ 36'-1"
4th LEVEL
+ 35'-5"
removable McNichols plank grating
3RD LEVEL
+ 23'-1"
2x3 hss steel tube structure, hot-dipped galvanized
2ND LEVEL
+ 12'-0"
removable window washing platform attachment point in-line with each vertical window mullion
1ST LEVEL
0'-11" ±
TOP OF BULKHEAD
0'-0"
Top of New Bulkhead = 12.6'
(Location based on FEMA Elevation Certificate)
10" thick bulkhead wall
approved free drainage fill
concrete box drain
existing grade
approved fully compacted fill
approved Enkamat drainage mat on all walls below grade
Bearing point
5/8" washed rock surrounded by approved Mirafi fiilter fabric both sides of wall
6" diameter perforated PVC drain pipe to approved location both sides of wall
steel bridge frame attachment to IHI wall panel
backfill drainage swale
approved waterproof membrane on all below grade concrete walls
furring wall, keep 1/2" from face of concrete wall
steel bridge frame attachment to IHI wall panel
existing stable bank cut
approved Enkamat drainage mat on all walls below grade
2-1/2" concrete topping
12" thick concrete retaining wall
5/8" washed rock surrounded by approved Mirafi filter fabric
6" diameter perforated PVC drain pipe to approved location
37'-1"
12'-4"
11'-0"
11'-0"
11'-2"
12'-10"

Fort Worth Block House

Fort Worth, Texas / 1996

A house addition for two artists and their children, this project asserts the independent character of this family while carefully fitting into a neighborhood of modest bungalows near the Kimbell Museum in Fort Worth, Texas. The owners, Cameron Schoepp and Terri Thornton, loved the original house and the scale and texture of the neighborhood, but they needed much more space as their family grew. Terri is a curator at the Fort Worth Modern Art Museum and a drafter of exquisitely detailed and small-scaled studies of seeds, nuts, stems, and other dried and shriveled organic detritus; Cam is a well-known sculptor and art professor in the Dallas–Fort Worth area, and has been a longtime and frequent collaborator with us on many public art and architecture projects around the country. With our close creative collaboration, we worked together to develop design concepts and approaches to materials and production methods that were integral to their interests and skills as artists and makers of public works, which they would employ as the self-builders of their home. Energized by both the opportunity to build a creative work on the scale of a house and the budgetary necessity to do as much as possible of the work themselves, we collectively sought out building methods and materials that would be compatible with these requirements and be related to our creative work in other areas. Cam's sculpture work in concrete, steel, stone, resins, and wood provided him with the skills and tools to produce many of the detail elements of the house in his well-equipped studio, and we selected a modular lightweight precast block structural system for the primary building mass to allow for incremental assembly according to their own self-build schedule.

While our experiments in temporary installations are important in testing the limits of current ideas and possibilities, when building a house for a family, whether one's own or someone else's, the issues and risks—financial and physical—are somewhat different, and in considerably closer focus. It is easy to conceive of various cutting-edge design prototypes, but for longterm inhabitable structures, especially houses, there is a natural conservatism based in risk aversion that is legally mandated in zoning and building codes as well as—in not entirely overlapping ways—a simple imperative for fundamental human concern in design and construction decision-making. While we are not averse to formal risk, conventional expectations and understandings in the real-estate industry view deviation from the norm as synonymous with considerable financial risk, regardless of whether such fears are based in logic or even in statistical market rationality. Community review boards, narrowly focused planning professionals, real estate appraisers, and bankers can have tremendously dampening effects on innovation, individuality, and site-specific fitness for purpose, causing multiple negative effects throughout the real estate, design, and construction industries. We mention these issues here not because this house has been particularly normalized through risk aversion—in fact it is relatively unusual in its form, siting, surfaces, and construction methods—but because it is a case that brings to the surface so many complex issues in balancing interests in experimentation and innovation with equally involving issues of community integration and the health and financial security concerns embodied in the family house.

One of the challenges was to double the size of the house, on a small lot, without markedly changing the appearance of the street facade and the overall rythym and spatial balance of the neighborhood streetfront as a whole. The new addition takes the form of a squat tower in the back yard rising up into a grove of mature pecan trees, which serve to screen it from view from the street. The original house is left essentially untouched, with new spaces built into the independent concrete block settled into the trees in the backyard. On the roof is a treetop sculpture court surrounded with a light steel bougainvillea trellis, providing shade and privacy and offering nighttime views of Fort Worth and the Texas sky. The building is constructed of structural insulated concrete panels and raw steel left to rust and to trail streaks of iron-oxide color over the course of time into the cement-plaster wall finish.

The prefabricated building system used here, based on the Rastra system developed in Austria after World War II, is both relatively inexpensive and somewhat unusually applied. In this case, the system is an autoclaved aerated concrete (AAC) panel system with encapsulated polystyrene beads for light weight and high insulating properties. The panels, which come in modules 15 inches (.38 m) high, 10 feet (3.05 m) long, and of varying widths according to structural and insulation requirements, can be stacked vertically or horizontally and are light enough to be handled and placed by two workers without cranes or other mechanical lifts. Round hollow cores run bi-directionally through the panels, and through the walls built with them, allowing reinforcing steel to be placed as needed in horizontal and vertical directions. Once the panels are stacked and the

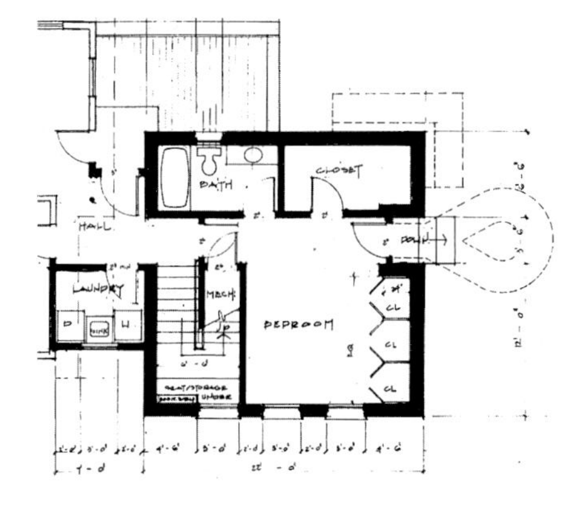

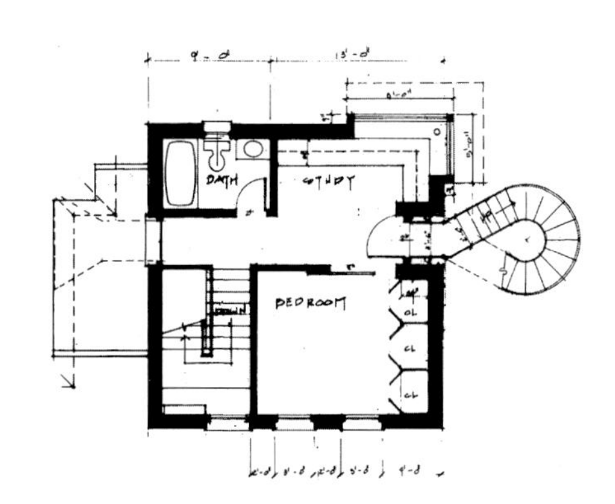

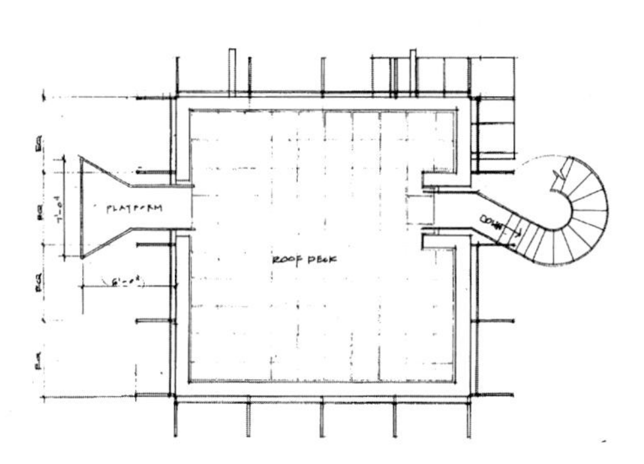

OPPOSITE
Schematic building plan studies: first floor (left), second floor (center) and roof terrace (right)

Top row: Lightweight, insulative concrete panels arrive by truck; excavation and footing construction; approximately 2 x 10 ft. panels can be saw-cut and fit on site

Middle row: Panels are erected vertically and glued together, with rebar placed horizontally and vertically through hollow cores; cores are pumped full of concrete grout, encapsulating rebar and creating strong structural grid

Bottom row: Site-cast concrete window box elements are structurally tied into rebar grid and cast along with grout cores in walls; porous wall panels are given a finish coat of cement plaster on interior and exterior surfaces

steel is placed, the cores are filled on site with a concrete grout that locks the whole system together. One of the great advantages of the system is that there is no formwork, other than the panels themselves, which are left in place and form the base for applications of interior and exterior finishes.

The Rastra system, and other proprietary variations of the same concept, have become relatively common in Europe and have begun to see some gradual increase in use in North America. This house was among the earliest uses in the United States, receiving some of the first panels from a new factory in Mexico. Among the many advantages of this system is the combination of high insulation values afforded by the air content and encapsulated polystyrene beads accompanied by the equally valuable attribute of thermal mass. This is an unusual combination, offering simultaneously the best of two material qualities that are most often mutually exclusive. High insulation values are necessary to minimize energy loss and to maximize indoor comfort, particularly in climates such as that of Fort Worth, where the weather can be bitterly cold and windy during the winter and fiercely hot and humid during the summer, requiring substantial energy consumption for both mechanical heating and cooling systems.

The Fort Worth site context is very typical of what might be found in many older, inner-city neighborhoods. This area of Fort Worth is known as the cultural district and has several major institutions of modern architectural interest, including Louis Kahn's Kimbell Museum, the Amon Carter Museum by Phillip Johnson, the new Fort Worth Modern Art Museum by Tadao Ando, and the former home of the modern art museum, designed by well-known Texas architect O'Neill Ford. Surrounding these institutions are neighborhoods of modest bungalow housing, with occasional concrete block mini strip malls, 7-Elevens, dry cleaners, and grocery stores. The typical pattern of the neighborhood houses consists of a one- or two-story bungalow with a porch near the street front and a side driveway leading to a detached chunk of a garage in the back yard. Originally low and single-story, in many cases the garages have grown as large or larger than the houses to accommodate more cars, a second-story apartment, or a backyard business such as a small welding shop or lawn mower repair. We love this kind of real neighborhood, the honest fabric of a uniquely American style of relatively compact, diverse, and culturally and economically vibrant city dwelling. Often, too much emphasis is placed by regulatory agencies and community design review boards on the visual appearance and traditional detailing of the archetypal house units on the one hand, and to the abstract uniformity of archetypal lot coverage ratios, height lines, and economic use and parking patterns on the other. Absent from the frequently resultant guidelines for rigid uniformity in style, pattern, and use is any site-specific and accurately perceptive recognition of the actual diversity and occasionally anomalous pattern-breaking that gives true architectural and social vibrancy to the traditional, jumbled neighborhoods that are so wonderful to live in and visit and that offer such variety in modes of living and means of earning a living on one's own urban lot. With this realistic understanding of the neighborhood in mind, this chunky backyard concrete box bumping up to a traditional bungalow is highly sensitive, appropriate, and contextual within its neighborhood, as well as an intentionally polemic statement of our views on appropriate development and densification in this common and important type of American urban neighborhood.

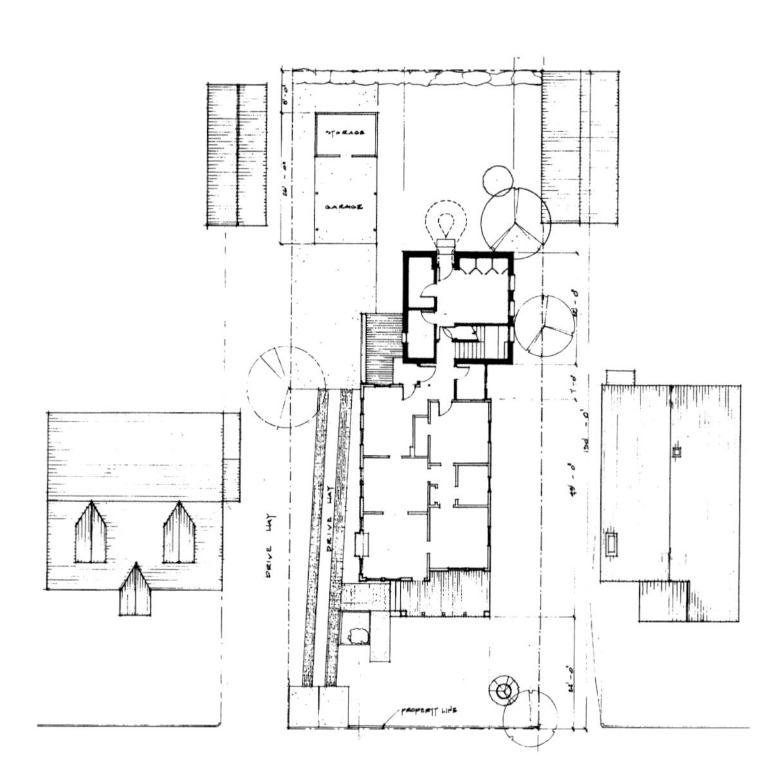

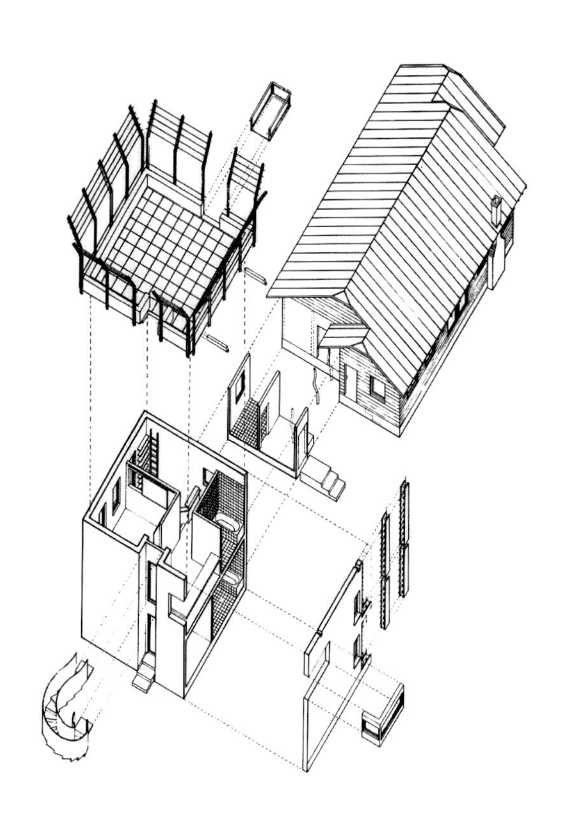

OPPOSITE
Left: Schematic site plan
Right: exploded axonometric view showing relationship of new addition to existing house

Above: Massive concrete and plastered walls with unpainted steelwork which will eventually decorate surface with weathering streaks of red iron-oxide rust

Orchard House

Sebastopol, California / 2005

When Ben Kinmont and Naomi Hupert first approached us about building a house for their family, they were living in New York but were looking for land in Sonoma, California, where Ben had grown up. Naomi had been raised in Manhattan, in the loft where they were raising their own children. As much as they all loved the city, the whole family was excited about moving to a place where they could live more connected to the land and to the cycles of nature and the seasons. We looked with them at several potential properties, suggesting the different design approaches that would be appropriate to each. When they chose to purchase five acres of century-old Gravenstein apple trees, we initially saw it as less interesting than some of the more dramatic hillside sites with spectacular views, but the more time we spent there the more we became fascinated by the structure of the orchard, and what it suggested for the buildings to be constructed there.

What has developed through a three-year collaboration with Ben, Naomi, their children Ian and Natasha, and the builder, Drew Allen, is a highly site-specific cast concrete construction, rationally prefabricated through the use of a limited set of repeated, modular formwork and prefabricated truss framing components. This approach allowed a high degree of adaptability to the landscape while keeping construction costs low. Initially conceived as a pre-cast concrete project, the system evolved instead toward a site-cast concrete building using a modular set of prefabricated formwork. The concrete modules follow the tree grid, emulating the existing rhythm of mass and clearing familiar to the spatial experience of the traditional orchard.

The house is built in conformity with the strict rectilinear geometry of the tree grid while equally exploiting the secondary diagonal surprises particular to human motion through an agricultural field. The site was intensely studied for the individual particularities of each unique tree within the orchard field, in turn lending the house design this same character of individual conditions within a predominantly regularized system. True to the character of the orchard, the house is laid out in long sequences of interior and exterior courtyards as defined by the adjacent trees, affording long, metered views along the rectilinear and diagonal axes of the field. The massive concrete walls align with the rows of tree trunks, while the open volumes of the rooms and exterior courts align with the open space between trees, resulting in a direct spatial continuity between house and landscape, figure and void.

The house is a low, single-story volume that is wheelchair accessible throughout and planned to provide easy access directly onto the surrounding orchard floor from every room. The roof is flat and low, well below the top limbs of the orchard, which swallows the house into its canopy. There is no vantage point from which one can see the building as a whole—it winds in and out of the trees, intimate in scale from any one vantage point, but with the limits of its extents always slightly out of reach, disappearing around a corner and into the orchard.

The range of materials is strictly limited—heated concrete slabs, raw concrete primary walls inside and out, with secondary walls and the indoor-outdoor ceiling plane clad in white plaster. In addition to the concrete, portions of the exterior walls and trim are clad in galvanized steel, which provides a dull reflection of the changing colors of the trees nearby. Minimal cabinetry and shelving for large numbers of books are made from simple modules of plywood, painted white to match the plastered walls and ceiling. There is no distinction between windows and doors, all of which run from floor slab to ceiling plane as hinged vertical panels. Made from lush, sensuous redwood planks reclaimed from the local winemaking industry's barrels, these elements, stained deeper in color from the Zinfandel they previously held, provide the only counterpoint of opulence to the house's otherwise simple palette of raw concrete, white plaster, and galvanized steel.

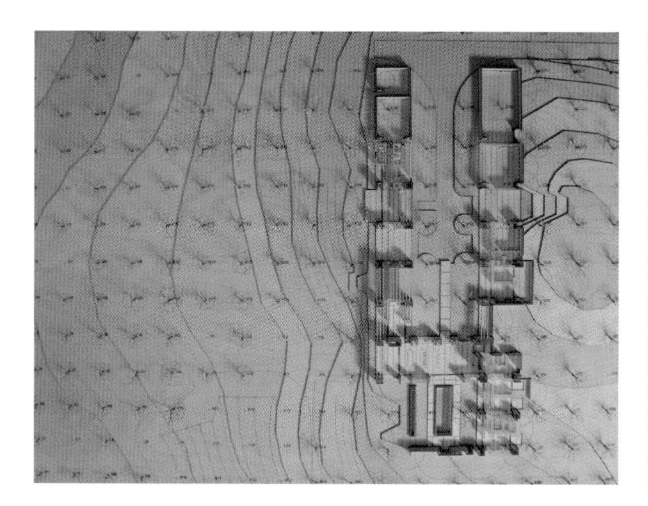

Site model

OPPOSITE
Top: View from orchard looking northeast
Bottom: Dining room doors line west- and south-facing afternoon dining terrace with massive concrete cooking fireplace

Orchard House

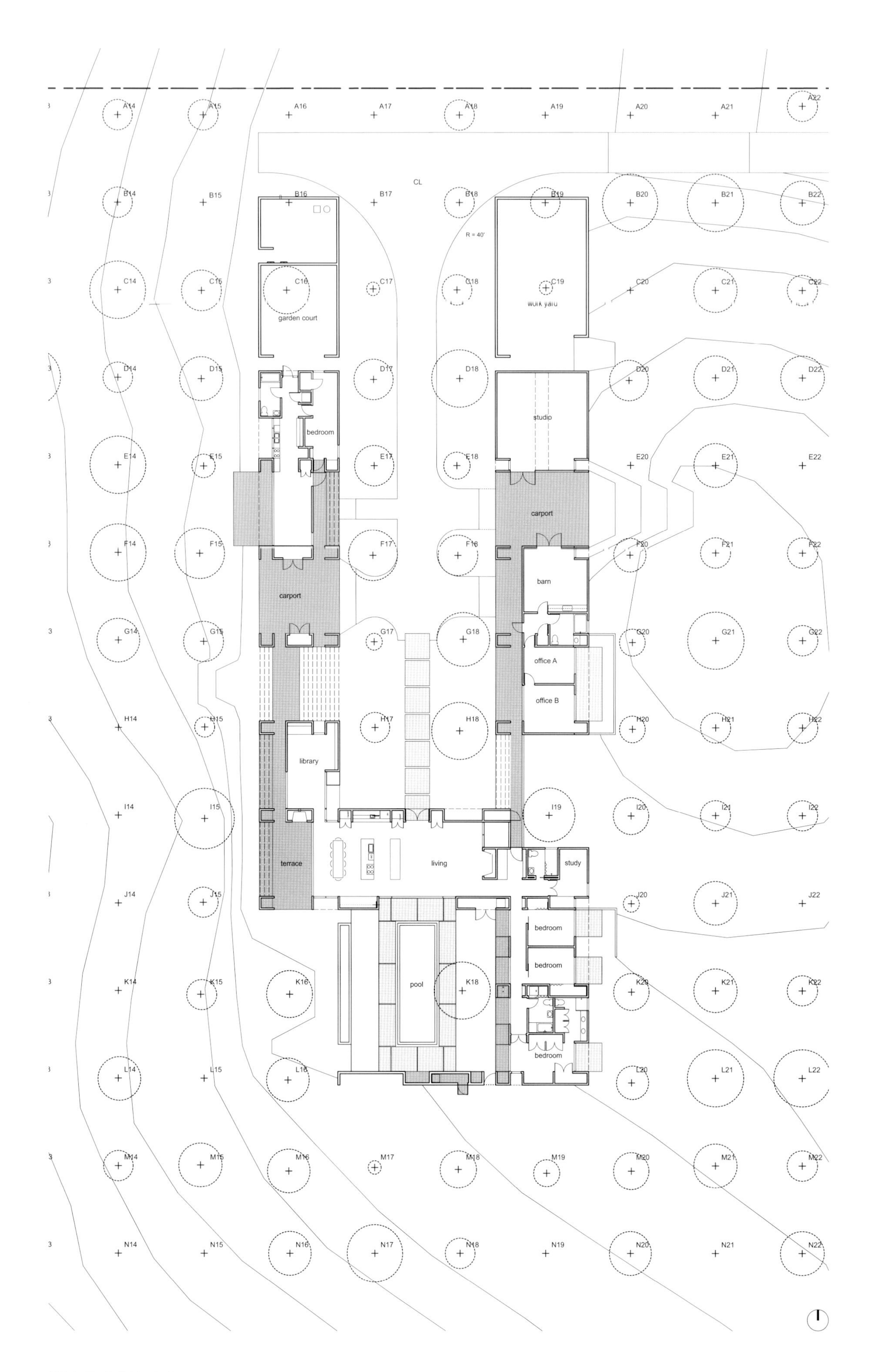

Building plan

OPPOSITE
Top: Page from orchard cataloging studies
Bottom: Aerial photograph and detail views of the orchard

OVERLEAF
During the initial design process, the orchard was documented as an overall grid of individual trees, each one photographed, described, and cataloged in order to develop a site-specific understanding of individual character and event within a modular system.

Concrete Systems

	1	2	3	4	5	6	7	8	9	10	11
A	A1 10"; mid age; nice	A2	A3 6"; Different species. Late bloomer.	A4 18"; survivor; pure; sprouting fr. nice big trunk	A5 10"; Different species. Late bloomer.	A6	A7 20"; scrub from stump	A8 18"; wide; nice	A9 4"; OK scrubby	A10 5"; young; small; weak but OK	A11
B	B1 10"; mid age; multi; nice	B2 8"; mid age; small; full; nice	B3 scrub [illegible]	B4 6"; Different species. Late bloomer.	B5 20"; gnarly; nice from stump	B6 scrub	B7 10"; Different species. Late bloomer.	B8 18"; many	B9	B10 16"; 6 major; full	B11
C	C1 10"; mid age; nice	C2 12"; sparse; whacked maybe OK	C3 16"; 2 or 3; sparse; gnarly; nice	C4 12"; survivor; OK	C5 14"; 5 limb; gnarly; nice	C6 scrub	C7 10"; mid age; full; OK	C8 8"; mid age; low; OK	C9 6"; mid age; small; gnarly; OK	C10	C11 10"; 7 major; very nice
D	D1 3/4"; new; nice	D2 medium; mid age; multi; OK	D3 12"; sparse; gnarly; OK	D4 medium; 2 major; nice - OK whacked	D5 12"; 4; pure; nice	D6 12"; 5; small; Nice - OK	D7 8"; mid age; full; OK	D8 14"; 3; full; pure; nice	D9 4" double; small; OK - weak	D10 12"; OK - weak shoots from a tall stump	D11 20"; pure; very nice
E	E1	E2 medium; mid age; multi; OK	E3 10"; straight; lush; pure; nice	E4	E5 15"; survivor; v. small; OK from stump	E6 12"; 3 major; wide; gnarl; nice - small	E7	E8 14"; 5 limb; pure; nice	E9 10"; 3 major; full; very nice	E10 10"; old; weak - not special	E11 8"; young; OK
F	F1 6"; Different species. Late bloomer.	F2 9"; small; full; round	F3 much scrub from stump	F4 scrub from tall stump	F5 20"; survivor; nice from stump	F6 8"; Different species. Late bloomer.	F7 8"; Different species. Late bloomer.	F8 small; Different species. Late bloomer.	F9 10"; Different species. Late bloomer.	F10 20"; survivor; 1 major; gnarly; very nice	F11 12"; old; 4 limb; OK weak
G	G1 multi; dead	G2 medium; sparse; very nice interesting	G3 3/5"; scrub maybe OK	G4 18"; OK [illegible]	G5 8"; Different species. Late bloomer.	G6 18"; gnarly; very nice [illegible]	G7 burn pile	G8 10"; mid age; OK	G9 5"; young; small; good	G10 20"; gnarly; nice	G11 8"; species. bloomer.
H	H1 12"; 2; full wide; low	H2 14"; very nice might leave	H3 14"; survivor; OK. From stump. Fallen stump is of intertest.	H4 20"; 3 major; sparse; gnarly; OK - Nice	H5 2"; Different species. Late bloomer.	H6 18"; survivor; from stump Maybe OK	H7 small; Different species. Late bloomer.	H8 8"; Different species. Late bloomer.	H9 24"; 3 major; gnarly; very nice	H10 20"; 4 major; gnarly; OK	H11 18"; 3 major; gnarly; nice
I	I1 12"; 3 limb; OK - whacked	I2 little; multi; Different species. Late bloomer.	I3 6"; Different species. Late bloomer.	I4 Dead stump	I5 4"; young; small; OK	I6 10"; Different species. Late bloomer.	I7 10"; mid age; many; full; OK	I8 12"; survivor; OK grown from stump	I9 24"; small; full; gnarly; nice with moss	I10 18"; 2 limb; OK	I11
J	J1 10"; [illegible]; nice whacked back	J2 18"; 3 major; very nice	J3 12"; nice low double	J4 18"; 2 major; thin; nice	J5 24"; 3; big; nice covered ivy	J6 8"; small; gnarly; OK	J7 12"; old; Different species. Late bloomer.	J8 12"; old; 3 major; thin; gnarly; OK - nice	J9 20"; 3 major; nice	J10 24"; new; low; [illegible]; OK - nice, strong, from stump	J11 10"; Different species. Late bloomer.
K	K1 small; Different species. Late bloomer.	K2 12"; 3 major; pure; nice	K3 8"; small; OK	K4 10"; little; nice twisty from old growth	K5 9"; mid age; many; small; OK	K6 12"; full; OK	K7 14"; 4 limbs; nice	K8 6"; small; scrubby; OK	K9 10"; old; gnarly; OK	K10 20"; 3 limb; tall; gnarly; very nice	K11 5"; Different species. Late bloomer.
L	L1 10"; pure; nice	L2 tiny; Different species. Late bloomer.	L3 small; survivor; low; OK. Scrub from stump.	L4 10"; old; sparse; gnarly	L5 9"; mid age; many; small; OK	L6 4"; Different species. Late bloomer.	L7 scrub	L8 2"; Different species. Late bloomer.	L9	L10 4"; Different species. Late bloomer.	L11 20"; little; gnarly; very nice. grows from stump.
M	M1 small; multi; OK - scrub	M2 12"; straight; pure; nice	M3	M4 18"; 3 major; very nice	M5 3"; OK - little. From big stump.	M6 2"; Different species. Late bloomer.	M7 2"; Different species. Late bloomer.	M8 16"; old; 4 major; nice	M9 3"; young; tiny; scrub from old stump	M10 20"; old; big; OK - weak. Survivor.	M11 5"; young; OK
N	N1	N2	N3 small; low; nice. Interesting scrub from stump.	N4 5"; young; little; OK	N5 22"; big; gnarly; nice	N6 small; low; OK. Grows from tall stump	N7 14"; full; pure; nice	N8 8"; mid age; small; nice	N9	N10 little; Different species. Late bloomer.	N11

	12	13	14	15	16	17	18	19	20	21	22
A	A12 12"; tall; weak but OK. No fruit.	A13	A14 3"; DEAD; No leaf; Different species. Late bloomer.	A15 10"; 3 major; little fruit; Different species. Late bloomer.	A16	A17 12"; OK; 4 major; OK but weak. Under oaks. Green Lichen	A18 12" + 8"; OK; 2 major; medium; OK. Double.	A19 12"; survivor; Great stump. Under oak.	A20	A21	A22 8"; not healthy; Different species. Late bloomer.
B	B12 mini; no fruit; scrub from stump	B13	B14 0"; small; OK. No fruit.	B15 stump	B16	B17 30"; [illegible]; Shoots from old stump.	B18 6"; Different [illegible] Late bloomer	B19 4"; tiny; OK. Healthy. Much fruit	B20 8"; OK; 3 major; medium; Different Late bloomer.	B21 12"; 5 major; wide; Vikki's favorite form. Healthy.	B22 16"; 3 major; Nice
C	C12 18"; 4 major; pure; nice	C13 8"; OK	C14 24"; 3 major; sparse; Nice	C15 18; 3 major; low; OK	C16 20; 4 major; OK	C17 4; new; small; Good. 8' height.	C18 6"; tiny; Different species. Late bloomer. 6' height	C19 5"; small; OK. Healthy. Scrubby.	C20	C21 12"; 3 major; Nice	C22 8"; 4 major; Different species. Late bloomer.
D	D12 large; 2 major; big; gnarly; OK	D13 20"; 5 major; OK to nice. Big trunk.	D14 6"; Different species. Late bloomer.	D15 14"; 4 major; Nice	D16 Shoots from stump	D17 18"; 3 major; sparse; Nice with interesting trunk. 24' canopy. Lots of fruit.	D18 24"; 4 major; Nice. Healthy. 30' canopy. 76" circum = stump.	D19 small; Different species. Late bloomer.	D20 12"; 4 major; OK. Weak	D21 16"; OK. Whacked sideways. 2 trunks living.	D22 14"; 3 major; Nice
E	E12 8"; stubby; OK - from 18" broad stump	E13 4"; Weak. Scrubby	E14 20"; survivor; gnarly; OK but not great. Grows fromold stump.	E15 12"; 2 major; Nice	E16 8"; young; many; low; OK	E17 18"; 2 major; OK	E18 8"; OK; Different species. Late bloomer. 10' height. Fruit.	E19	E20	E21 12"; OK to nice. Whacked.	E22
F	F12 8"; Different species. Late bloomer.	F13 small; Different species. Late bloomer.	F14 20"; thick; Nice. Whacked sideways. 1/2 Living.	F15 24"; 3 major; sparse; gnarly; Nice	F16 8"; older; small; OK. Dwarfish. 11' Hieght with small canopy. 22" circum = stump.	F17 24"; 4 major; lot of fruit; Very nice. Branches are trunk-like. 25' canopy.	F18 medium; 1 major; odd; OK with nice form at sky.	F19 8"; Different species. Late bloomer.	F20 20"; 2 major; gnarly; Nice	F21 small; little; Different species. Late bloomer.	F22 10"; Different species. Late bloomer.
G	G12 12"; old; small; gnarly; OK	G13 4"; small; OK	G14 12"; [illegible]; low; Nice	G15 18"; 4 major; Nice	G16 30"; 2 major; wide; screwy; OK. Hollow. Medium fruit	G17 18"; regrowth; hollow; not wide; pure; Ok. Not special. Grows from tall stump. 53" circum = trunk.	G18 20"; 1 major; Nice. Hollow. 70" canopy.	G19 24"; survivor; 2 major; gnarly	G20 10"; low; OK health. not bad.	G21 20"; Weak. Whacked. Grows from old stump.	G22 10"; OK. scrubby.
H	H12 16"; 4 major; tall; nice	H13 2"; Different species. Late bloomer.	H14 18"; 5 major; Nice	H15 4"; young; low; OK	H16 8"; young; many; sparse; OK	H17 6"; young; many; Nice. 24" Circum = trunk. Much fruit.	H18 20"; 4 major; wide; Very nice. Picnic tree. 70" circum = trunk. 20' canopy.	H19 8"; young; low; OK. Not special. Scrubby and weaker.	H20 10"; Different species. Late bloomer.	H21 6"; new; Regowth from 20" rotted stump	H22 survivor; small; OK health. Whacked. Double tree.
I	I12 18"; 2 major; thick; pure; nice	I13 12"; survivor; small; OK but not special. Grows from 12" trunk.	I14 12"; 6 major; sparse; OK	I15 18"; 5 major; Nice	I16 4"; young; OK	I17 6"; young; many; craggy; Nice. Short 8' Height	I18 24"; 5 major; odd; Very nice. Special. Lot of fruit. TRY TO SAVE.	I19 24"; 2 major; gnarly; Nice. Heavy to top. Crooked. Hollow trunk. Lots of fruit.	I20 10"; Different species. Late bloomer.	I21 10"; 5 major; Different species. Late bloomer.	I22 6"; new; small; OK
J	J12 18"; 2 major; pure; nice	J13 8"; mid age; OK. Not special. Scrubby.	J14 20"; sparse; Nice. Whacky. Broken with strong new limbs.	J15 10"; low; OK. Nice Form.	J16 12"; old; 4 major; small; Nice	J17 20"; full; tall; gnarly; Nice. 1/2 left of double trunk. Whacked. Medium fruit.	J18 20"; full; OK. Scrubby. Sprouts from old stump. No fruit.	J19 20"; 4 major; wide; Very nice. Lots of fruit.	J20 Very weak and scrubby	J21 14"; 3 major; Very nice	J22 Sprouts from stump
K	K12 6"; young; small; OK	K13 8"; Different species. Late bloomer.	K14 Shoots from low stump	K15 18"; survivor; gnarly; OK. Nice growth from old stump.	K16 24"; survivor; many; gnarly; OK. Dead main with good shoots. KEEP. 16' canopy	K17 Burn pile.	K18 20"; 2 major; Nice	K19 6"; Different species. Late bloomer.	K20 small; Different species. Late bloomer.	K21 12"; 2 major; OK to nice	K22 10"; 3 major; Nice form
L	L12 20"; 3 major; very nice	L13 Fallen; OK. Interesting leaner. Not special.	L14 18"; survivor; OK. Grows from old stump.	L15	L16 20"; survivor; OK. Lotso fo fruit and leaf. 17' canopy.	L17	L18 14"; 5 major; small; OK. Healthy.	L19	L20 20"; low; gnarly; OK. Weak	L21 20"; Ok to nice. Not strong. Regrowth.	L22 24"; OK. Grows from old stump. Whacked.
M	M12 10"; Different species. Late bloomer.	M13	M14 small; Different species. Late bloomer.	M15 14"; 3 major; v.sparse; OK	M16 12"; 5 major; dense; Nice. Flattop. Healthy	M17 8"; small; OK. Grows from old stump.	M18 10"; Different species. Late bloomer.	M19 6"; young; OK. Healthy. Not special.	M20 14"; 3+major; Nice	M21 14"; Different species. Late bloomer.	M22 8"; small; OK. Weak
N	N12 6"; young; small	N13 Burn Pile	N14	N15	N16 12"; 3 major; Different species. Late bloomer.	N17 20"; 7 major; Very nice	N18 3"; small; Scrubby	N19	N20 18"; 3 major; wide; Very nice	N21	N22 16"; 4 major; Nice

Orchard House

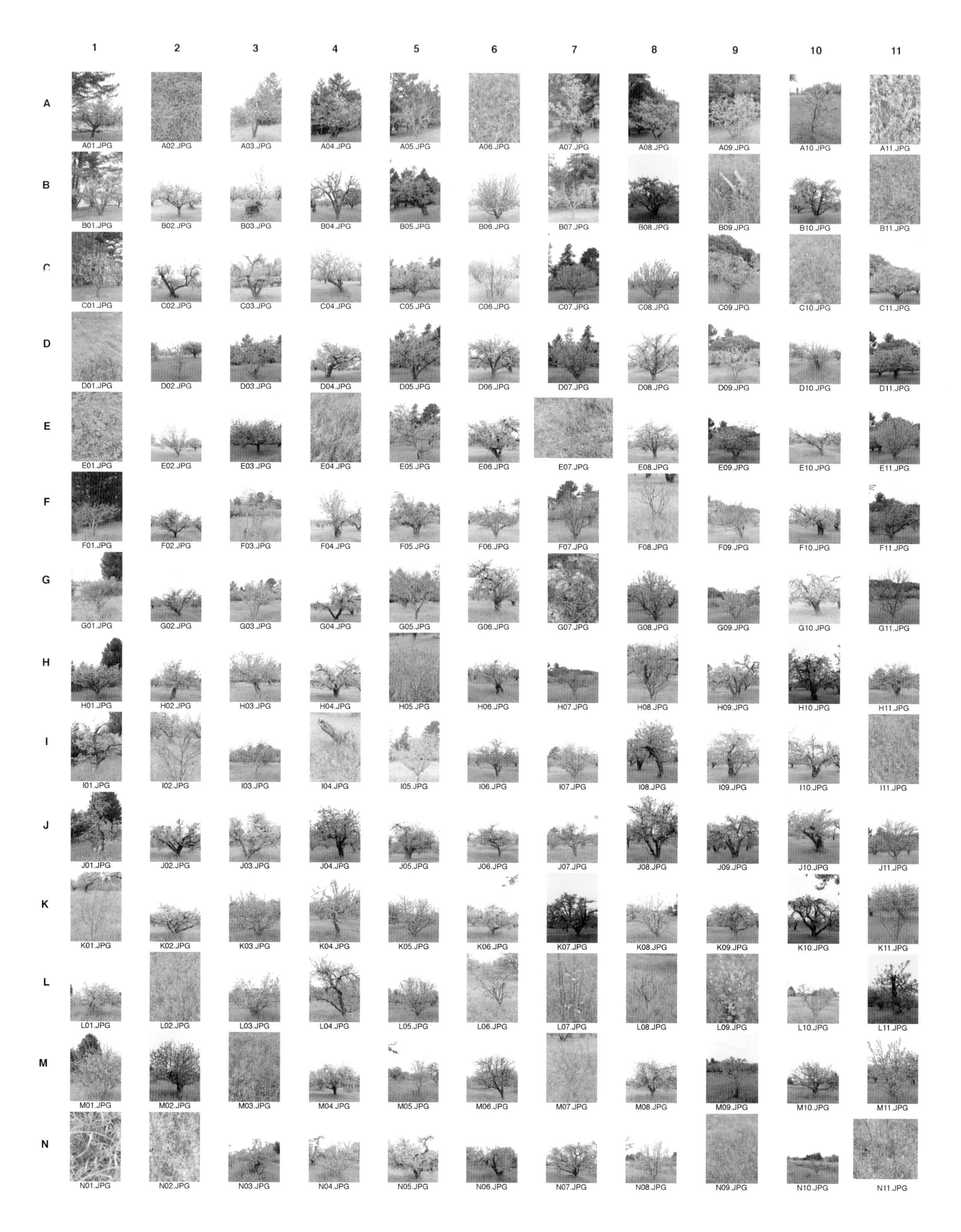

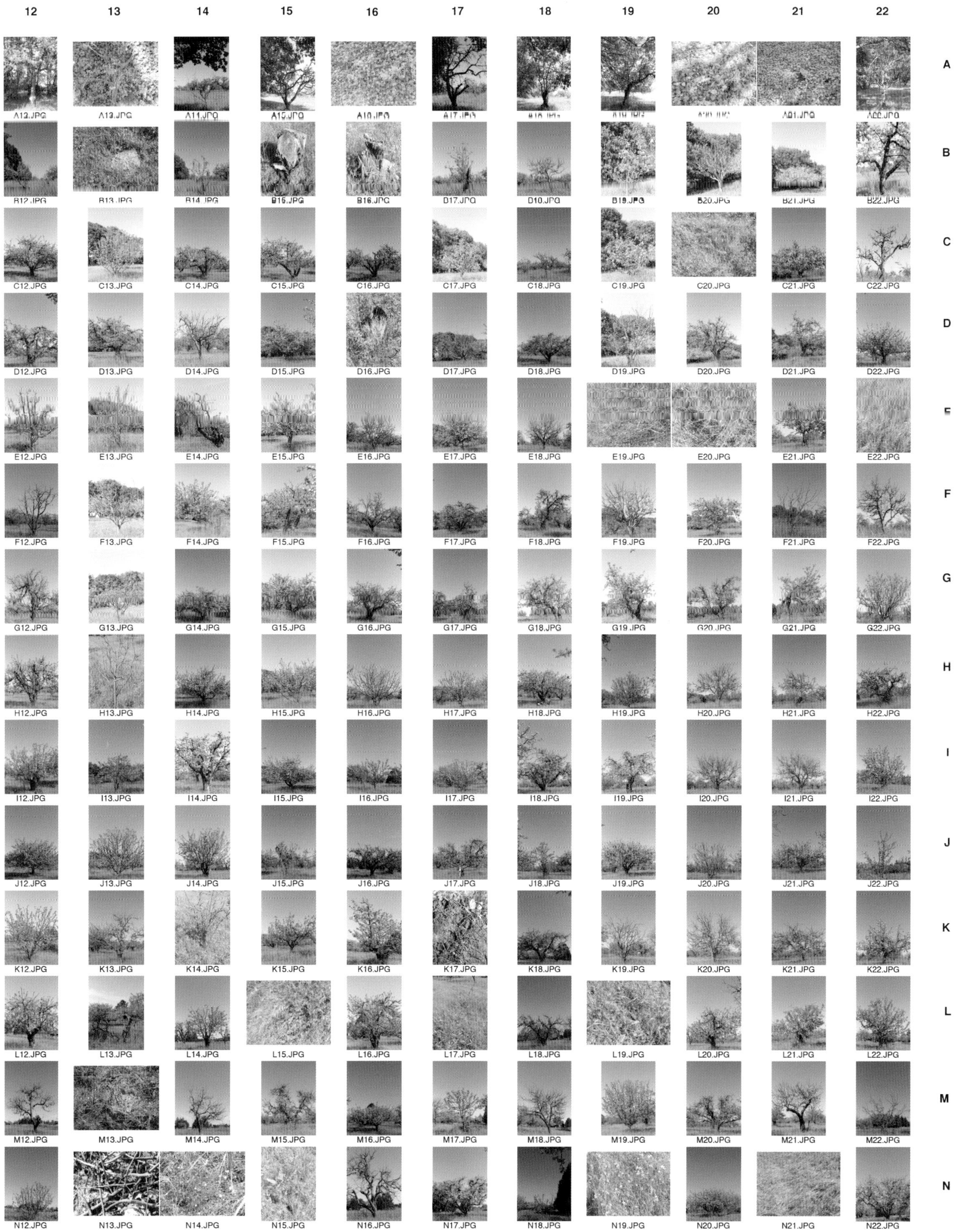

12	13	14	15	16	17	18	19	20	21	22	
A12.JPG	A13.JPG	A14.JPG	A15.JPG	A16.JPG	A17.JPG	A18.JPG	A19.JPG	A20.JPG	A21.JPG	A22.JPG	A
B12.JPG	B13.JPG	B14.JPG	B15.JPG	B16.JPG	B17.JPG	B18.JPG	B19.JPG	B20.JPG	B21.JPG	B22.JPG	B
C12.JPG	C13.JPG	C14.JPG	C15.JPG	C16.JPG	C17.JPG	C18.JPG	C19.JPG	C20.JPG	C21.JPG	C22.JPG	C
D12.JPG	D13.JPG	D14.JPG	D15.JPG	D16.JPG	D17.JPG	D18.JPG	D19.JPG	D20.JPG	D21.JPG	D22.JPG	D
E12.JPG	E13.JPG	E14.JPG	E15.JPG	E16.JPG	E17.JPG	E18.JPG	E19.JPG	E20.JPG	E21.JPG	E22.JPG	E
F12.JPG	F13.JPG	F14.JPG	F15.JPG	F16.JPG	F17.JPG	F18.JPG	F19.JPG	F20.JPG	F21.JPG	F22.JPG	F
G12.JPG	G13.JPG	G14.JPG	G15.JPG	G16.JPG	G17.JPG	G18.JPG	G19.JPG	G20.JPG	G21.JPG	G22.JPG	G
H12.JPG	H13.JPG	H14.JPG	H15.JPG	H16.JPG	H17.JPG	H18.JPG	H19.JPG	H20.JPG	H21.JPG	H22.JPG	H
I12.JPG	I13.JPG	I14.JPG	I15.JPG	I16.JPG	I17.JPG	I18.JPG	I19.JPG	I20.JPG	I21.JPG	I22.JPG	I
J12.JPG	J13.JPG	J14.JPG	J15.JPG	J16.JPG	J17.JPG	J18.JPG	J19.JPG	J20.JPG	J21.JPG	J22.JPG	J
K12.JPG	K13.JPG	K14.JPG	K15.JPG	K16.JPG	K17.JPG	K18.JPG	K19.JPG	K20.JPG	K21.JPG	K22.JPG	K
L12.JPG	L13.JPG	L14.JPG	L15.JPG	L16.JPG	L17.JPG	L18.JPG	L19.JPG	L20.JPG	L21.JPG	L22.JPG	L
M12.JPG	M13.JPG	M14.JPG	M15.JPG	M16.JPG	M17.JPG	M18.JPG	M19.JPG	M20.JPG	M21.JPG	M22.JPG	M
N12.JPG	N13.JPG	N14.JPG	N15.JPG	N16.JPG	N17.JPG	N18.JPG	N19.JPG	N20.JPG	N21.JPG	N22.JPG	N

Orchard House

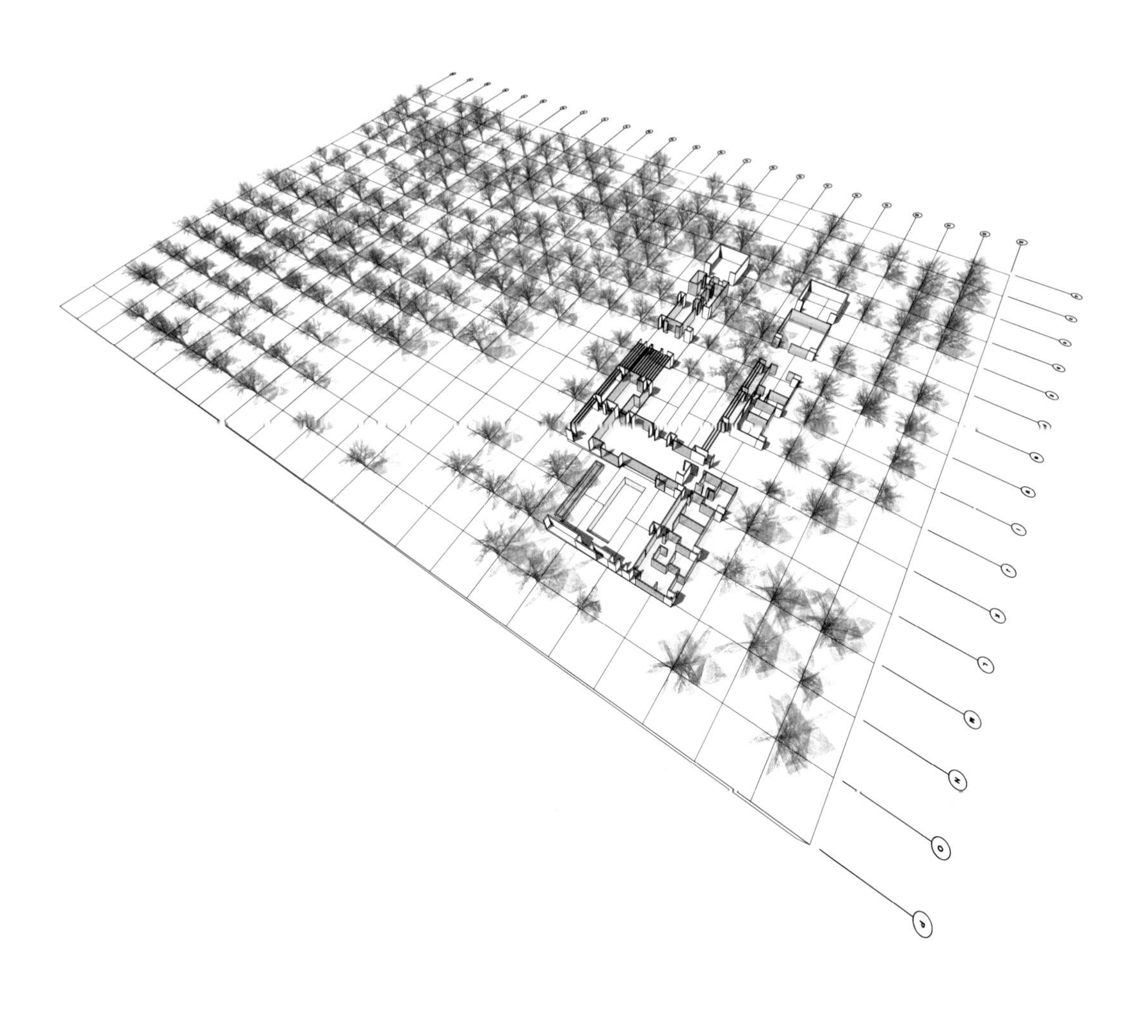

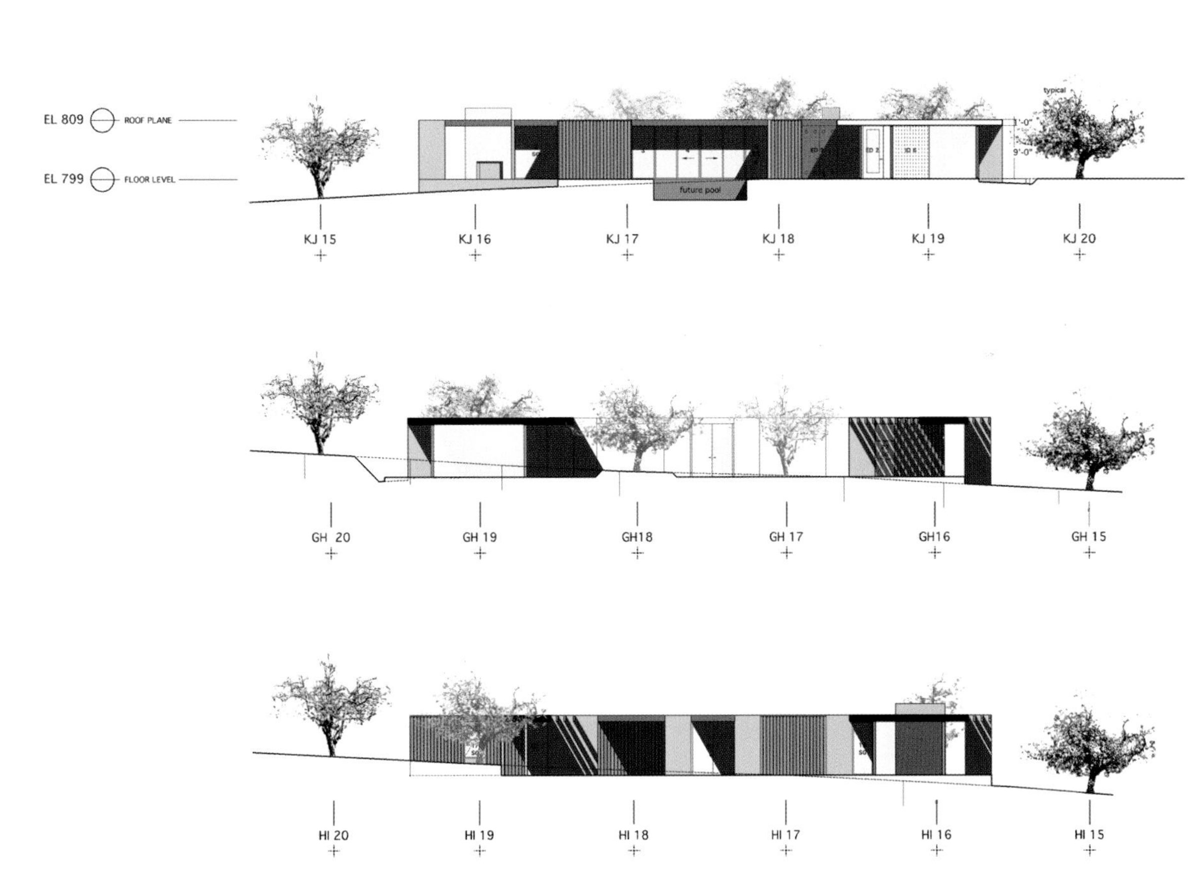

Top: Aerial perspective of house and orchard grid. Construction document grid and notation is laid out in relation to orchard rows.
Bottom: Rendered building elevation/sections as viewed from in between specific rows of trees

Top: Aerial perspective looking from southeast
Bottom: Rendered building elevation/sections

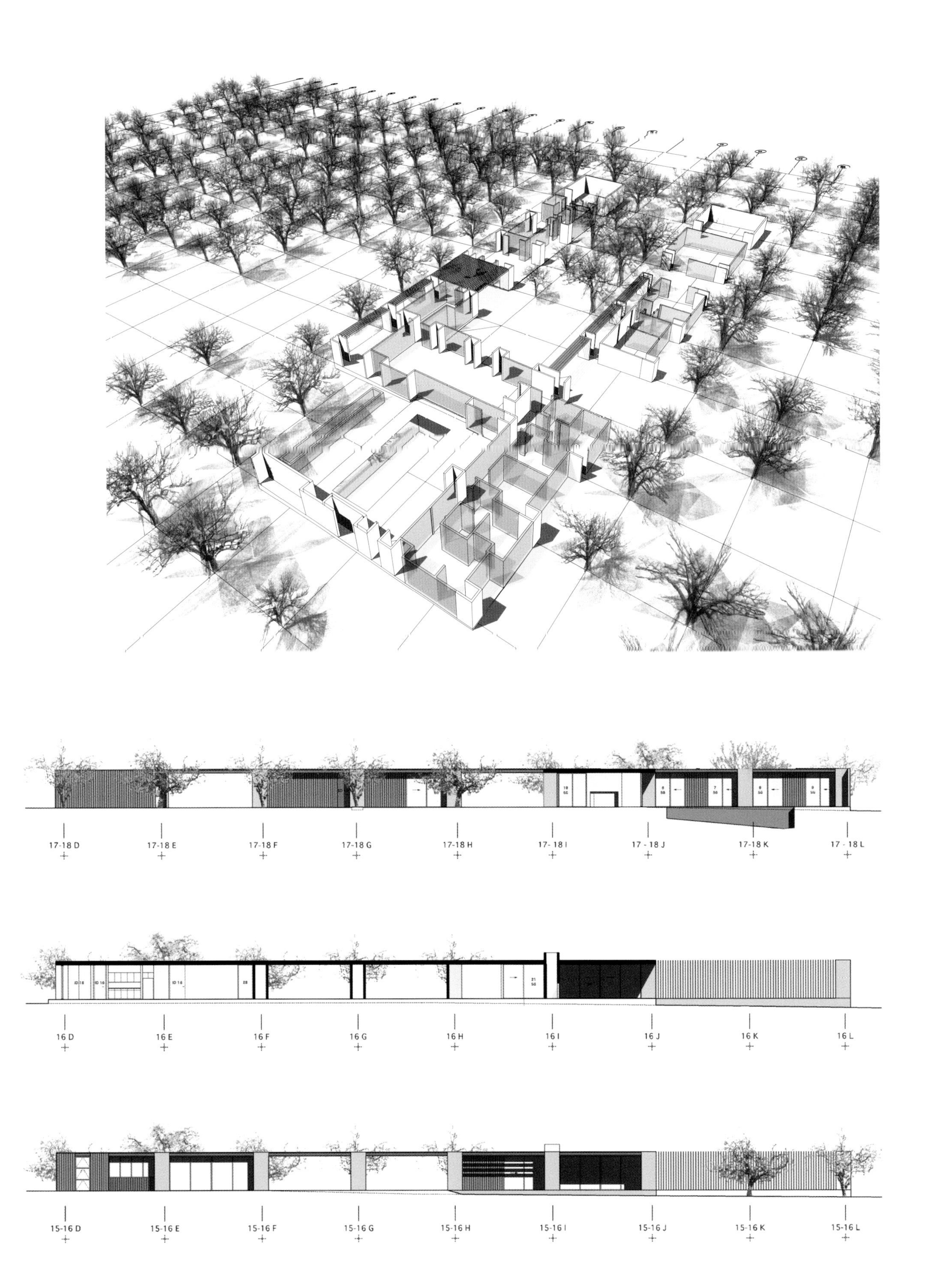

Orchard House

Top: Interior view of dining room looking southwest

Bottom left: Primary living space looking east from dining room, across kitchen, and into sitting room with massive interior cooking fireplace

Bottom right: View from kitchen looking west across dining room and terrace and out into the orchard row occupied by the primary living space

Top left: Interior cooking fireplace. The house and site are all designed to be wheelchair accessible without interruption from interior to exterior.

Bottom left: Exterior view from entry courtyard looking into intersection of library, dining room, and dining terrace

Right: Rhythm of openings along gallery corridor maintains the continuity of north-south orchard rows through the house

Orchard House

Top: View along library doors, looking south toward dining terrace
Bottom: View southward through entry courtyard toward main entrance

Top: Night view looking east across south courtyard
Bottom: View down orchard rows along entry sequence during early morning fog

Chapter 4

Steel Framing

The introduction of steel into building systems can be seen as somewhat incremental, starting with nails and other fasteners and more recently including extensive hardware connectors that help strengthen and connect wood framing elements. As primary structural systems, light-gauge steel framing and heavy structural steel systems can be applied separately or together in highly efficient offsite building situations. The strength-to-weight ratios of steel make possible light and strong structures appropriate to difficult building sites, long cantilevers, and programs requiring large expanses of openings or long spans.

The introduction of the use of steel has transformed architecture and the shape of cities in the last century, allowing taller and more open structures of all sizes and types. All-steel houses are still rare, since wood framing or masonry construction are typically less costly for most types of low-rise buildings, but steel homes have been long predicted as the wave of the future. There were many innovative and experimental projects done in the 1920s and 1930s in both Europe and in the U.S. that explored the qualities of steel for its ability to separate structure from enclosure and create open glass-walled buildings. Mies van der Rohe's German Pavilion in Barcelona (1929) and Tugendhat House in Brno, Czechoslavakia, (1930) were built at about the same time as Richard Neutra's Lovell House in Los Angeles (1929) and Pierre Chareau's Maison de Verre in Paris (1932), all using steel as a primary structural system as well as an integral part of the design aesthetic. Later well-known examples of houses using similar structures of rectilinear grids of steel columns and exposed steel beams and decking include work by Charles and Ray Eames, Raphael Soriano, Pierre Koenig, and Craig Ellwood, all in California.

Beginning at this same time, and continuing through the war years and the 1950s, other groups of architects, engineers, and inventors were more explicitly exploring the potential for industrializing housing production using pre-engineered building systems derived from machine manufacturing techniques. Buckminster Fuller's work on the Dymaxion House project began in 1927, and he continued refining the concept through several prototypes built through the 1940s, although financing for larger-scale production was never obtained. Other brave ventures sought to harness the great buildup of industrial capacity during the war for postwar housing needs, most notably the Lustron Houses, designed by Swedish-born inventor Carl Strandlund, which were built at a large factory in Columbus, Ohio, from 1948 to 1950. This project was the most successful of the manufactured steel house projects, with almost 2,500 units built before the company shut down in 1953 with great financial losses for

Construction and aerial axonometric views of Ishida Ferrari Gallery in Tokyo, Japan, in which we combined prefabricated steel frames from Japan with prefabricated panelized wood framing infill shipped from the United States.

Strandlund and other investors. Despite the alluring promise of tapping the resources of the steel manufacturing industry, which had been greatly built up during the war years, it was precisely the large scale required for cost-effective production that made it difficult for the houses to be widely accepted into the market. The Dymaxion and Lustron houses were conceived as products to be mass produced, which was at odds with the varying needs of different sites and different owners. Lack of flexibility and adaptability is often cited as a chief failing of these designs, and in the end they could not compete with the more flexible and inexpensive wood and masonry houses that were the norm.

Although light-gauge steel stud construction has been available in North America since the 1930s, it has until recently been mainly used for non-load-bearing partition walls in commercial construction. Carpenters and homebuyers were slow to adopt it for residential building, but in the last ten or twenty years it has become increasingly common. This is due in part to the declining quality of wood available for building, as well as price volatility in the wood markets as supplies have become less reliable. Aware of the expanding opportunity, the steel industry has made a coordinated effort to produce components in sizes and configurations appropriate to residential construction, and have created complete building solutions based on their products.

Because most light-gauge steel framing follows the same modules and construction sequences as wood framing, it is relatively easy to interchange or even mix the two systems without affecting the overall design of the building, and it is common for builders to bid their projects both ways before deciding which system to use.

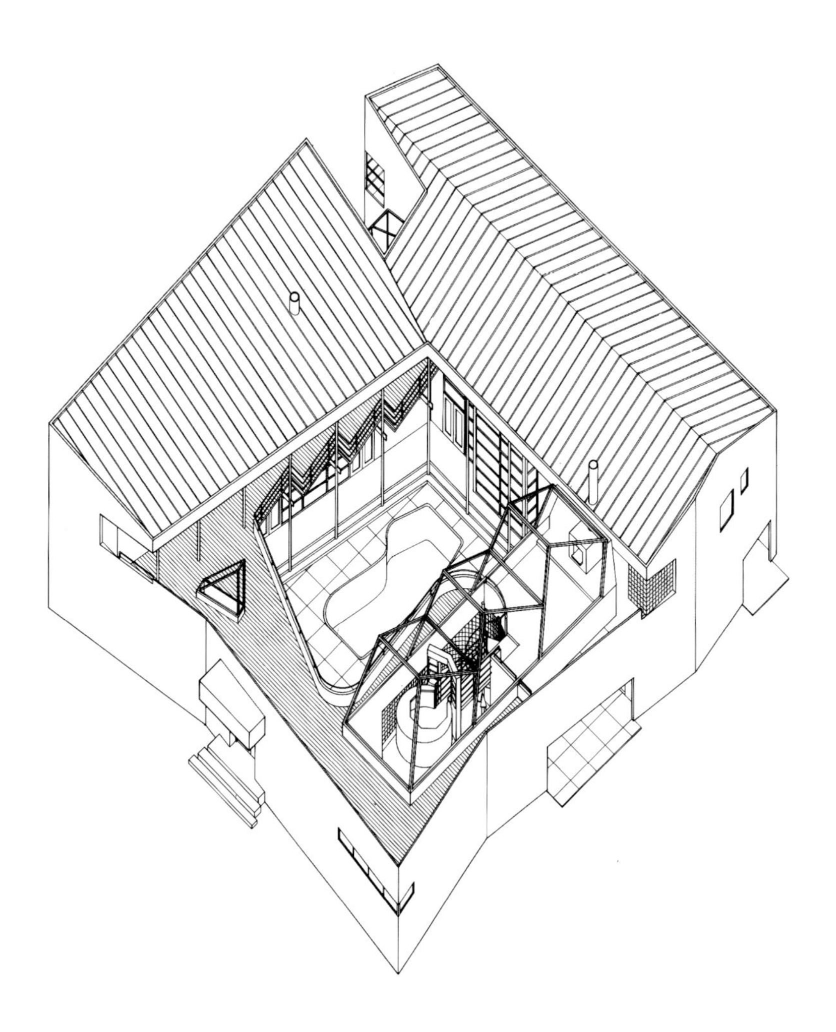

Steel Framing

Exploded component views of experimental folded, perforated sheet steel stair assemblies

Parts List:

a. Folded railings
b. Folded stair/landing
c. Folded structural railing inserts
d. Perforated sheet steel CAD/CAM cut with waterjet
e. Handrail and handrail support system
f. Steel mounting plates

Assembly Sequence:

1. Stair system patterns cut with waterjet
2. Brake-formed stairs
3. Brake-formed railings
4. Brake-formed structural railing inserts
5. Handrail and handrail support system
6. Steel mounting plates

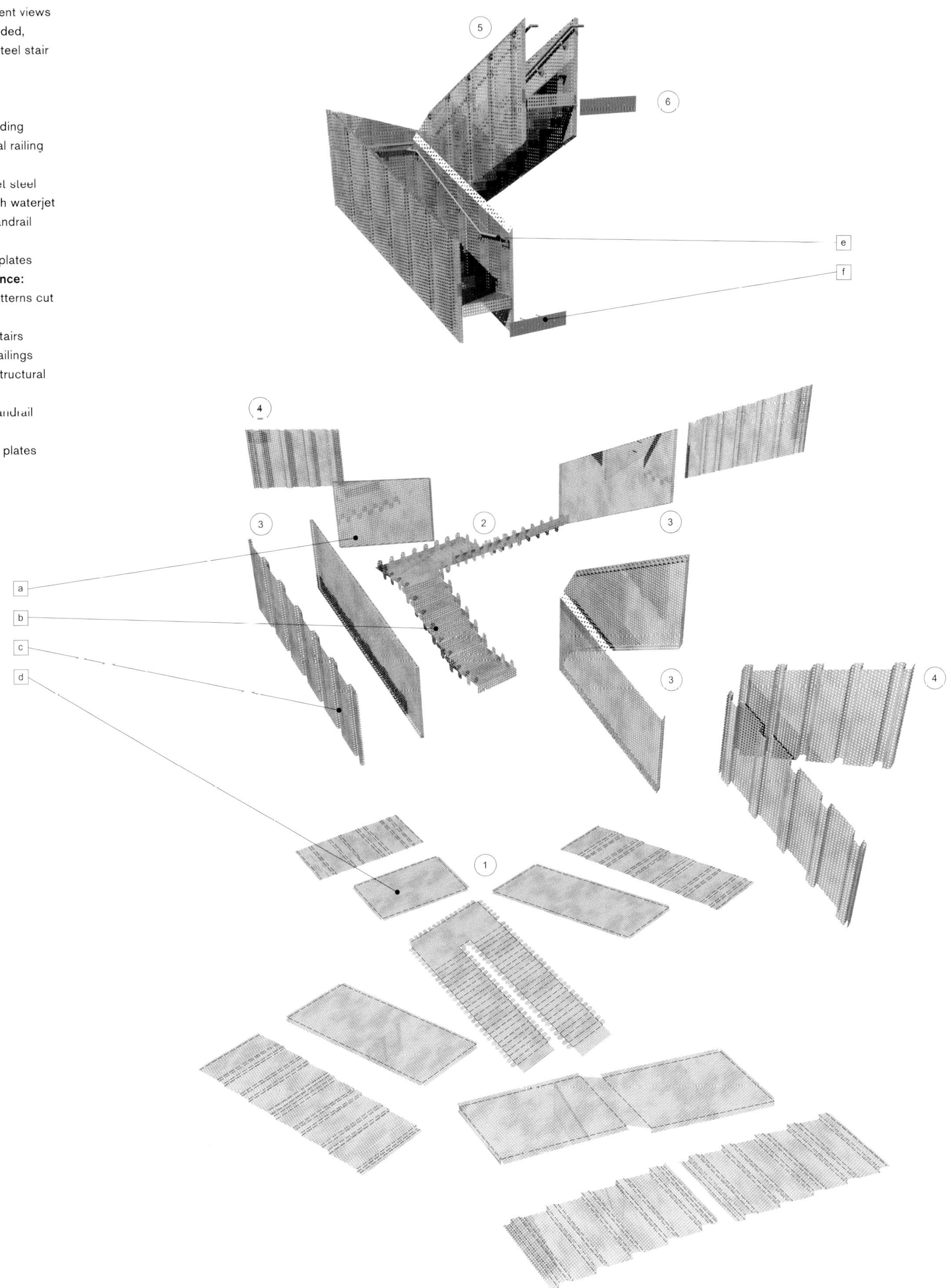

Cantilever House

Granite Falls, Washington / 2004

The impetus for the development of a prefabricated building series based on extensive use of structural steel frames originated with the desire to build a house on a beautiful rocky promontory in the Cascade Mountains north of Seattle. While the owner's 360 acres of land provided many excellent and more sensible locations to build, we were immediately drawn to a narrow rock ridge that provided the best near and distant views in all directions of mountains, lakes, streams, cliffs, and the islands and bays of Puget Sound in the distance, as well as a spectacular view across a narrow valley of twin year-round waterfalls spouting from the opposite cliff face. The site conditions required a completely different approach to the design than anything we had done before, and though it initially proved to be an exceptionally individual situation, we instead chose to use it as an extreme test for a more broadly applicable modular system that could be deployed for a wide variety of site conditions. A second prototype two-family configuration is in design for a relatively flat urban site in San Diego, and several other projects are in development for sites where the small footprint and larger upper stories are advantageous. Despite the fairly unrestrictive zoning constraints for the Granite Falls site, the challenging topography and geotechnical conditions are an example of factors that define the overall design strategy of the system. The small ground-floor building footprint/foundation reduces the cost of this expensive area of the house and allows the points of attachment to adapt to varying slope and soil conditions.

The Cantilever Series of projects explores the combination of multiple prefabrication techniques to build low-cost, high-quality, site-adaptable and program-adaptable manufactured buildings, taking advantage of the best qualities of each system and overcoming the limitations associated with using any single system on its own. The reliance on structural steel as a primary frame allows great flexibility in the relationship of the building to the site itself, as well as in the configuration of the building enclosure relative to the frame.

The concept of hybridizing multiple material and structural systems facilitates the utilization of relatively common, non-proprietary products available from multiple manufacturers, which encourages competitive pricing and enhanced availability of components. Coordinating and combining existing manufactured systems distributes capital investment in system detailing, inventory, and prototyping among a group of sophisticated manufacturers with already established, project-specific customization and component packaging and delivery systems. By creatively combining a small number of manufactured systems into one comprehensive approach, there is reduced need to undertake the costly design of an entirely new system, and there is increased certainty of long-term applicability and widespread industry support. Because the system components are adapted from readily available, multiple-manufacturer products, the buildings can tap into the economies of large scale that result from the research and production capacities of large manufacturers. The small number of relatively simple raw materials used in the products mean that the manufacturers are in turn purchasing in large quantities from competitive sources. The customization for specific sites and programs is a result of the specific configuration of pre-designed modular components within a system of typical assemblies designed by the architects in collaboration with the manufacturers.

The building system used in the first two prototypes is a marriage of two common mass-produced building elements—prefabricated steel structural frame, of the type commonly used for commercial and high-rise structures, and a structural insulated sandwich panel (SIPs) system that provides all non-glazed building envelope areas. The enclosure system is designed around using the same interchangeable, rearrangeable, low-labor structural panels for the walls, floors, and roof, thereby minimizing unpredictable site labor costs, speeding construction, and reducing disturbance to the existing site context. The construction period from foundation through exterior enclosure could be completed in as little as four weeks, with an additional six to eight weeks required for installing building systems, fittings, and finishes. No interior walls are essential to the structural integrity of the building, so the interior build-out of the shell can be completed by the owners or less skilled labor without structurally dictated quality or scheduling requirements.

Although the materials and methods of construction for this system are chosen for efficiency and affordability, particularly when faced with difficult site conditions, the underlying design principles guiding the Cantilever Series include the larger goals of producing affordable, high-quality buildings that offer variety, adaptability, convertibility, strength, simplicity, spatial richness, and optimized access to views and light.

Construction views looking south (left) and up from top of cliff face below (right)

OPPOSITE
Dusk view looking south across rocks

Cantilever House

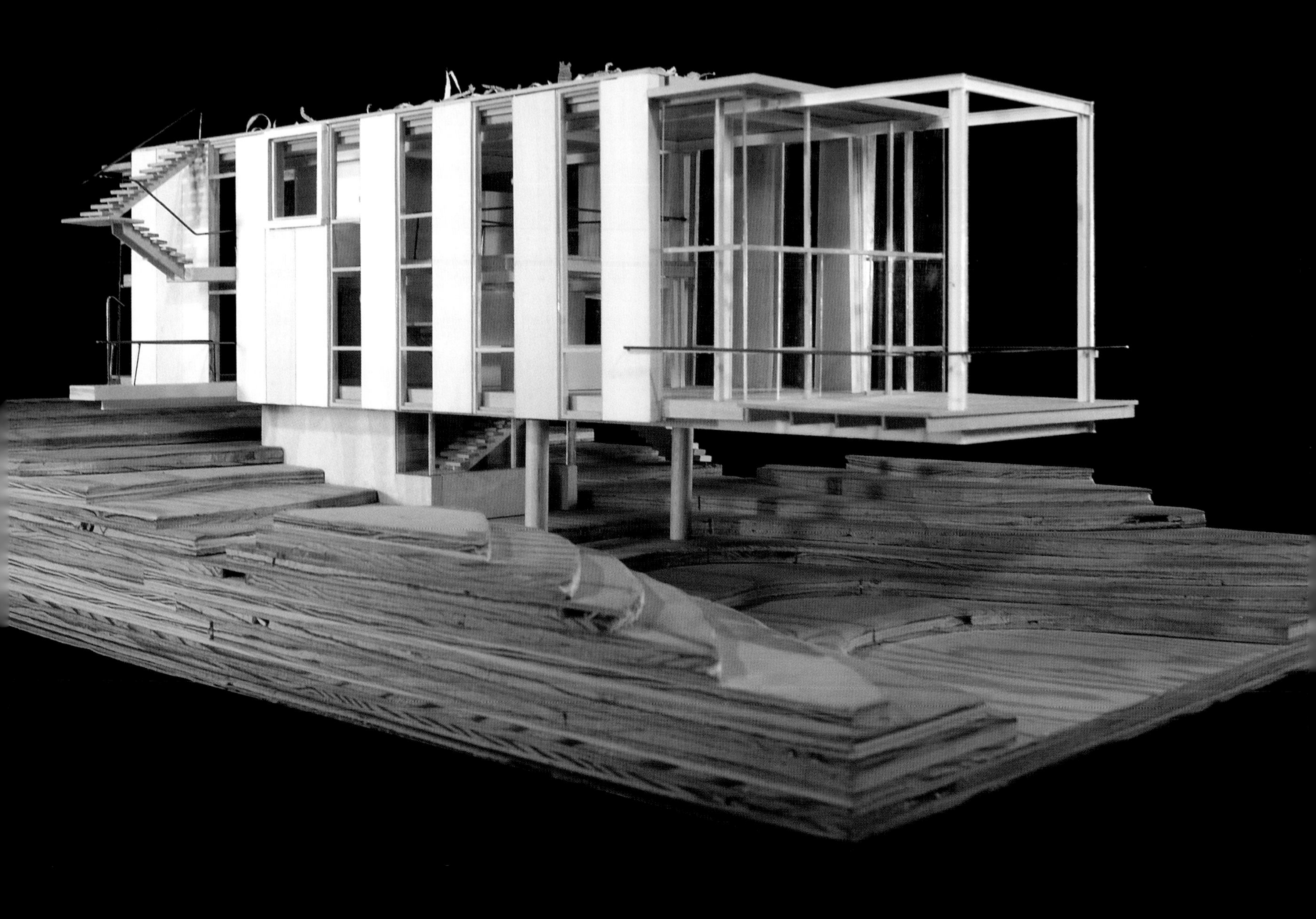

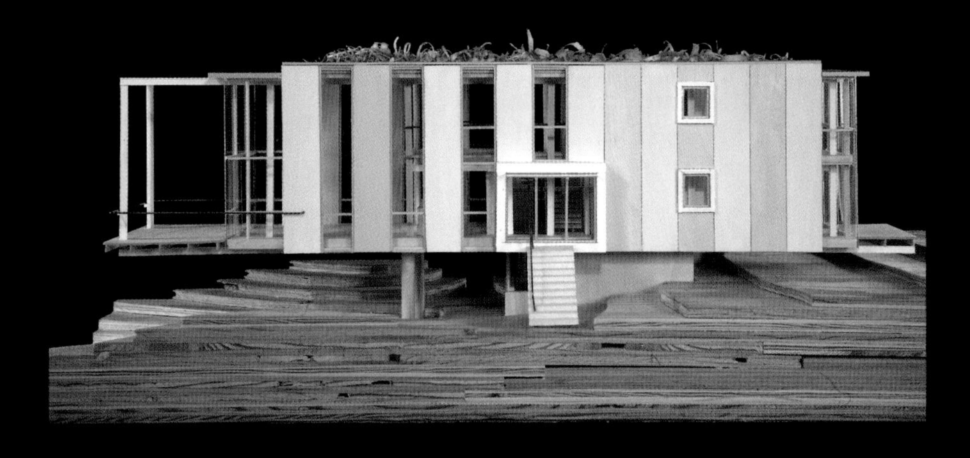

Steel Framing

OPPOSITE
Study model, South elevation view (top) and north entry elevation view (bottom)

Below: Vicinity site plan

The cantilever truss structure is particularly well-suited to hillsides or unusual geotechnical conditions as it requires a minimal number and area of foundation points, preserving the landscape while greatly reducing site development and costly onsite foundation costs. The modular steel frame structure can span between supporting columns or foundation walls located below truss bay connections, and can cantilever up to thirty-two feet (9.75 m) at the ends of the building. In addition to providing great flexibility for placement and configuration of the foundations, the two-way Vierendeel truss creates a complete, self-contained structural system independent of the building enclosure and interior walls. All gravity and shear forces are resisted solely by the moment-frame truss structure, and no shear walls are required, providing complete freedom in the placement of windows and other wall openings, and in the positioning and later modification of interior or exterior walls.

In the Granite Falls prototype, the secondary structural system and primary insulated enclosure system utilizes manufactured sandwich panels composed of expanded polystyrene foam glued between skins of Oriented-Strand Board (OSB), and is designed to integrate with a range of window or curtain wall systems by various manufacturers. These panels have great structural rigidity in spanning between the steel supports as roof, floor, and wall systems and can cantilever beyond the frame support points to allow different building widths without modifying the pre-engineered structural frame. The overall dimensions of the building are configured so that each four-foot-wide (1.2 m) panel is shipped to the site cut to the full width of the floor or roof, or to the full height of the walls, reducing the handling of small parts and accelerating the onsite erection process.

In the case of the two current prototypes and in many other urban and natural hillside locations, the roofscape is an important visual element, both as seen from the building itself and as a part of the larger built or natural environment. The roof structure is designed to have a sod roof covering, or it can accommodate a rooftop terrace living space surfaced with a variety of materials. The waterproof roof membrane is a one-piece, prefabricated pan rolled out and clipped into place in less than one day. The building assembly process allows the roof membrane to be installed early on in construction—even before wall and floor assembly—greatly reducing labor costs and unpredictable weather-related costs and delays.

The modular system is designed to accommodate a range of plug-in accessory items such as exterior decks and stairs, window boxes, sunscreens, daylighting shelves, and solar water heating or electrical generation panels. The intention is to have a basic building structure that allows for extensive customization over time, attaching accessory elements from many manufacturers without the need for custom attachment systems or costly modifications to the primary structure.

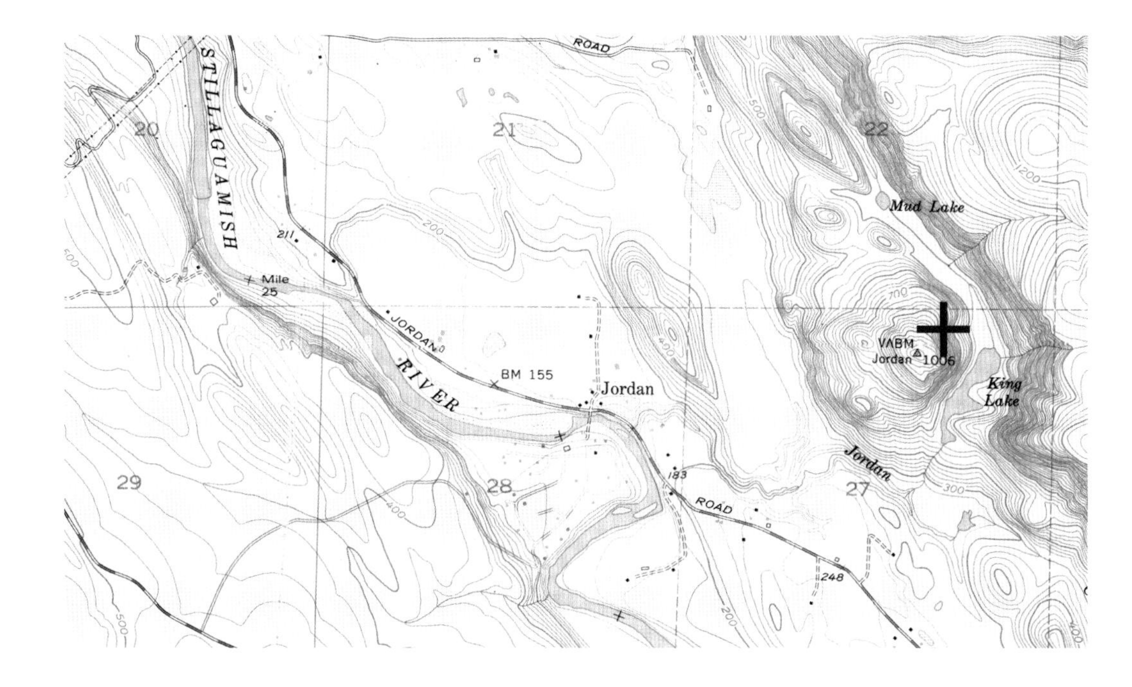

Cantilever House

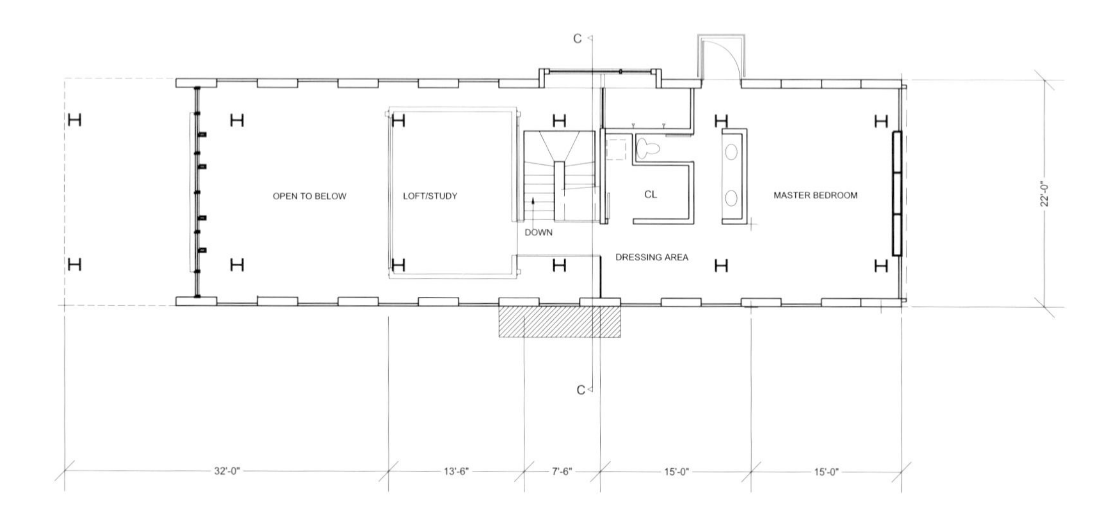

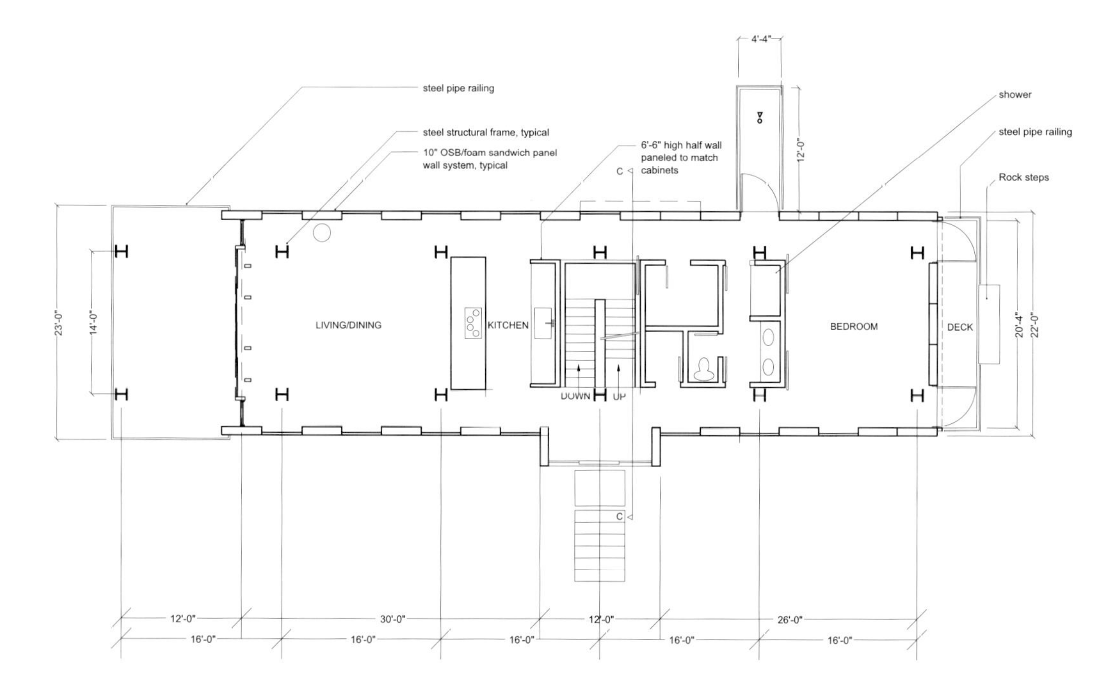

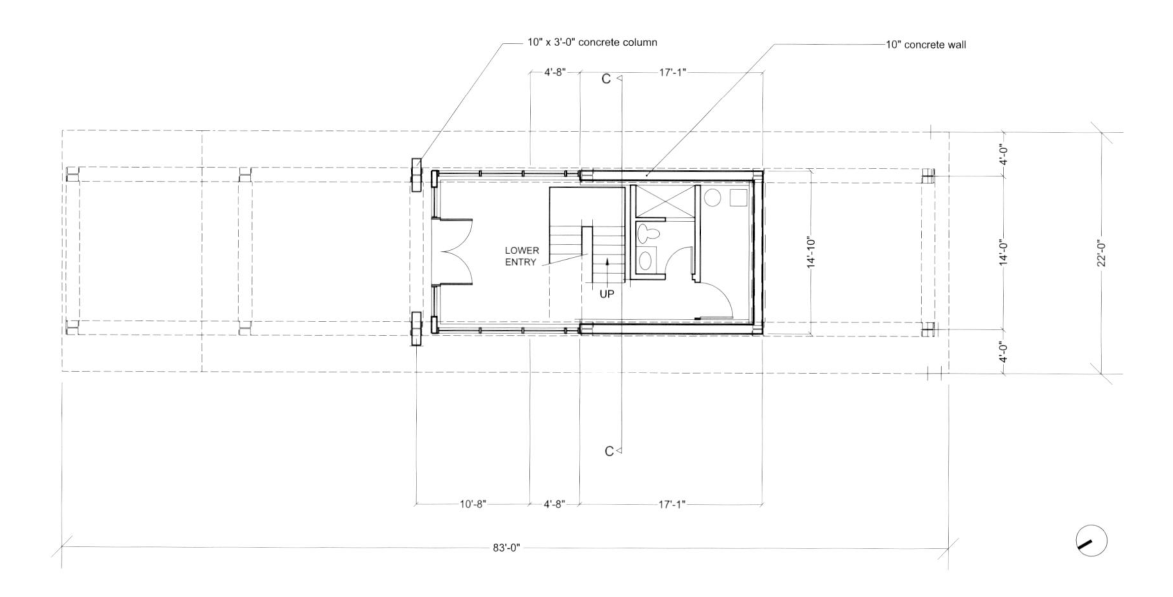

Floor plans. Top to bottom: master bedroom and loft floor; main floor with entry stair; foundation level with service entry

Steel Framing

Top: South elevation
Middle left: East elevation
Middle right: Cross-section through stair and floating loft
Bottom: Typical SIPs panel layout showing main floor panels that span within steel frame

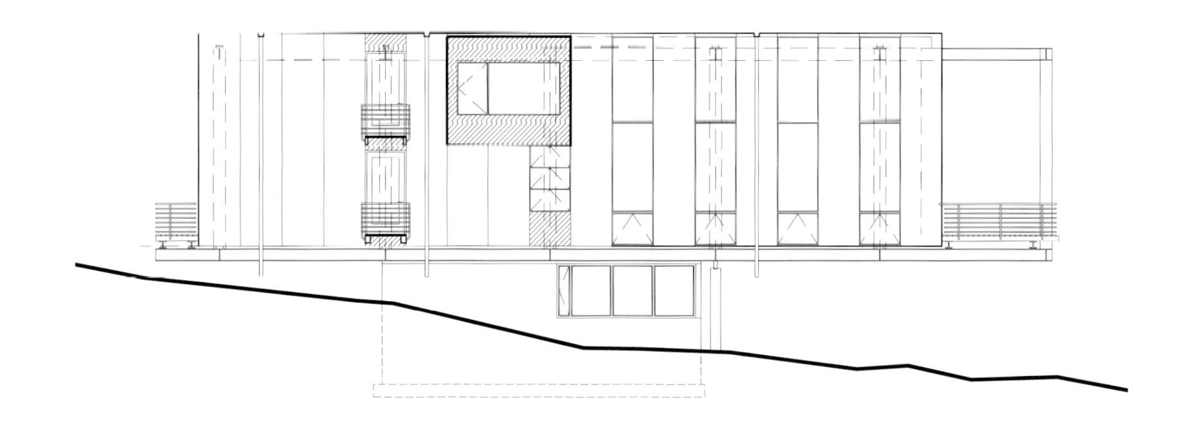

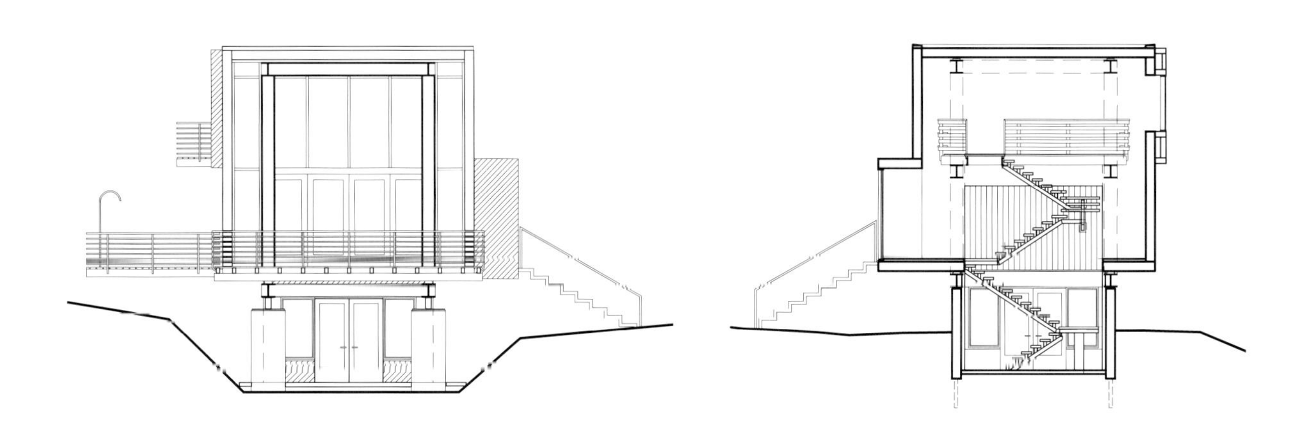

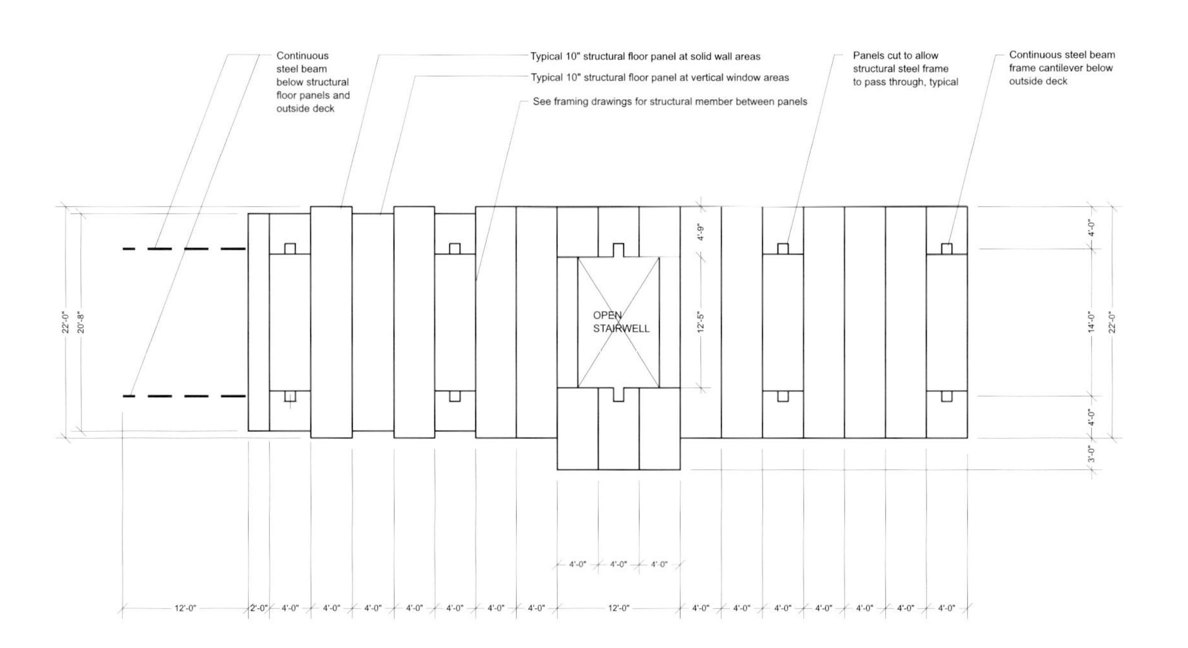

Steel Framing

OPPOSITE
Primary steel frame, secondary structure, and insulation enclosure system of SIPs panels

Right. Perspective view with prefabricated structural steel moment-frame truss lifted above to explain cantilever system

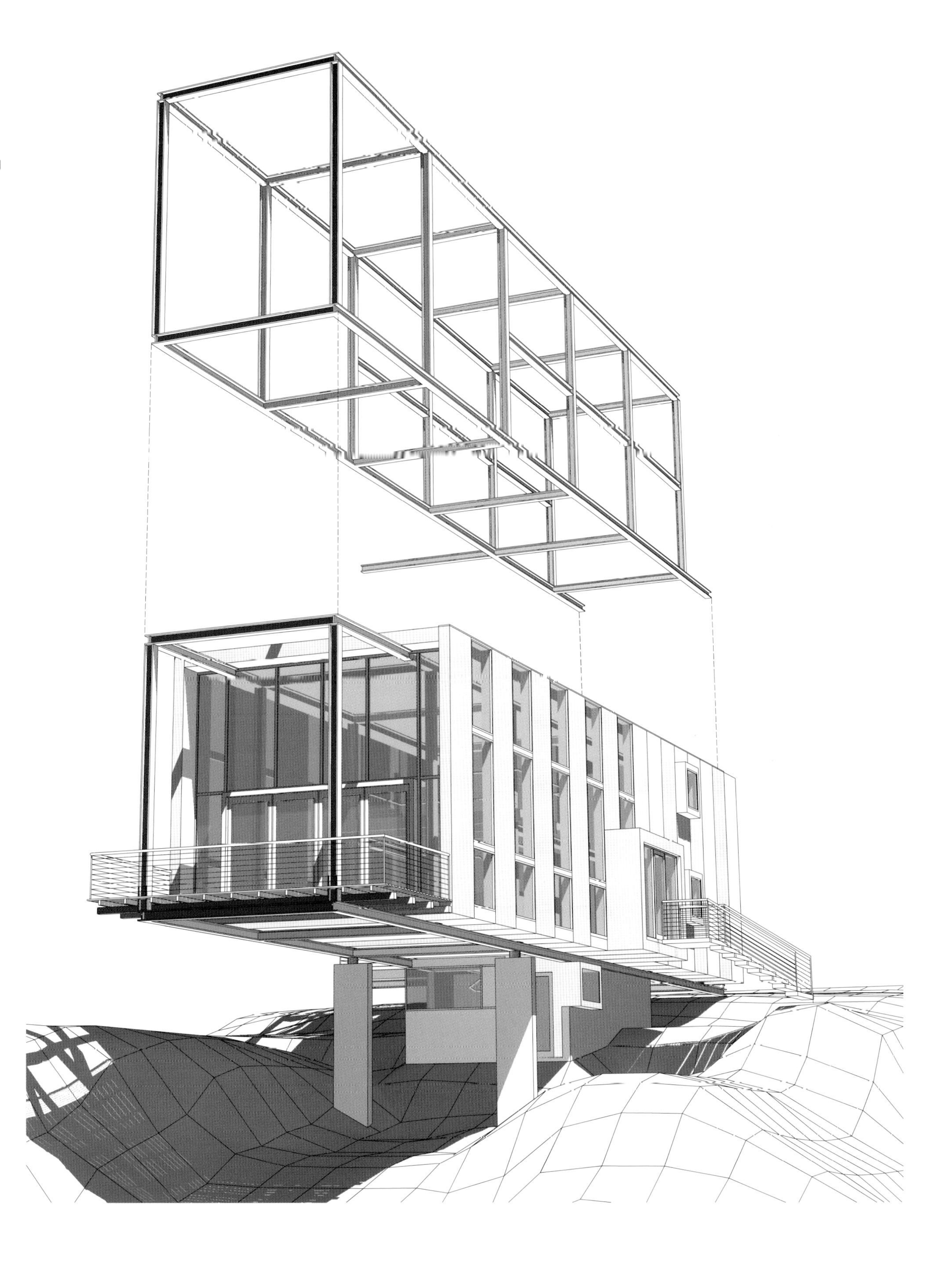

Cantilever House

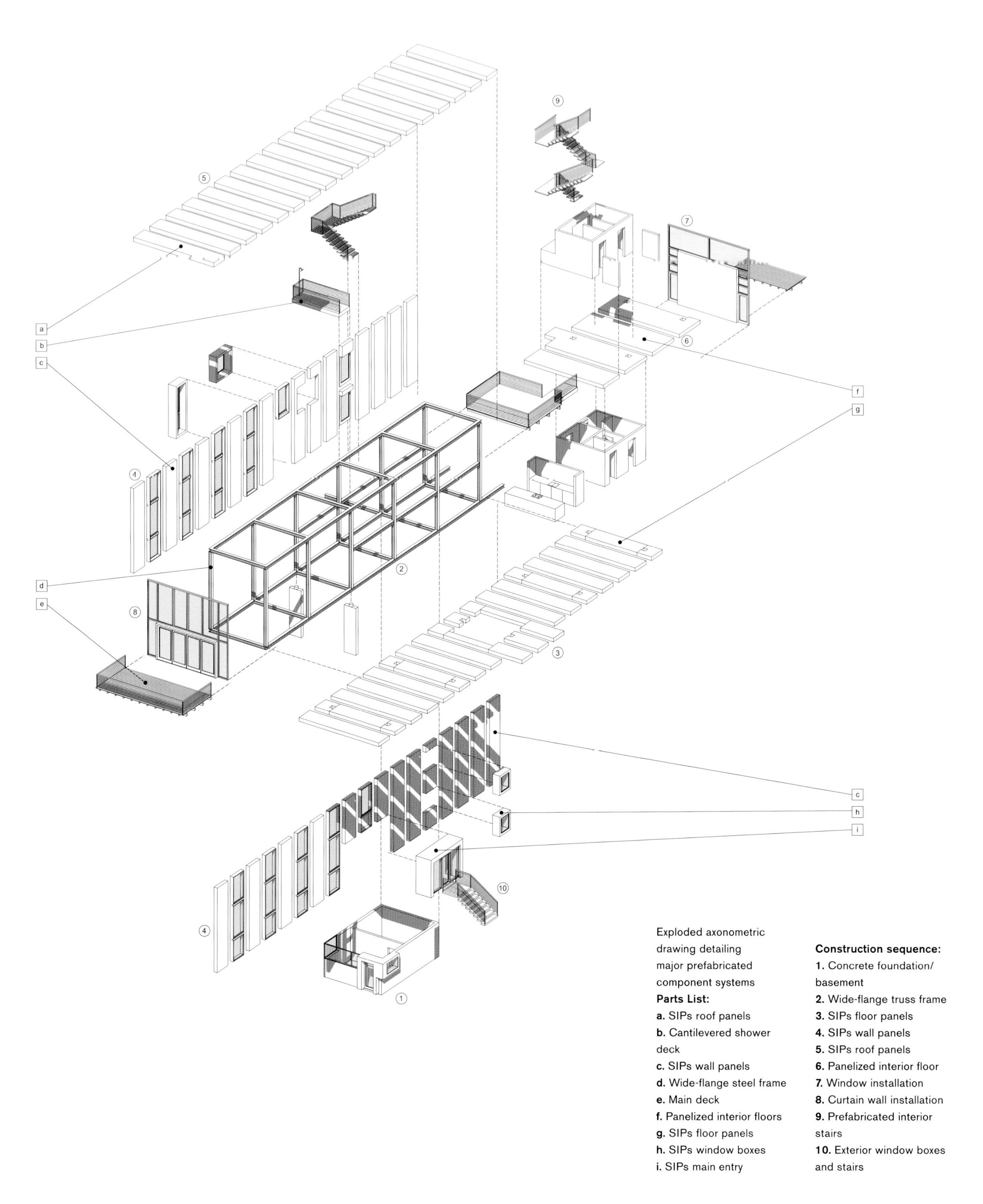

Exploded axonometric drawing detailing major prefabricated component systems

Parts List:

a. SIPs roof panels
b. Cantilevered shower deck
c. SIPs wall panels
d. Wide-flange steel frame
e. Main deck
f. Panelized interior floors
g. SIPs floor panels
h. SIPs window boxes
i. SIPs main entry

Construction sequence:

1. Concrete foundation/ basement
2. Wide-flange truss frame
3. SIPs floor panels
4. SIPs wall panels
5. SIPs roof panels
6. Panelized interior floor
7. Window installation
8. Curtain wall installation
9. Prefabricated interior stairs
10. Exterior window boxes and stairs

Interior detail views of finished construction, showing inset relationship of steel frame to cantilevered secondary framing and enclosure walls

Everglades Swamp House

Arcadia, Florida / 2005

Sited on a large tract of overgrown swamp and former ranch land fronting the Peace River in central Florida, the Everglades Swamp House is the first in a coming series of tropical stilt houses, largely prefabricated with a panelized light-gauge steel system. During the summer wet season this property can be flooded with five or six feet of water for weeks at a time. During the rains, neighbors of this site commute by canoe to their cars left on high ground by the nearest highway. By logic—and now also by law—houses in this area are required to be elevated on stilts well above the potential high-water mark. This provides a wonderful opportunity to lift the house up into the trees, away from the snakes and alligators and into the cooler breezes.

In contrast to the Cantilever Series, this situation allowed for an economical use of many points of support from the land. Instead of the more costly heavy structural steel-frame system, we chose to use a light-gauge steel-panel system that provides structure and building enclosure in a single component. It was not difficult to plan the building to have sufficient solid panels that would develop the required shear strength through the panels themselves, thus further reducing the need for heavy steel members. In this case the construction system is prefabricated by a steel building manufacturer in Georgia, who fabricates all the steel elements in their factory and ships them to the site, where they are lifted into place with a mobile crane by the local building contractor. The primary components are all galvanized light-gauge steel, with a small number of heavier sections of wide flange beams that collect and distribute the loads to hot-dip galvanized mild steel pipe columns for vertical support.

By elevating the main living floor fourteen feet (4.26 m) above the floodplain, the house is brought into the junglelike tree canopy of the Everglades swamp forest, well shaded by the leaves and open to the breezes during the nine months of the year when air conditioning is not required. The horizontal building plan is open to the jungle view in all four directions, with house functions programmed to allow zoned privacy as well as through-house views and ventilation. In keeping with the requirements for a low-maintenance, hurricane- and termite-resistant solution to the entire building, metal products are used throughout, with aluminum windows and alternating galvanized and painted sheet steel siding, intended to act as a geometric camouflage dissolving the building mass into the dripping tree canopy.

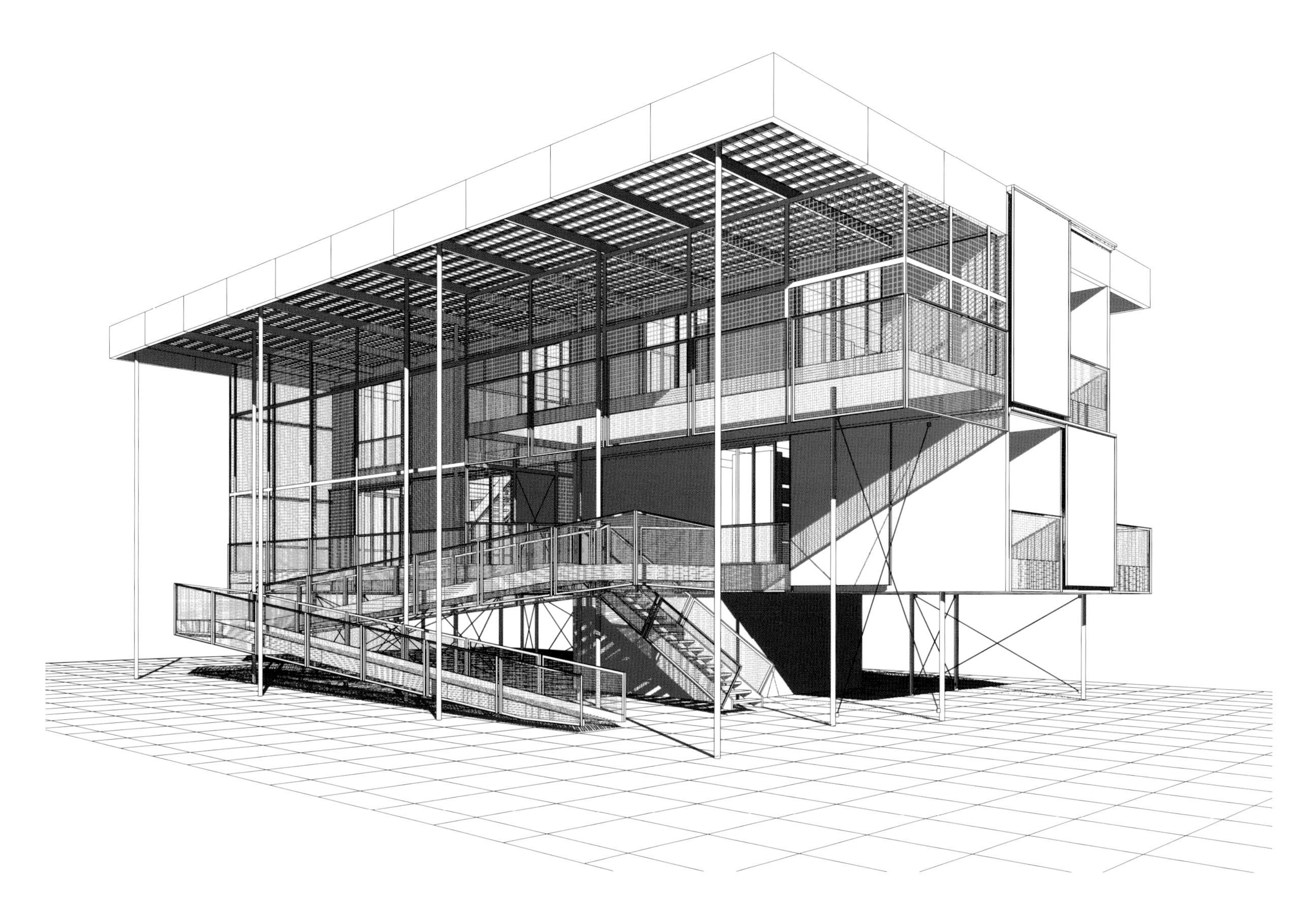

OPPOSITE
Entry perspective view of Swamp House. Note large shade trellis overhang, screen porches, hurricane shutters, and living spaces high above flood water

Below: Site plan and views of property

Everglades Swamp House

Left: (top to bottom) top-, main-, and ground-floor plans

Right: Inset perspective view of main-floor screened terrace on south face of house, looking east toward entry stair

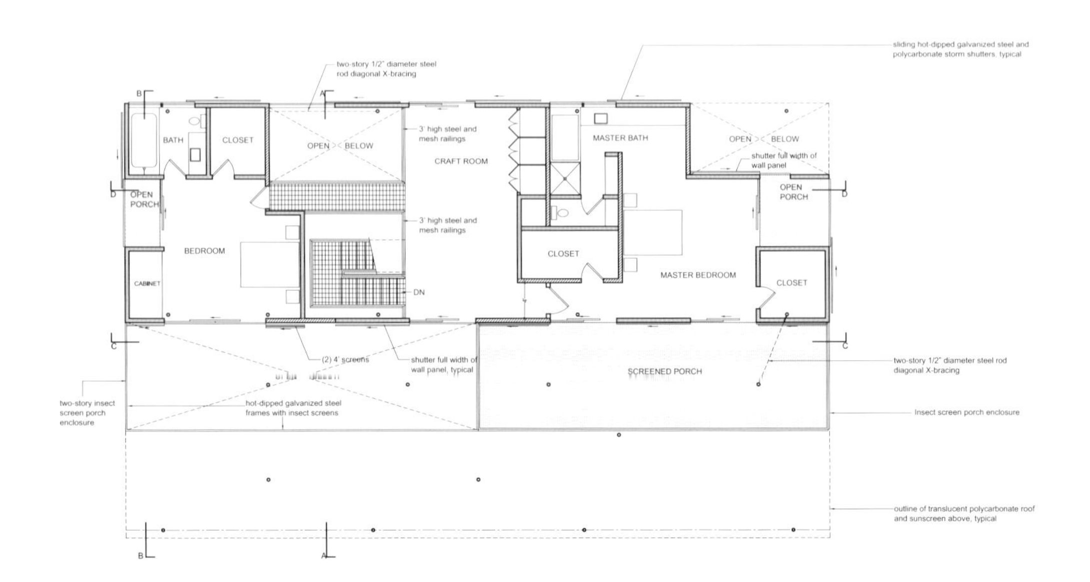

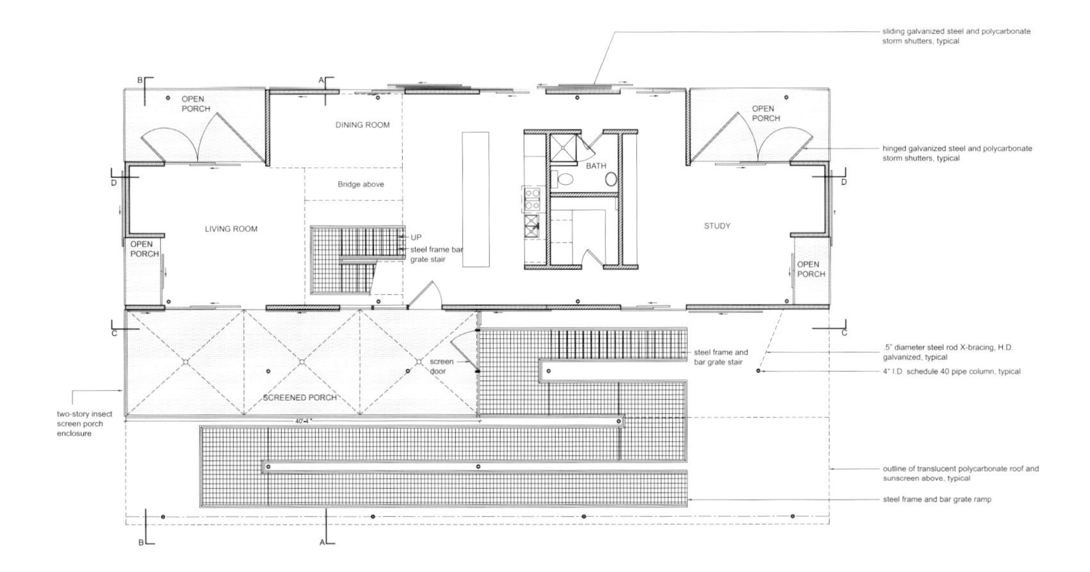

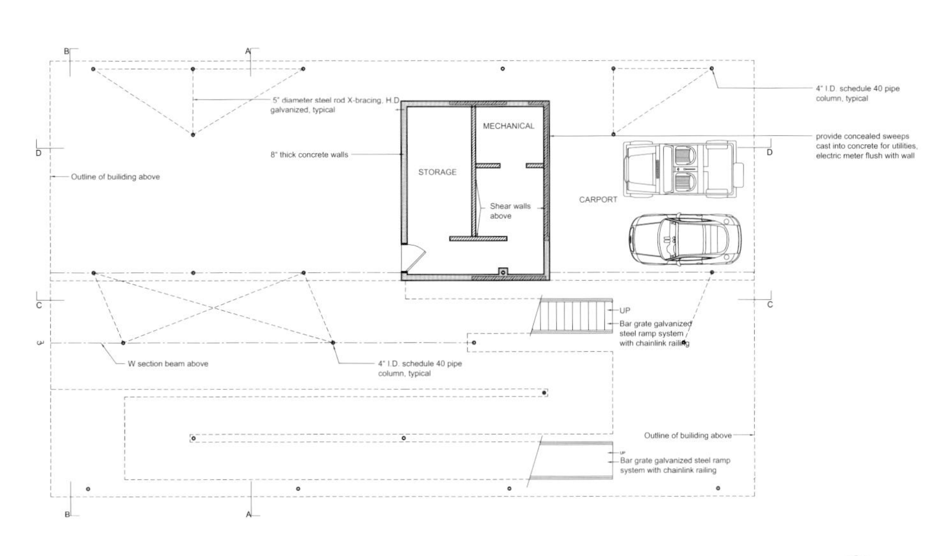

Steel Framing

Exploded perspective view of primary construction system components

Parts List

a. Polycarbonate roofing
b. 2x4 cedar purlin sunshade
c. Parapet sunshade system
d. Prefabricated steel/ bar-grate stairs
e. Insect screen deck enclosure
f. Prefabricated steel/ bar-grate ramp
g. Panelized steel/foam roof system
h. Panelized steel/foam wall system
i. Panelized steel/foam floor system
j. 4 in. diameter steel pipe columns
k. Steel/polycarbonate storm shutters
l. Steel/chain link railings

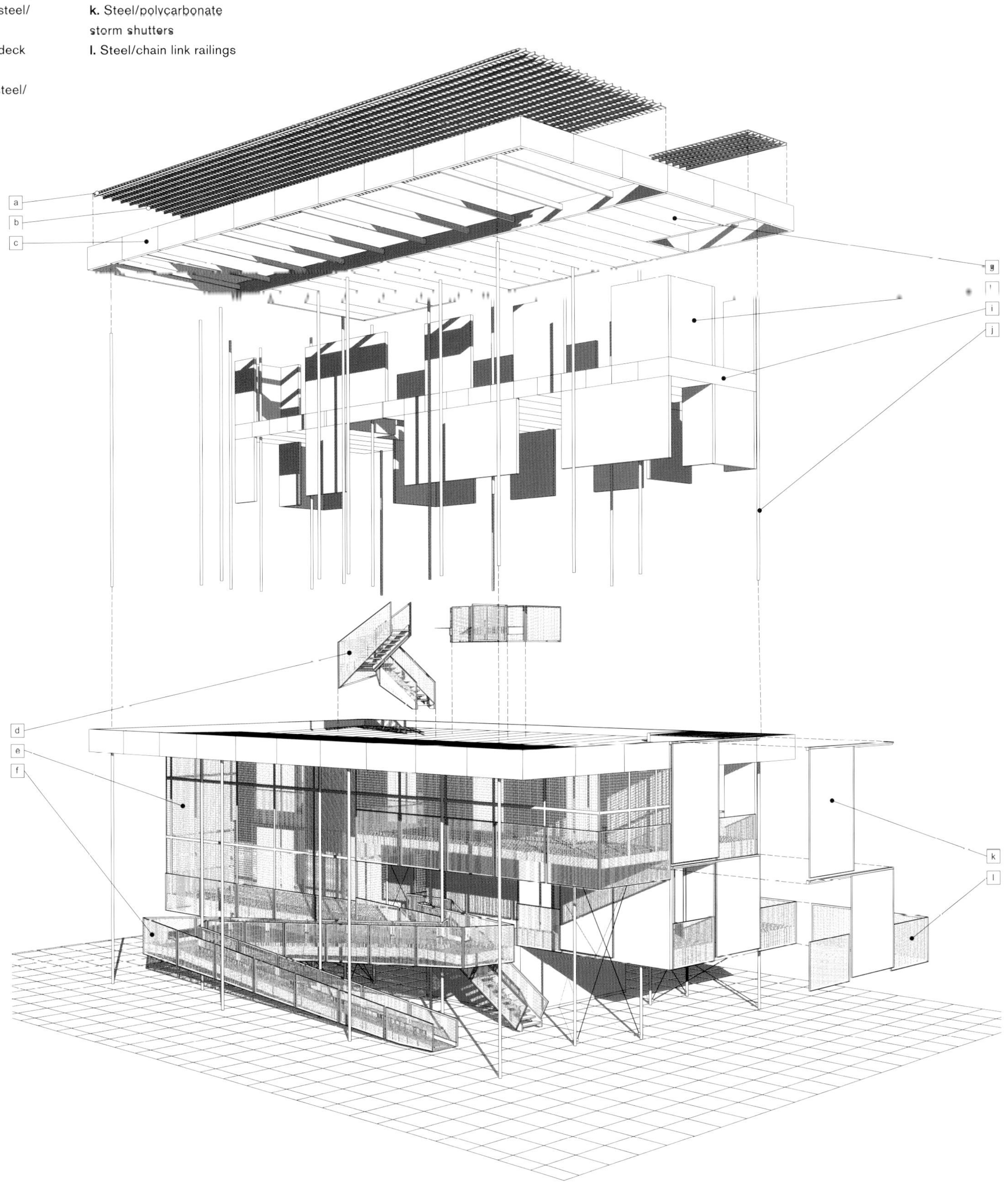

Everglades Swamp House

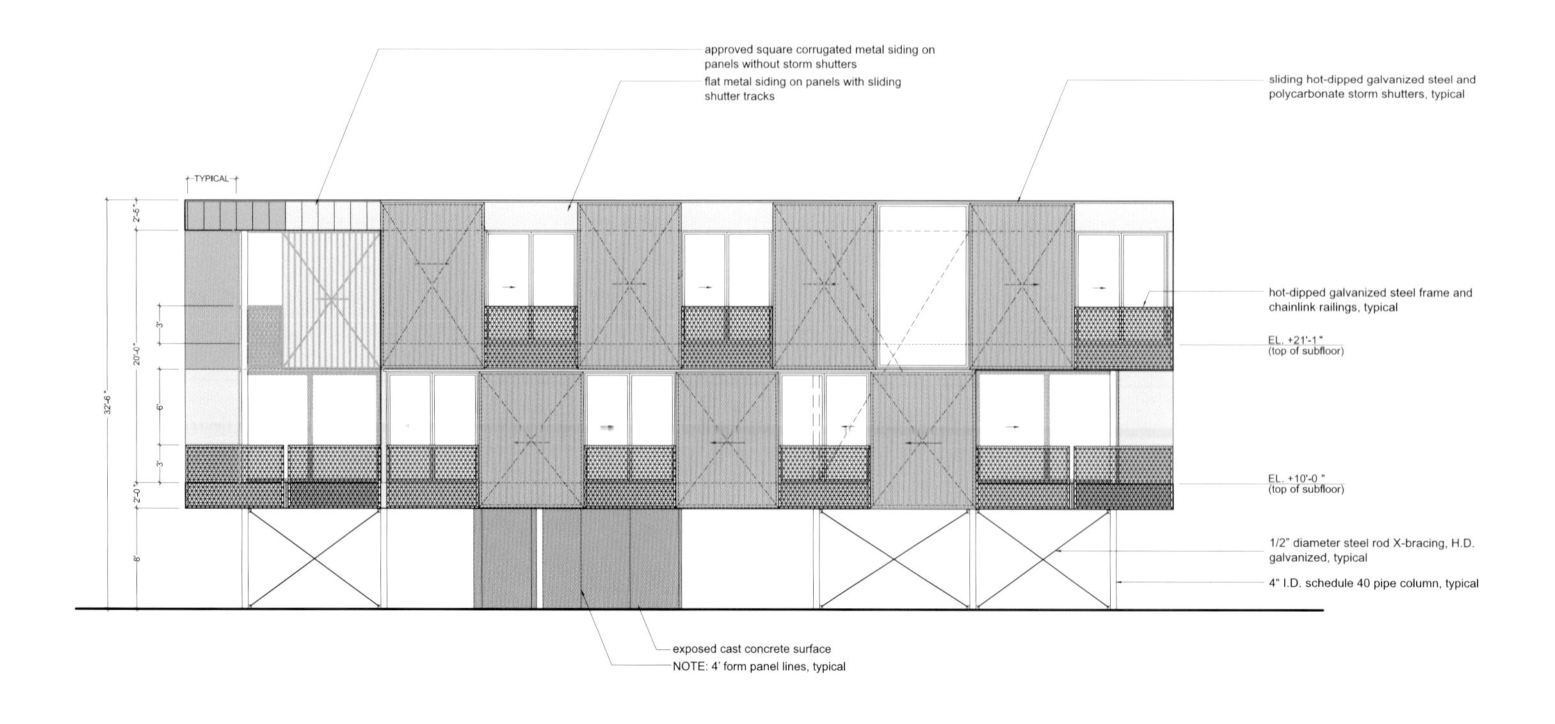

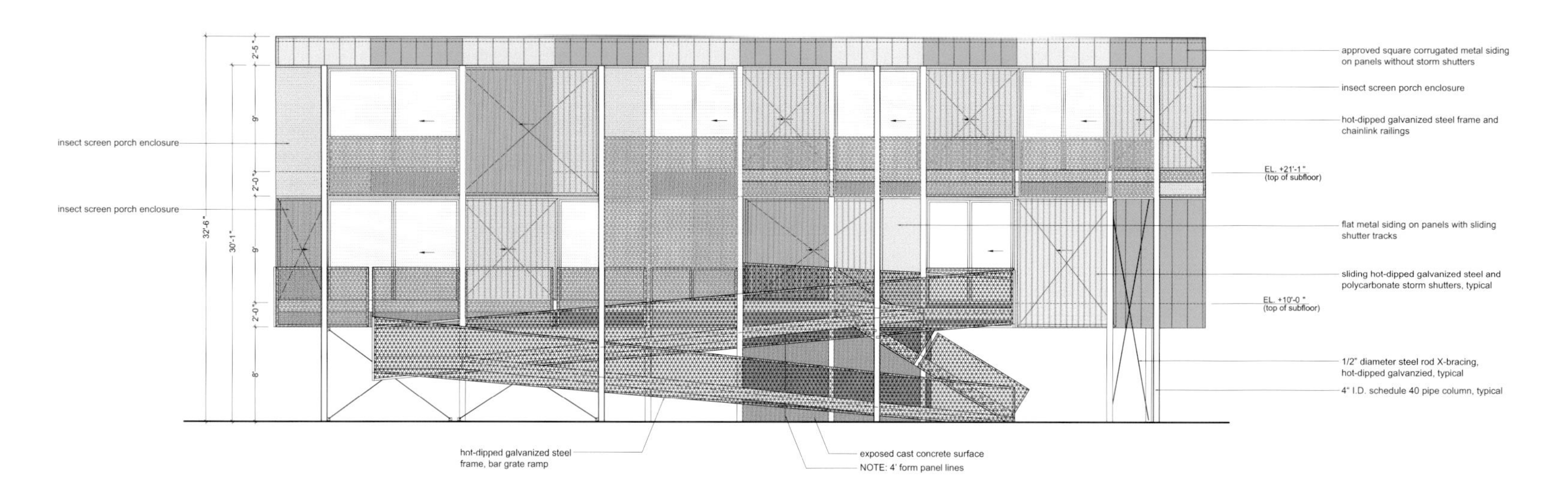

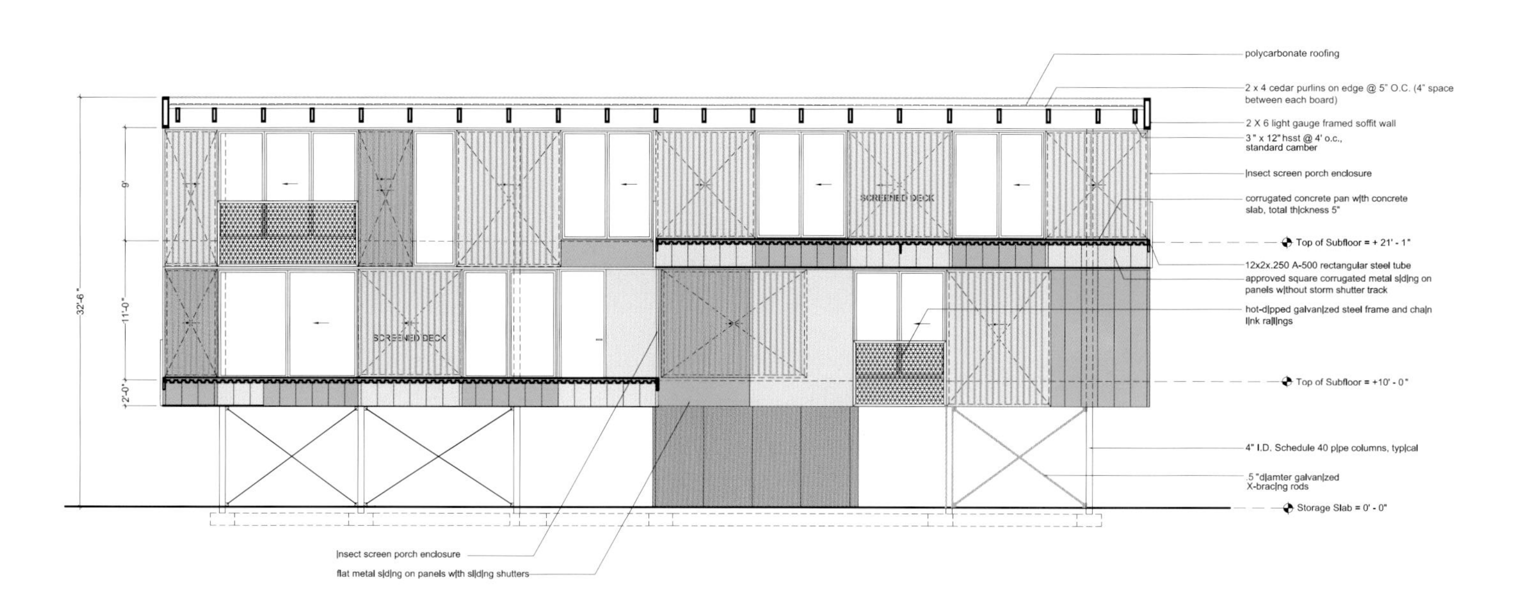

Steel Framing

OPPOSITE
Top: North elevation
Middle: South elevation
Bottom: Longitudinal section showing south elevation enclosure panels and hurricane shutter system

Top: West elevation
Middle: East elevation
Bottom: Cross-section looking east through entry and primary living areas

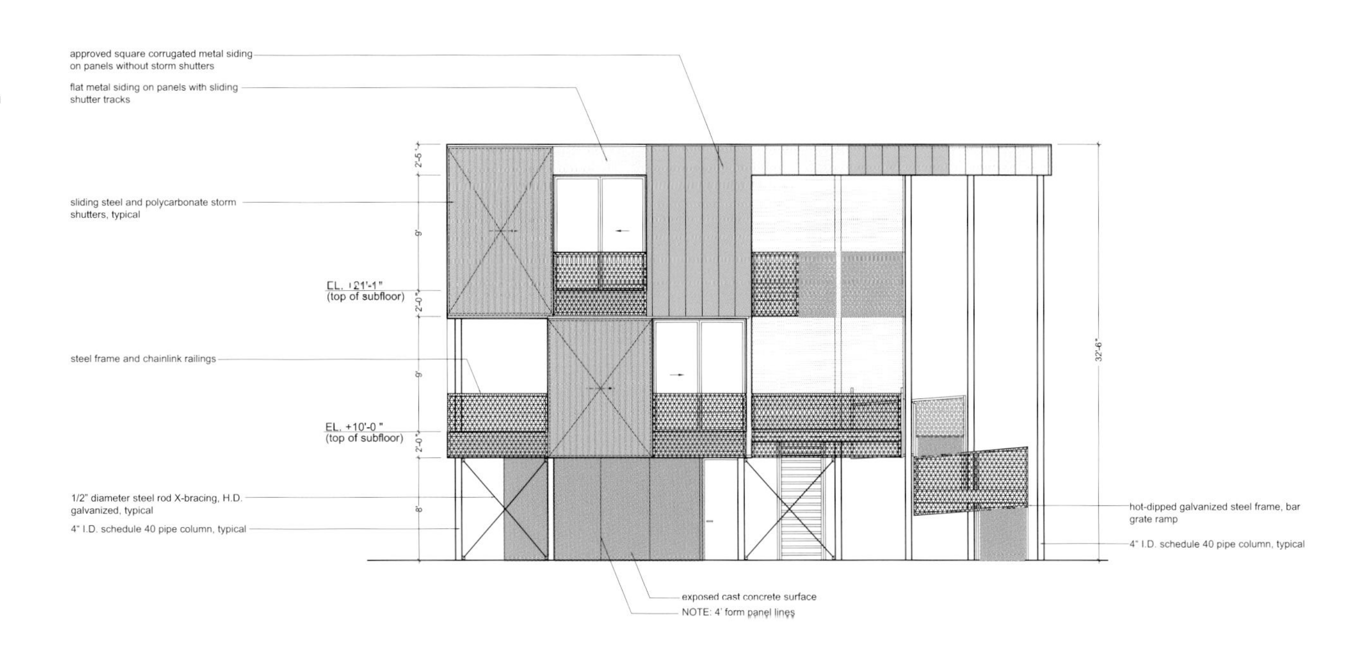

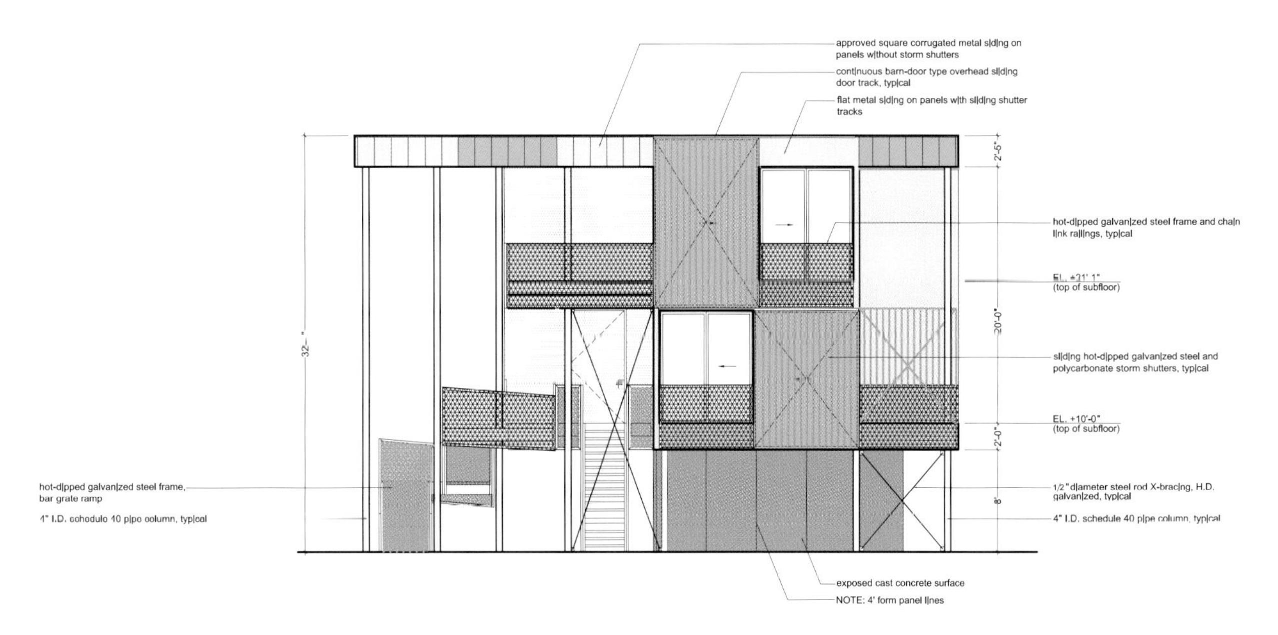

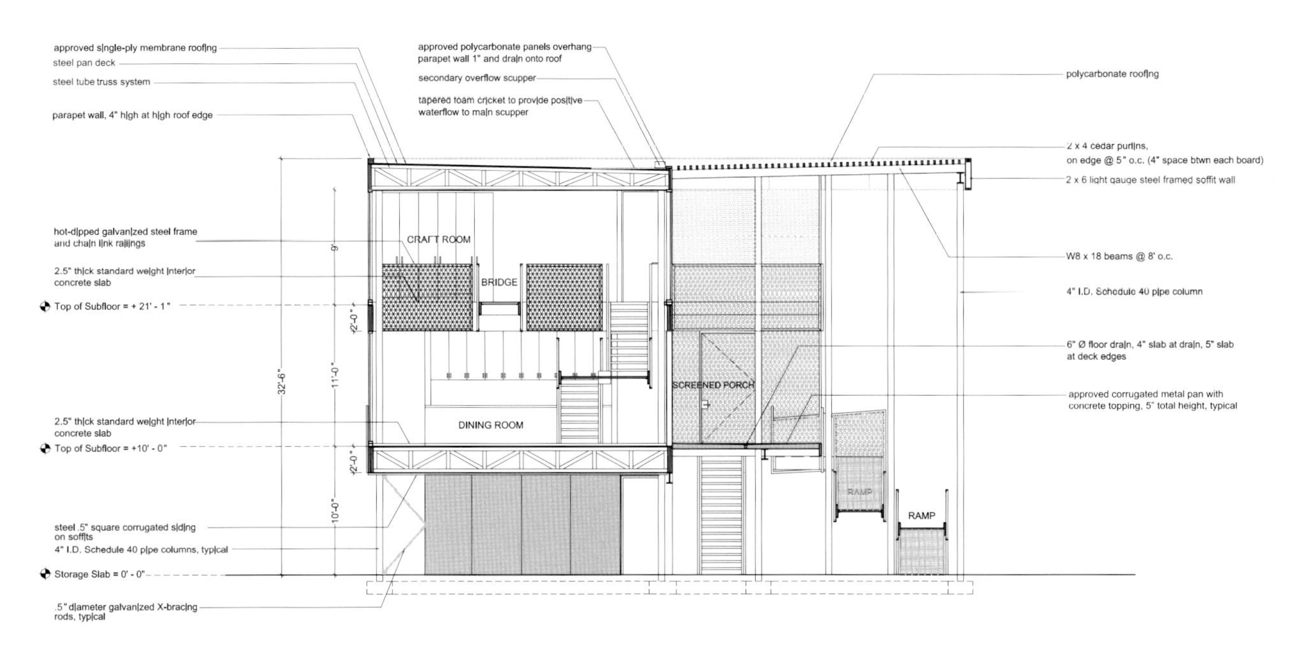

Airstrip Hangar House

Durango, Colorado / 2005

Chosen as the ideal site for a grass airstrip in a microclimate of optimum wind for small planes and gliders, this expansive hayfield on a broad plateau between the snowcapped La Plata range of the Rocky Mountains to the east and the desolate desert ridgeline of Mesa Verde to the west presented an exceedingly complex design and construction challenge. As the owners come and go to the site almost exclusively by air, the site has been designed as a giant environmental tableau, best viewed at high altitude and at high speed, at various angles from the sky. The airstrip itself is the most prominent feature, carefully excavated, graded, and planted with a fine, closely cropped green grass, appearing as a primitive hieroglyph carved out of the waving field of golden hay. In composition with this long slash across the land was the need for a large hangar and machine shop for six airplanes, a home, gardens, and several ponds required for water run-off control, irrigation, and desirable wetland habitat. Rather than scattering these elements across the land, the majority of the hayfield was preserved as working agriculture, the house and hangar were joined as a single dense structure defining the geometry of a surrounding orchard, and the ponds were placed in functional and aerially perceivable composition with the building figure and airstrip.

The form of the building was dictated by complicated siting issues: requiring the placement of the hangar door, for example, on the side away from drifting winter snows, using the hangar end of the structure to shield the house from winter winds and the rumble of a distant highway, creating a south-facing entry court and private gardens on the roadway side of the building, and giving simultaneous east-west, through-building views of the La Platas and Mesa Verde from nearly every room. To accommodate these competing requirements, while maintaining an affordably simple construction rationality that would allow primarily offsite construction using standard industrial steel frames, precast concrete panels, and commercial glass curtain walls, the form of the building became a gently arcing line of repeated, clear span structural frames. The sectional shape derives from the standard prefabricated airplane hangar frame, while the plan form arc is required to place each of the rooms into required position while maintaining a repeated structural element. This approach minimizes construction cost, time, and complexity through repetition, but it offers an unfolding differentiation of interior spaces and outside courtyards defined by the curve of the building. By continuing the geometry of the building form out into the landscape with a field of trees and gardens, the building, which might otherwise be a small dot next to the airstrip, now becomes a strong, dense, emblematic figure in its own right. From an airplane miles distant, even at a low oblique angle, the entire composition offers unique clarity, guiding the fliers home.

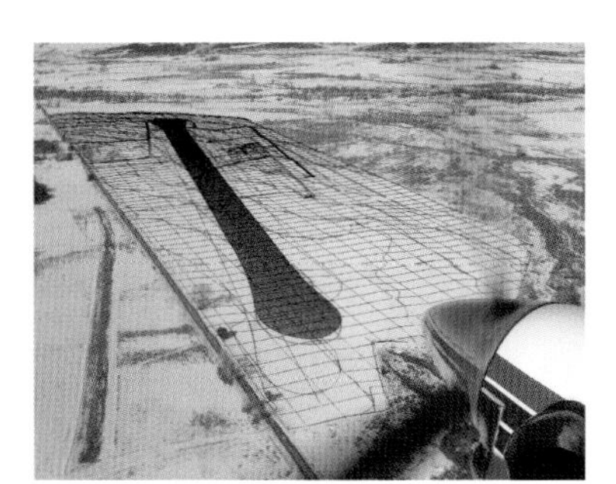

Left: Aerial perspective montage renderings of airplane approach from east, with Mesa Verde looming at end of new airstrip
Below: Site plan

OPPOSITE
Building detail and site study models: hangar, office, and workshops (top); west elevation (middle); aerial view of site model (bottom)

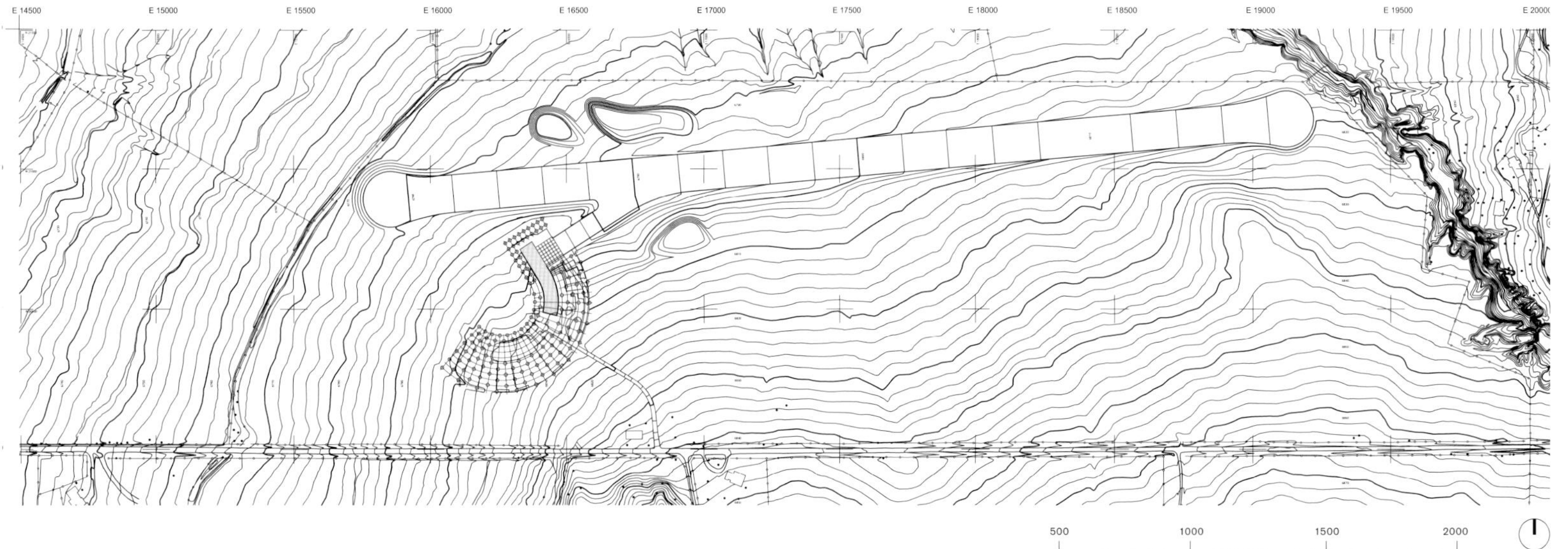

Steel Framing

OPPOSITE
Top row: Aerial view looking southeast (left); airplane landing view looking east (center); low flyby view of building and hangar apron, looking southwest (right)
Left. Hangar apron and landscape terracing, looking west (bottom); detail aerial view of model with roof removed, focusing on main living areas, terraces, and sunshade louvers (middle); west elevation of hangar area showing office space above, workshops below, and main hangar space at right (top)
Right: Aerial view of detail model with roof removed, showing repeated steel moment frames

Right: Ground-floor plan of hangar and living areas
Below: East elevation (top); West elevation (bottom)

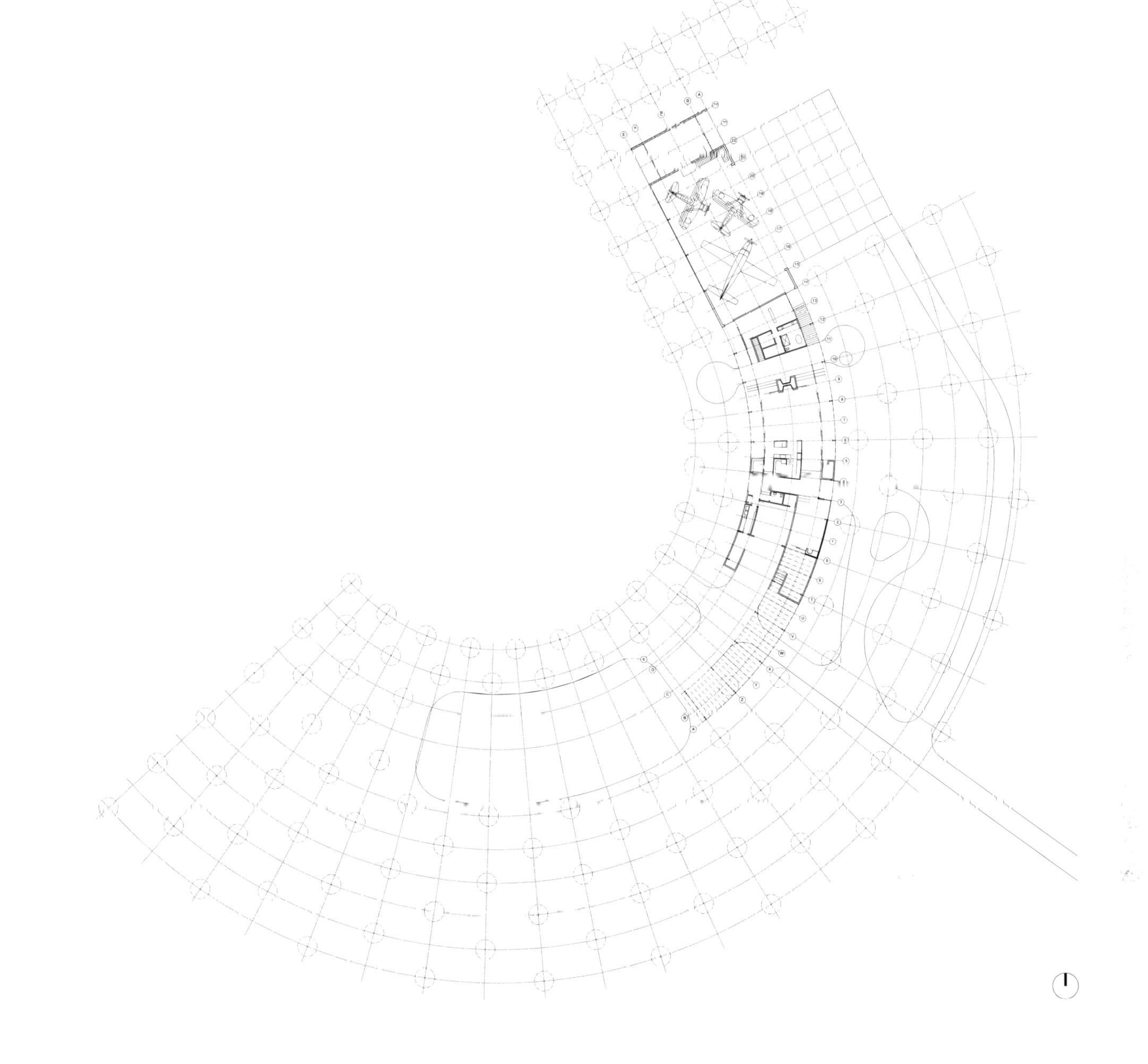

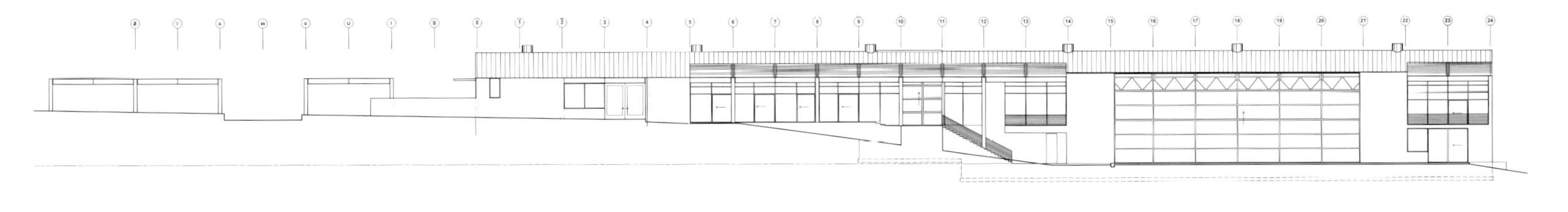

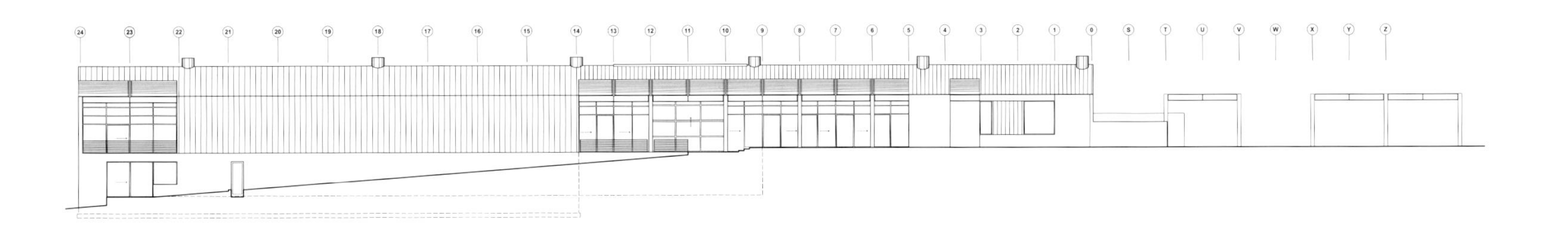

Wurster Workshop

Berkeley, California / 2002

One of the greatest advantages of working with steel as a primary structural system is the ease with which it can be integrated with other systems. In the Cantilever House at Granite Falls, the SIPs panel enclosure system was chosen for its insulative properties and its ability to provide a self-supporting secondary structural system. In the Wurster Workshop project, the steel frame is again expressed as the primary structure but used in connection with a non-structural plastic enclosure system that is designed to encourage the passage of air and light while providing a security barrier and rain protection for the outdoor workspace. Commissioned to serve as both an experimental production facility and as a showcase for new material applications and computer-controlled fabrication technologies, this building addition, interior renovation, and courtyard landscape ramp focus on the minimal definition of large, flexible spaces in order to allow for a wide range of activities and continual updating of the teaching program and fabrication tools.

The original Joseph Esherick–designed building, completed in 1964, is a solid achievement of simple cellular functionality, with didactically exposed construction and mechanical systems and an impressive use of pre-cast concrete as a primary structural and finish material. The new shop addition follows the structural geometry of the existing building frames but employs translucent plastic materials and computer-aided design and fabrication technologies to produce a multilayered rectilinear form with an environmentally active structural skin. The primary work area is defined by a ventilating roof and wall system that collects and outspouts the rain while allowing hot air and fumes to exhaust through a matrix of large roof apertures, functioning similar to a Dorade vent on a ship, which permits free flow of ventilation without water infiltration.

The double skin of site-cast fiberglass-reinforced polyurethane panels forms a dense field of thick translucent roof volumes serving as gutters and ventilator shafts hovering within a deceptively simple box that follows outward from the structural bays of the existing building and acts as a lanternlike pavilion within the large building courtyard. The courtyard is proposed to gain a new multipurpose functionality as an experimental construction space and informal amphitheater for outdoor lectures and performances and will tie together three levels of building space opening onto this newly unifying area. A broad concrete-supported ramp would rise upward as a rectangular lawn to gain the full sunlight otherwise escaping the shaded courtyard, symbolically drawing the campus ground through the two-story lobby space and into the landscape architecture studios on the building's third floor. With these additions to the courtyard, this previously underutilized outdoor space becomes an activated work area for design/build construction activities that integrate students from both the architecture and landscape architecture programs.

The workshop addition continues our interest in creating a rich, spatial density produced within a deceptively simple armature. The structure is rigidly rectilinear in external geometry, but the simple form is the result of folding a complex multifunctional skin of weatherproofing, daylighting, and air exchange into a diagonally cross-braced and bow-tensioned frame, then overpulling this tension into a supertaut rectilinear enclosure. What appears from the outside to be a box-form in repose is revealed on the interior to be a tautly counterbalanced matrix of tension and compression forces swelling within the steel frame, just constrained from an outward explosion of internal pressures. This physical compaction and live, internal tension is reinforced through the translucent projection of human activity and external environmental forces—sun shadows, rainwater, wind ripples—onto the multilayered, complex geometries of the internal skins. The bright flashes of welding, cutting torches, and grinding sparks will further light the skin with shadows and highlights of student fabricators dancing before artificial suns.

We have worked extensively with student construction projects in the past, and we have planned this space for future projects to itself involve direct student participation in the construction. The primary steel-frame assembly is intended to be erected by professional contractors, but the cleanly separated elements of substructure and skin will be independently fabricated and installed by students, who will be involved both as class participants and hired as project employees in construction management and fabrication supervisory roles. The flexible, translucent skin will be developed in collaboration with student researchers and is based on our previous experimental work in flat-cast, reinforced, UV-resistant, flexible polyurethane skin structures.

Top: Transparent perspective view detailing primary steel structural systems
Parts List:
a. Spray booth fume stack
b. Steel moment frame
c. Steel trough frame
d. Chain hoist beam
e. Polycarbonate roof system
f. Polycarbonate skin system
g. Polycarbonate trough skin
h. Sliding door rail
i. Concrete sliding door
Bottom: Contextual models of building addition and ramped courtyard terrace leading up to landscape department studios

Steel Framing

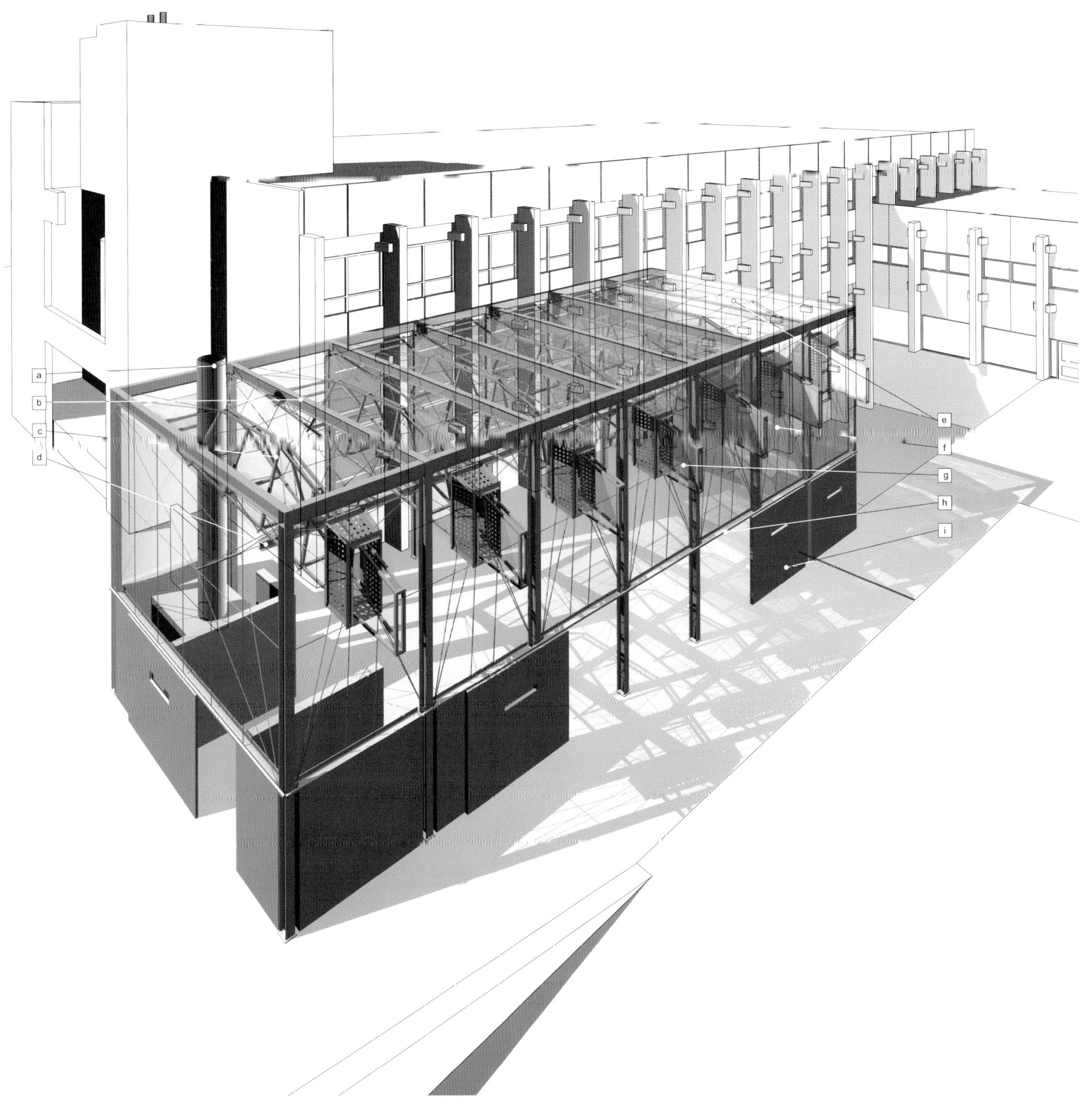

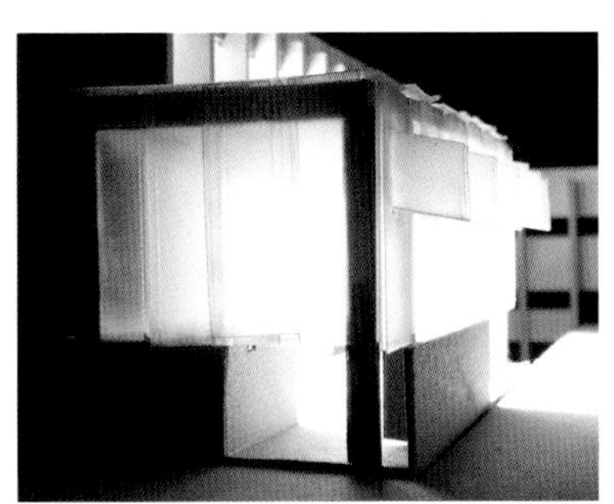

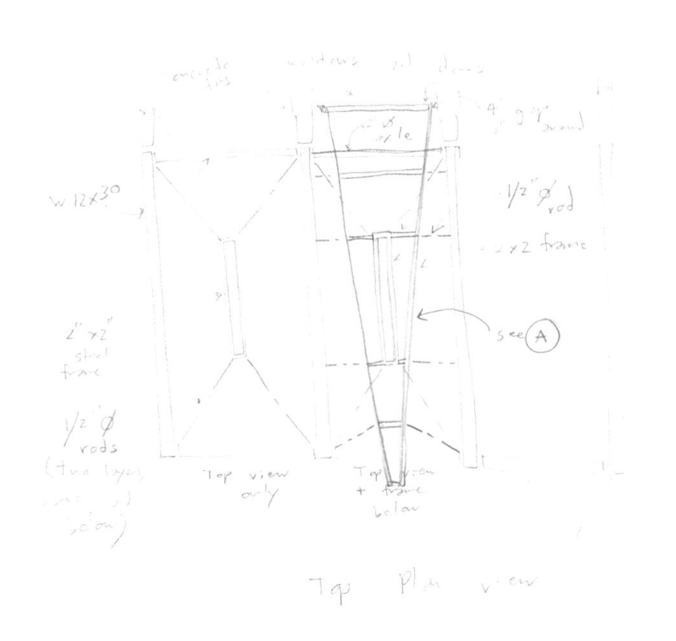
1/2" Ø rod
2x2 frame
see A
Top view only
Top plan view
1

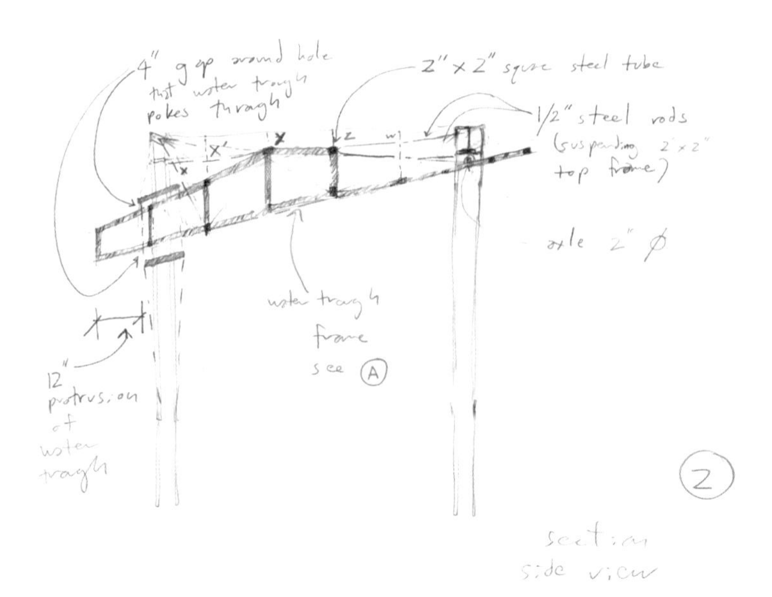
4" gap around hole that water trough pokes through
2" x 2" square steel tube
1/2" steel rods (suspending 2" x 2" top frame)
axle 2" Ø
water trough frame see A
12" protrusion of water trough
2
section side view

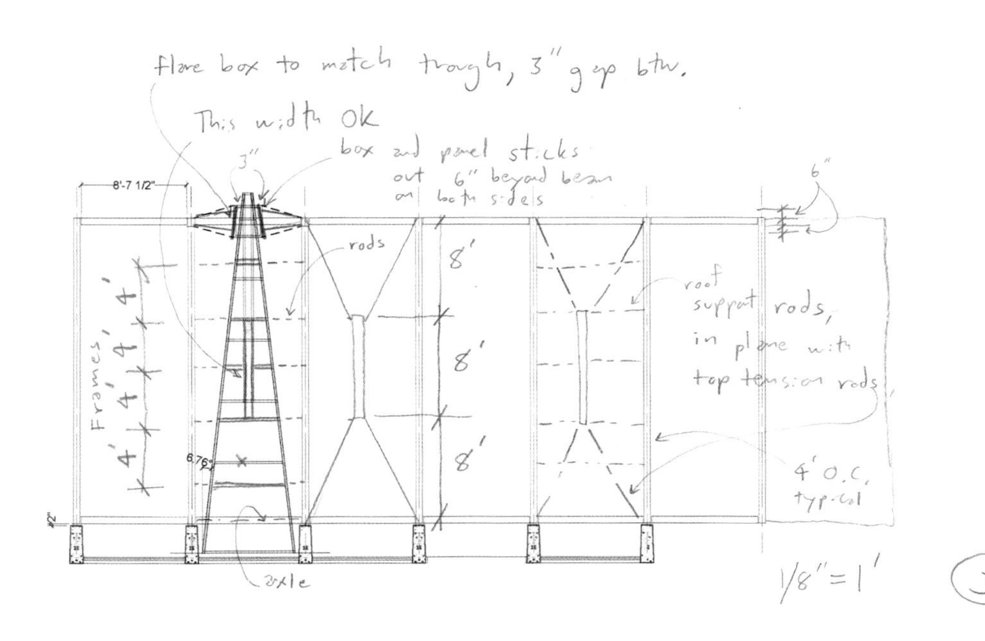
flare box to match trough, 3" gap btw.
This width OK
box and panel sticks out 6" beyond beam on both sides
3"
8'-7 1/2"
6"
rods
8'
8'
8'
roof support rods, in plane with top tension rods
4' o.c. typ. col
axle
1/8" = 1'
3

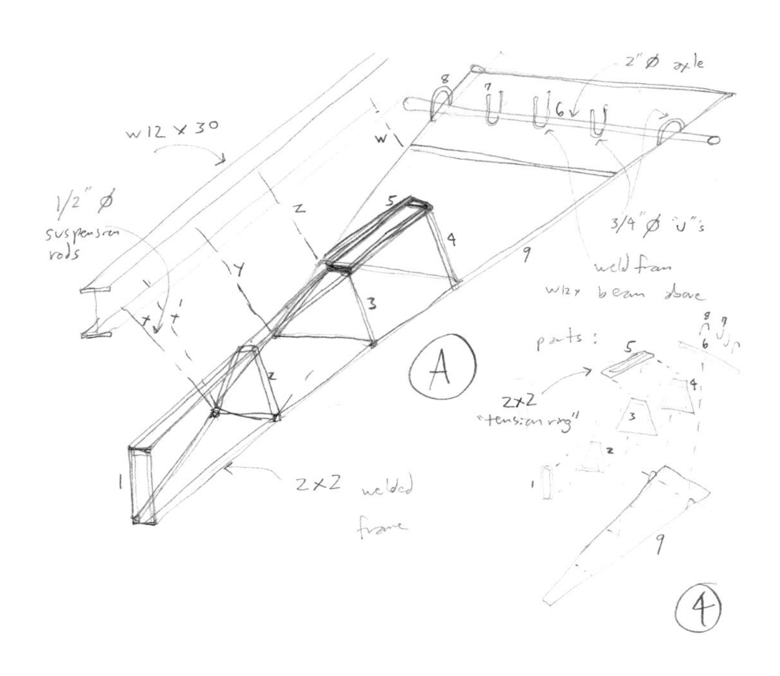
w12 x 30
1/2" Ø suspension rods
2" Ø axle
3/4" Ø "U"s
weld fan w12x beam above
parts:
2x2 "tensioning ring"
2x2 welded frame
A
4

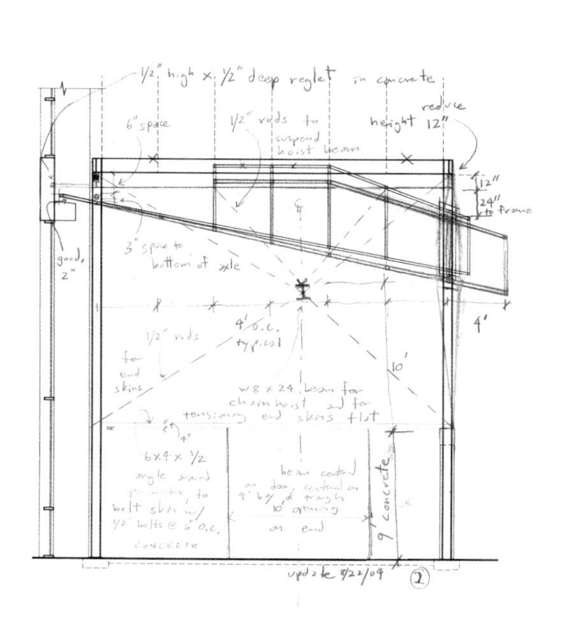
1/2" high x 1/2" deep reglet in concrete
reduce height 12"
1/2" rods
4' o.c. typ. col
4'

Steel Framing

"Redline" design sketches detailing major component fabrications during design process

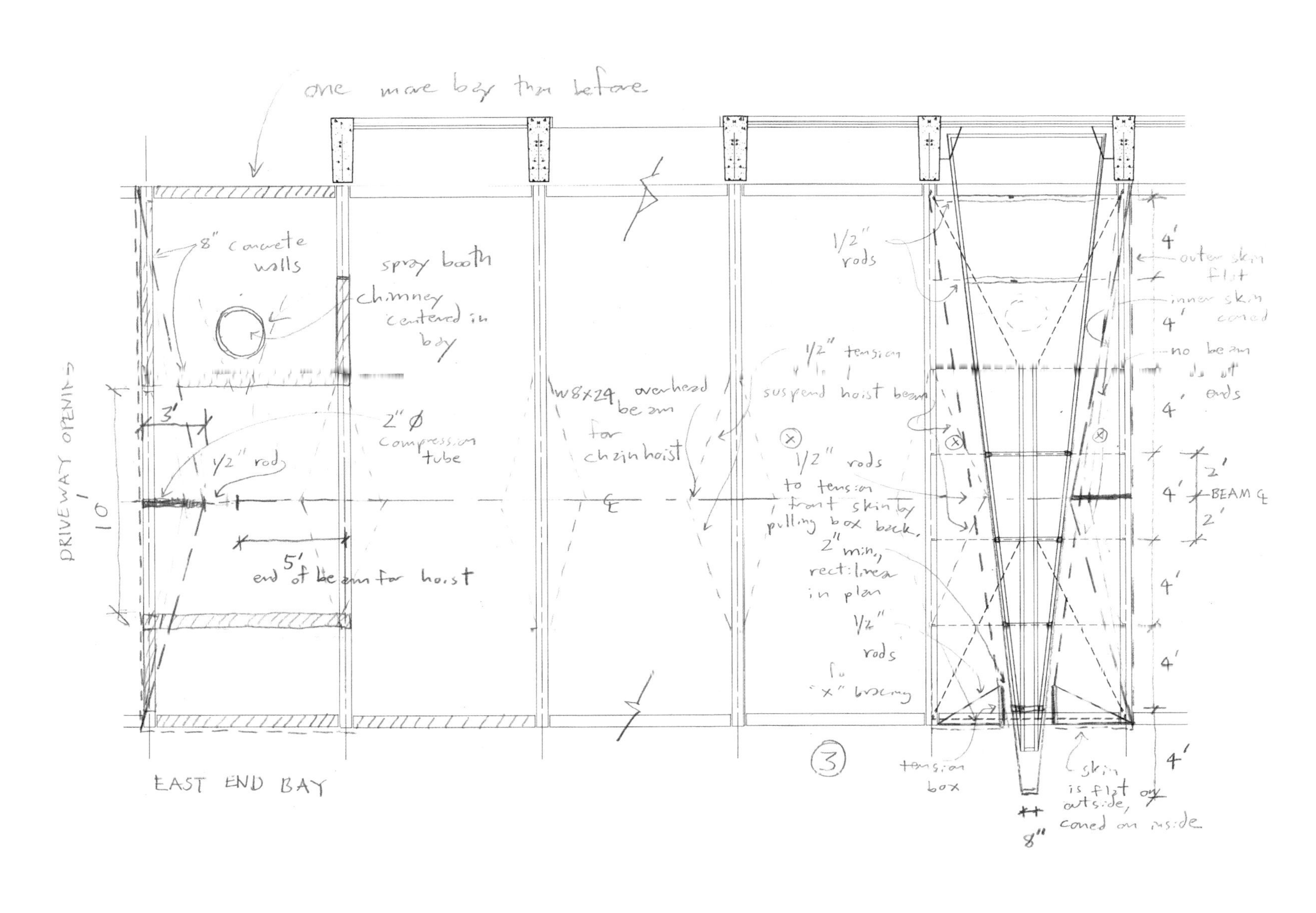

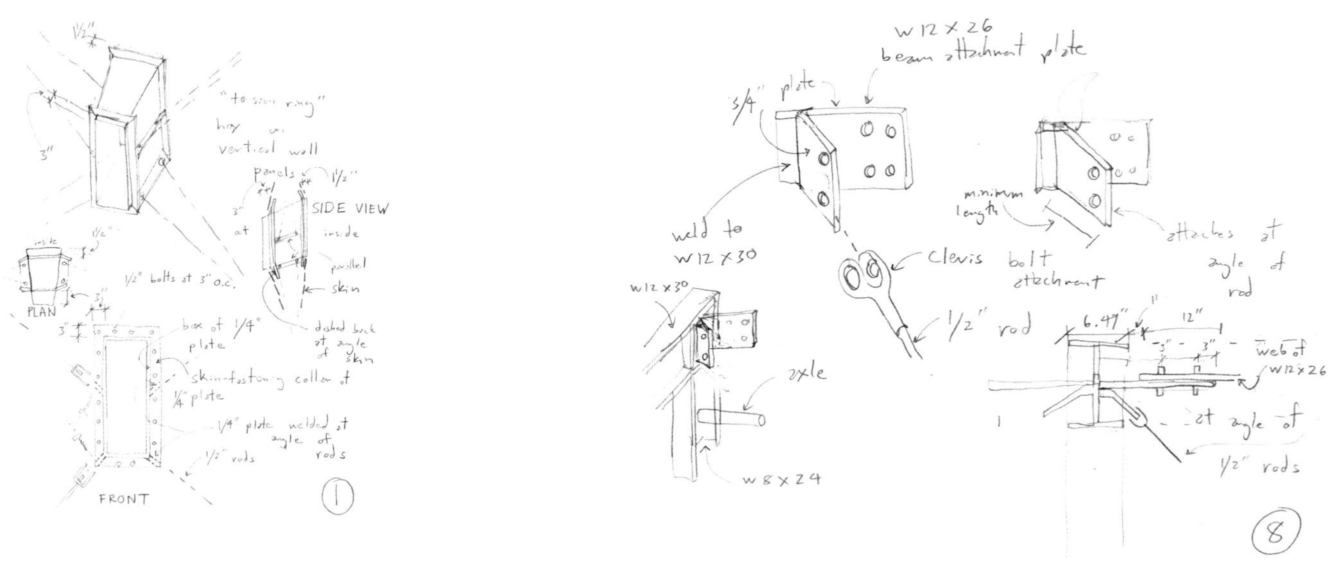

Wurster Workshop

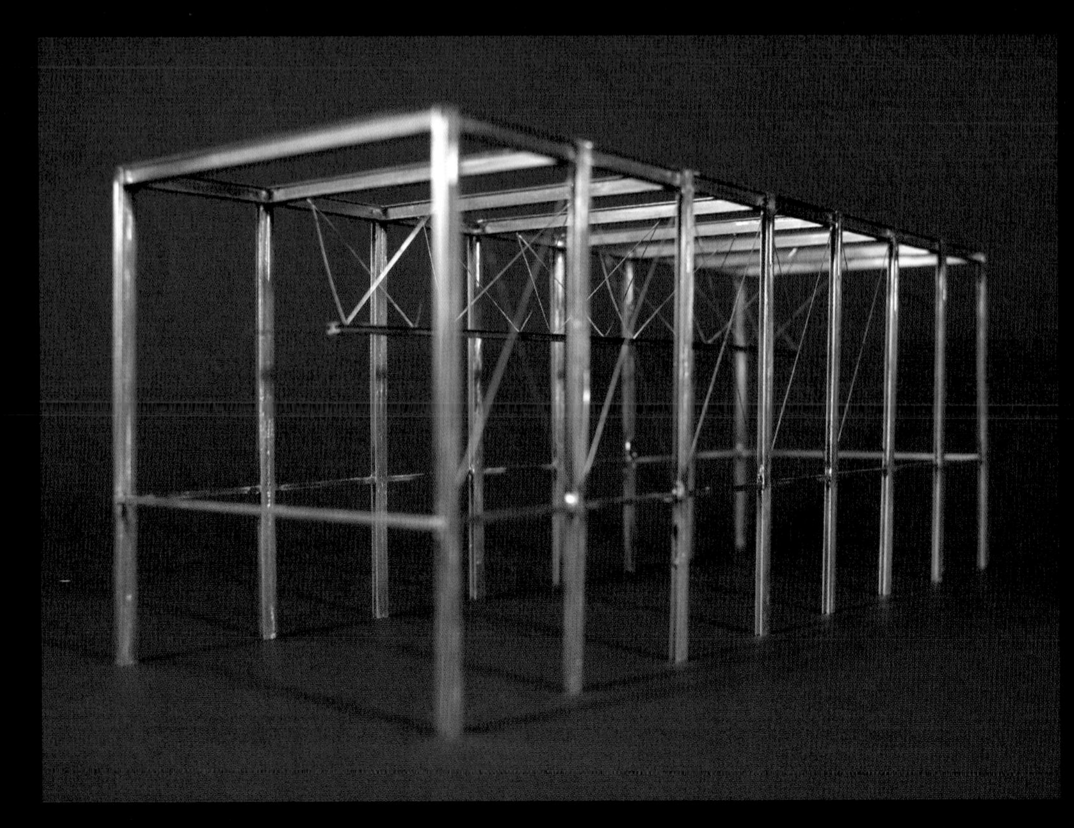

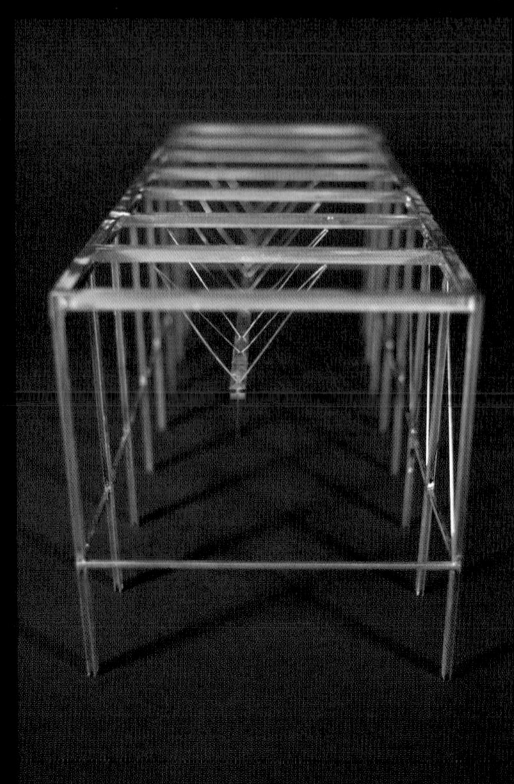

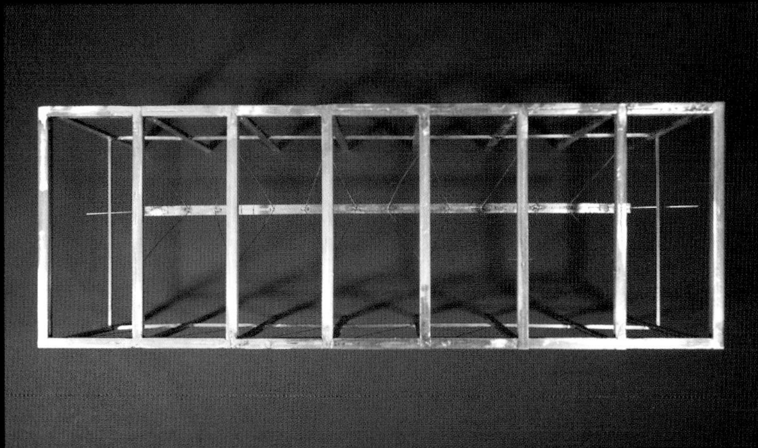

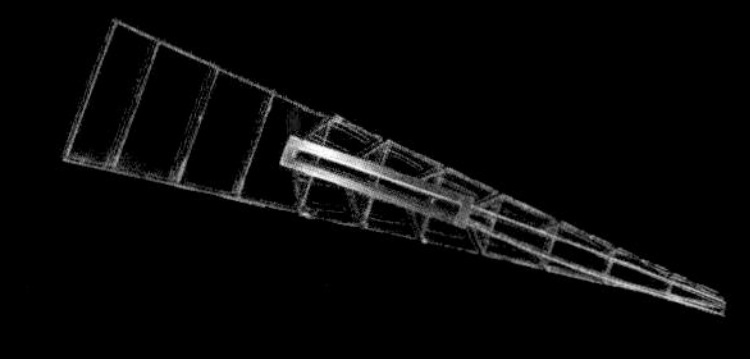

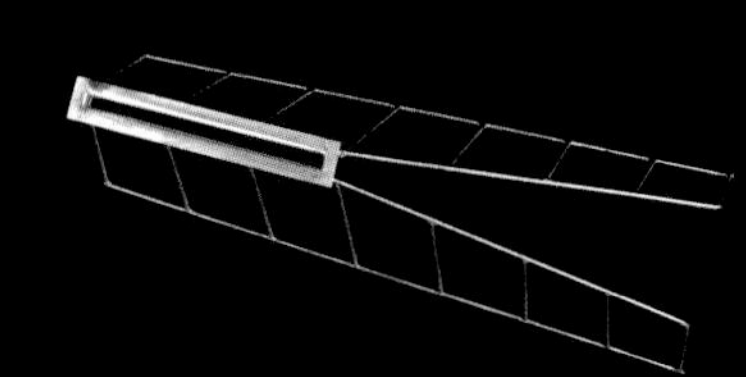

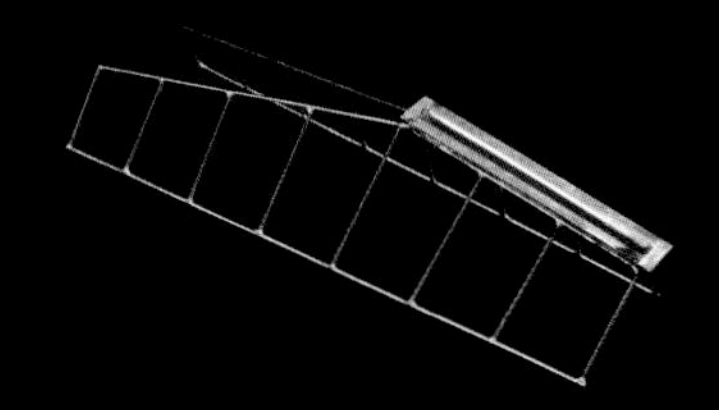

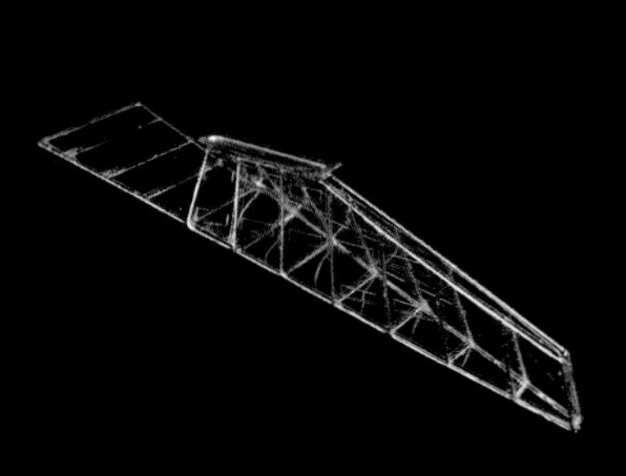

Steel Framing

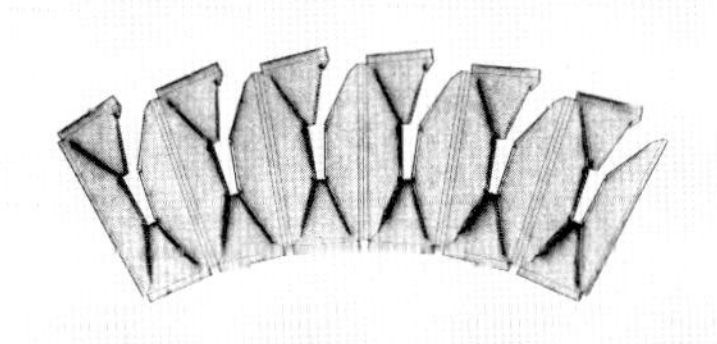
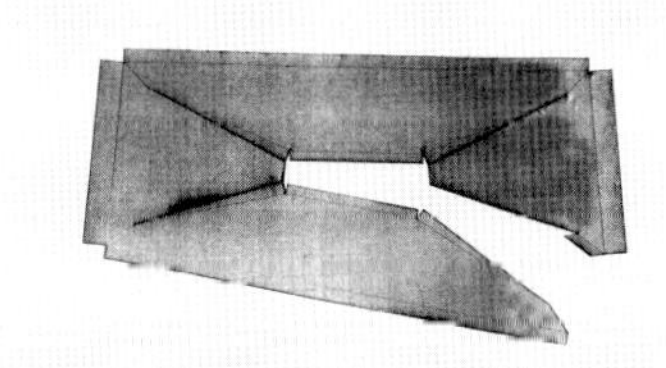
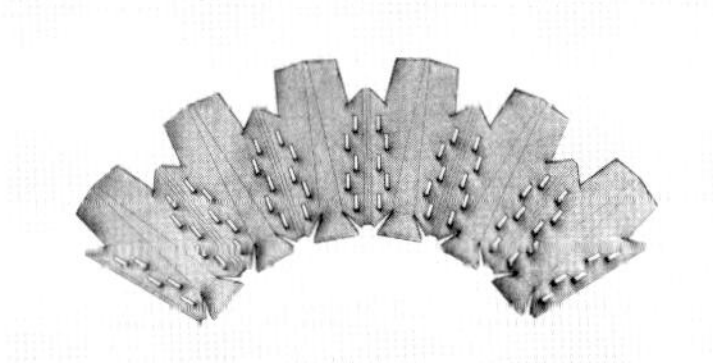

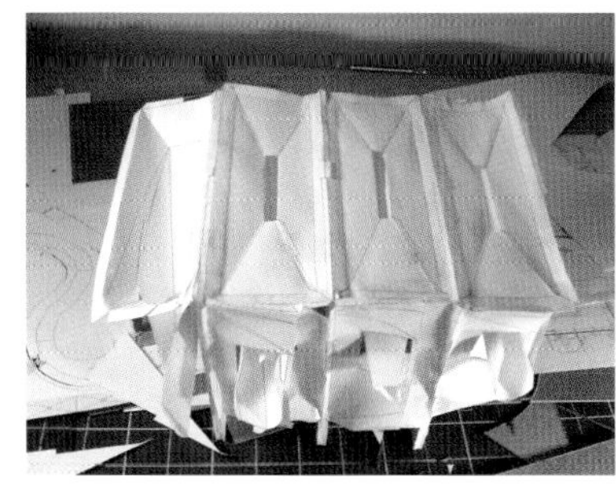
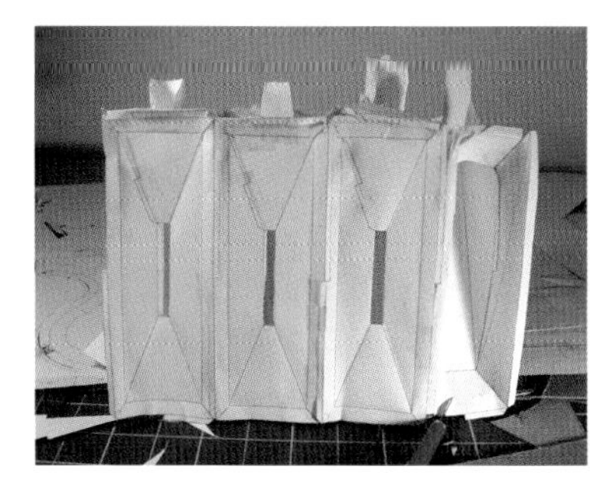
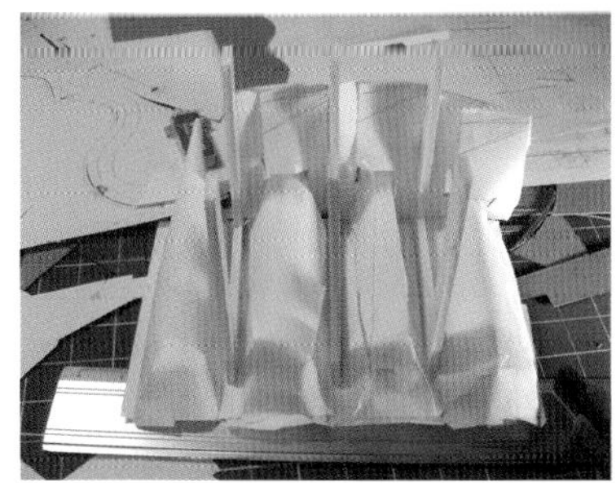

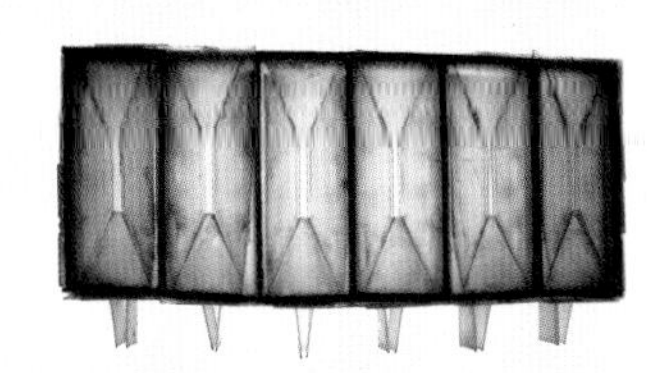
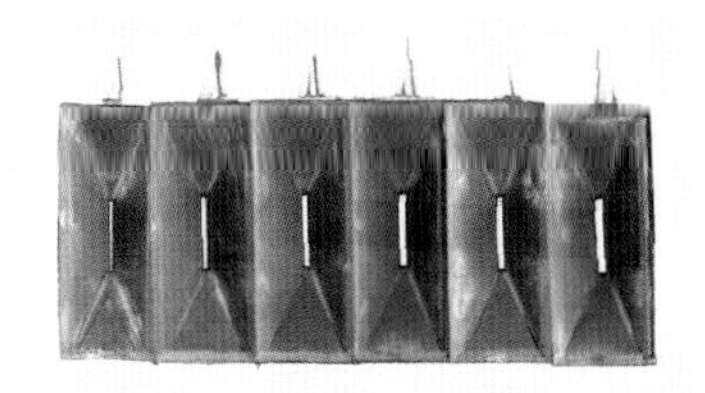
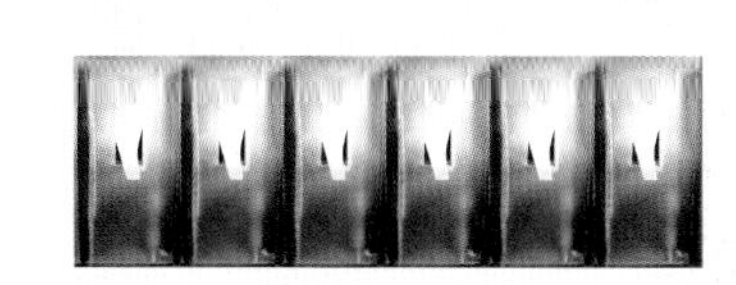

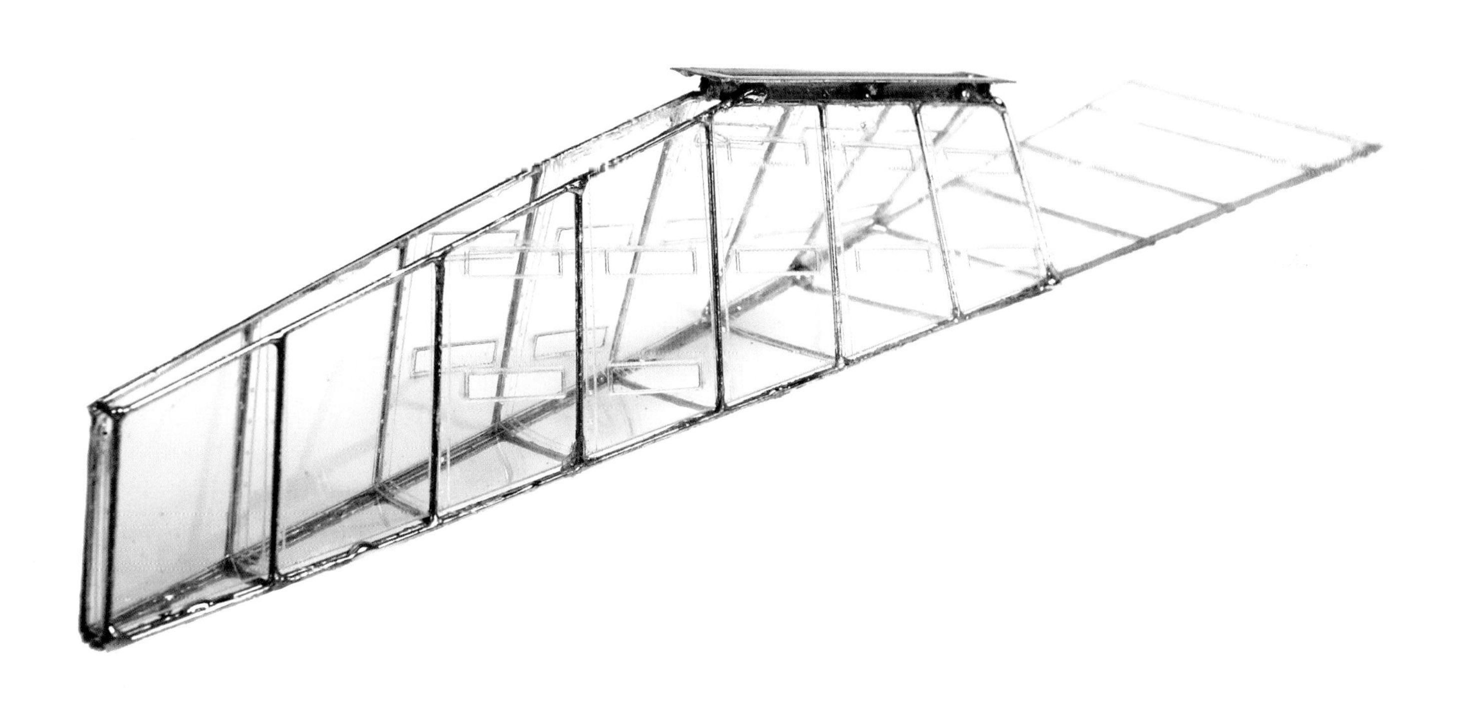

OPPOSITE
Top and middle: Detail study models of primary steel frame structure
Bottom: Detail study models of secondary frame structure creating ventilation venturis, water catchment troughs, and roof enclosure structure

Top row: Unfolded laser-cut studies of principal enclosure and water trough panels
Middle row: Early paper study model exploring spatially rich, folded-skin system of enclosure

Bottom row: Laser-cut polycarbonate and paper studies of transparent multilayered modular skin system

Above: Study model of hanging water trough roof structure that doubles as a ventilating venturi, drawing fumes upward through slotted side walls and exhausting out through the same hole in the roof that captures rainwater into the trough and spout

Wurster Workshop

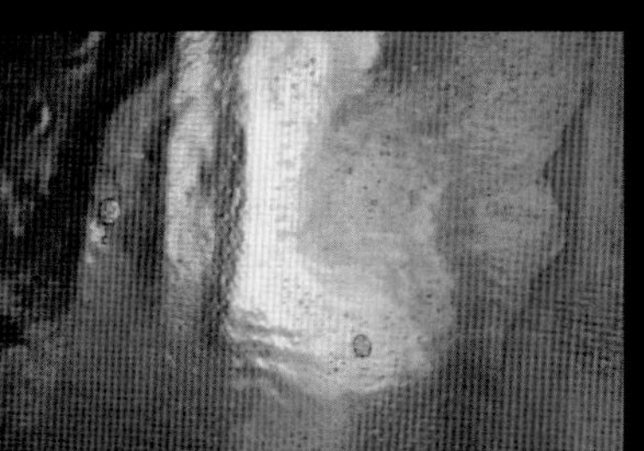

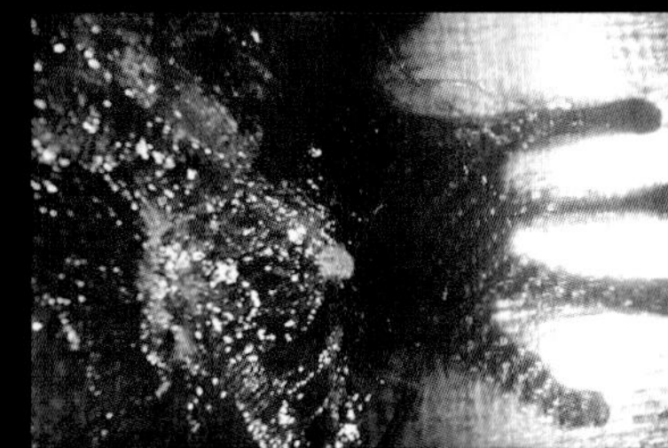

Top left: North elevation view of workshop addition as seen from main courtyard **Top right:** Interior view as seen through double-height window wall of existing shop interior **Second row:** Interior view showing deeply pleated, transparent ventilating skin and roof structures with suspended chain hoist beam serving as a primary tension element drawing the exterior skins into a flat, rectangular geometry **Third row:** Transparent ventilating roof system viewed from interior below (left) and from exterior above (right) **Bottom row:** Early site-cast translucent skin studies

Steel Framing

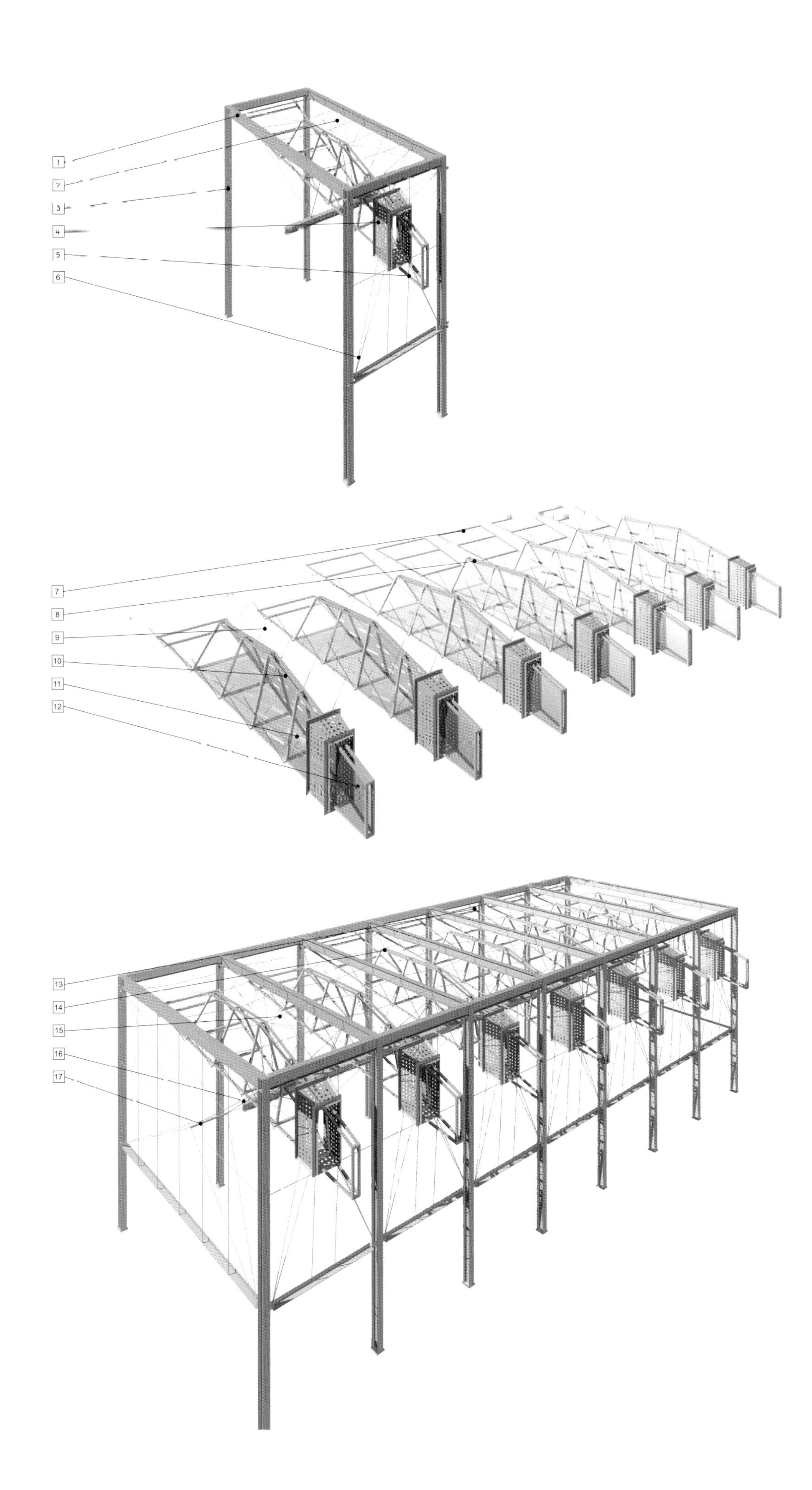

Exploded perspective detail views showing the primary and secondary structural framing and suspended water trough/ventilator roof system

Parts List:

1. W12 x 30 galvanized steel beams
2. .5 in. diameter steel rod cross-bracing galvanized steel
3. W8 x 24 galvanized steel column
4. Perforated steel plate tension ring
5. 2 x 2 x .25 in. square tube water trough, galvanized
6. .5 in. diameter steel tension rods
7. Water trough pan fits between concrete fins on existing building
8. Water inlet through interior trough skin
9. .5 in. diameter steel rod suspension structure
10. 2 x 2 x .25 in. square steel tube water trough frame
11. Exchange air inlets through interior trough skin
12. Translucent polycarbonate skin panels
13. 2 x 2 x .25 in. square steel tube frame extended to waterproof between concrete columns
14. 2 x 2 x .25 in. square steel tube roof water inlet frame and roof beam tension ring
15. .5 in. diameter steel rods, diagonal tension bracing
16. W8 x 24 overhead chain hoist track
17. 2 in. schedule 40 steel pipe skin-tensioning compression member

Wurster Workshop

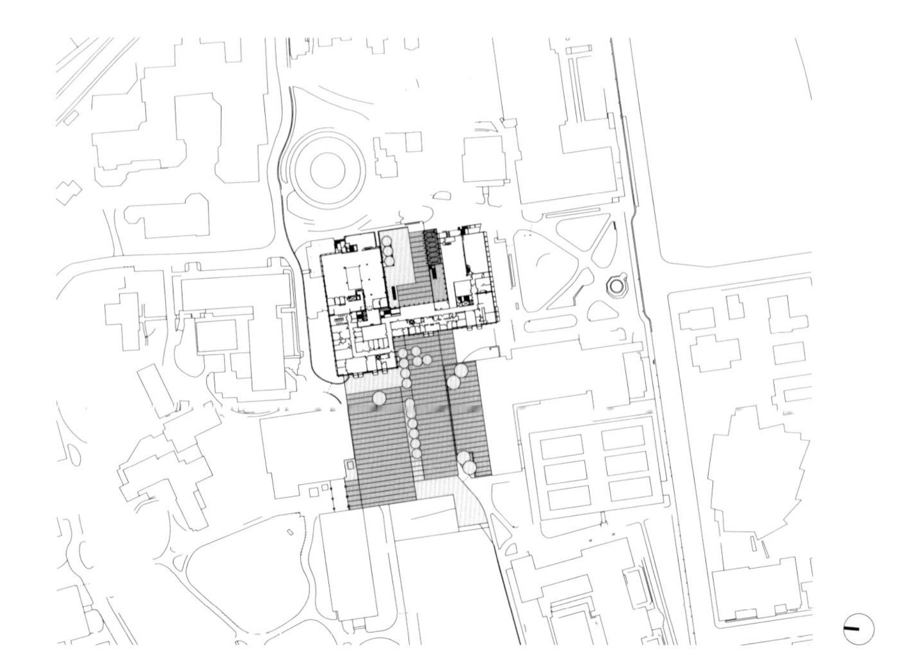

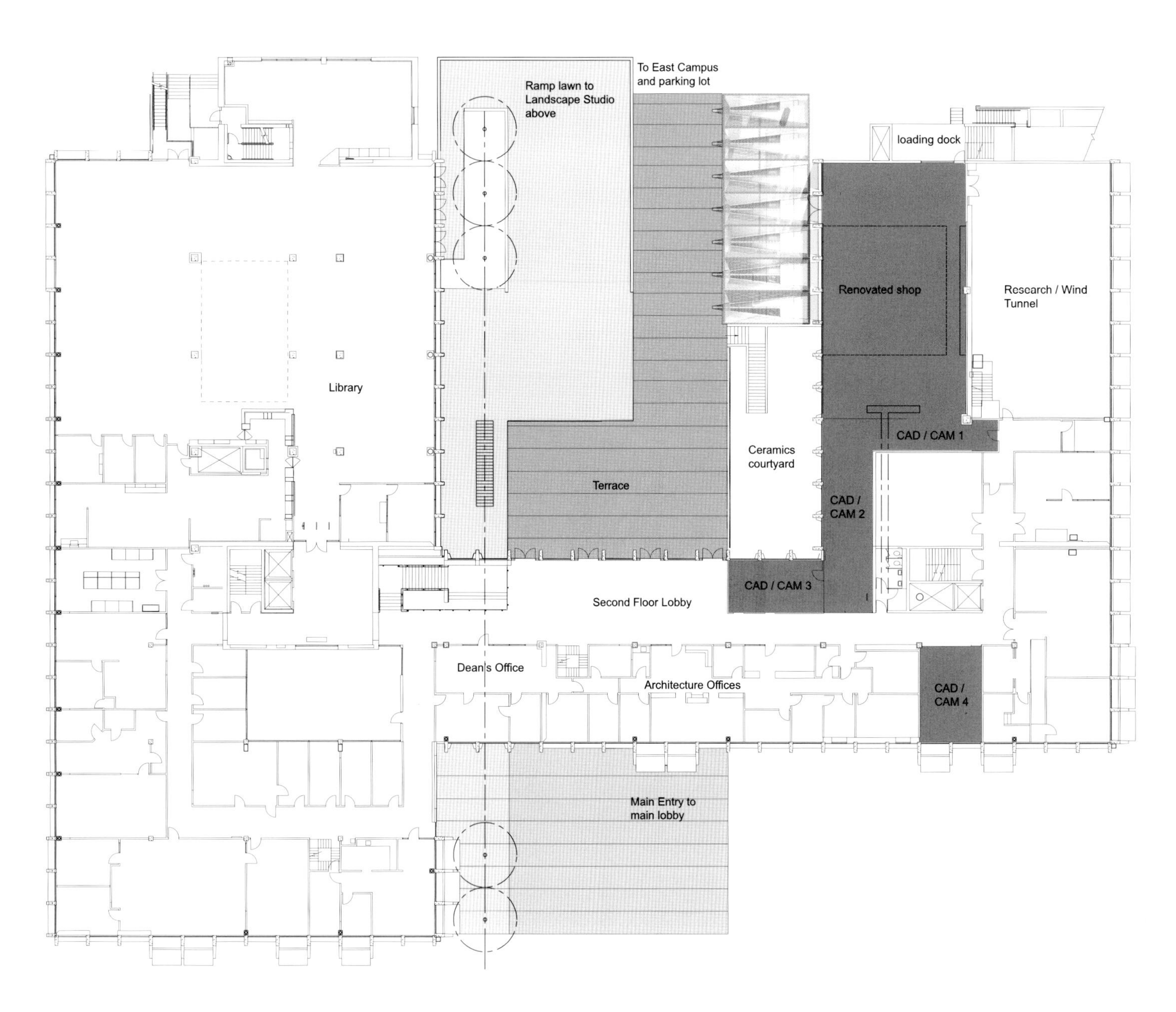

Top: Site plan showing workshop, ramp, and courtyard additions within overall campus plan
Bottom: Detail showing new additions and courtyard design inserted into existing Wurster Hall courtyard space

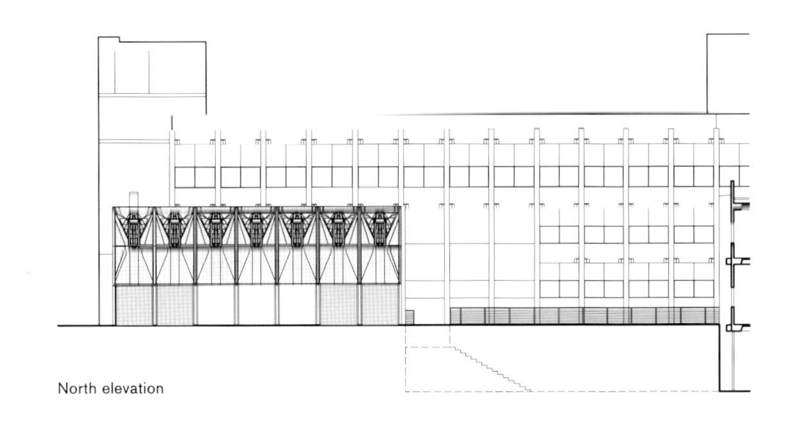

North elevation

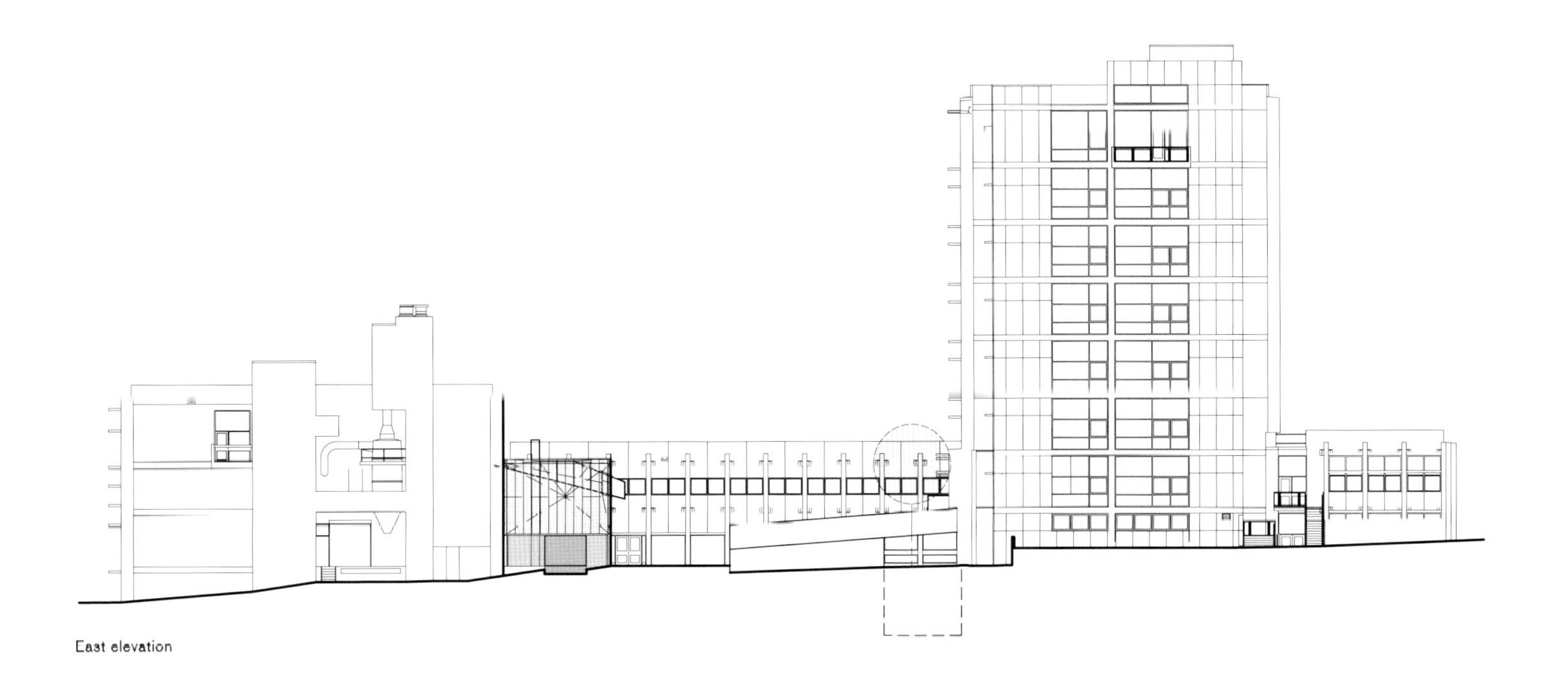

East elevation

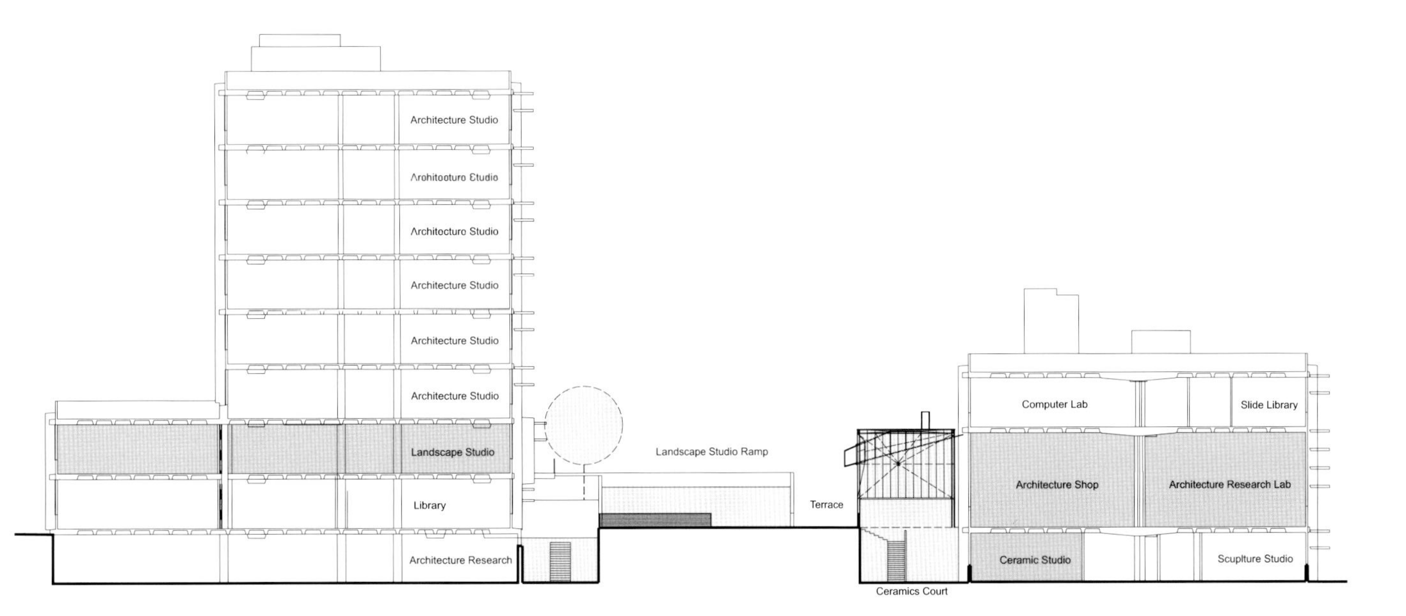

Section looking east

Chapter 5

Sandwich Panels

There are quite a few approaches to panelizing building systems, from simple panelization of 2x4 wood framing to innovative proprietary systems using steel, plastics, concrete, and other materials. This chapter deals with a more advanced system of panelized construction, using engineered structural panels that are composed of a sandwich of multiple materials with different properties. Frequently combining high levels of insulation properties with structural loading and spanning capabilities that significantly exceed those of conventional or panelized wood framing, this system provides unique design opportunities for producing large, open spaces with less need for exposed structural supports

Frank Lloyd Wright is often credited with some of the earliest applications of sandwich panels in the building industry, but the principle of laminating multiple layers of dissimilar materials can be found in many industries going far back in history, such as the use of layers of different varieties of steel in tools and weapons, where some layers are chosen for hardness and ability to hold a sharp edge, and others chosen for ductility. More than 1,000 years ago, Japanese swordsmiths were fabricating sandwiched steel swords in this fashion, with many thousand thin layers making up the finished blade. Sandwich techniques have also played an extremely important role in the aerospace industry since its earliest days, in the search for ways to make lightweight yet strong structural elements.

Wright's application of sandwich panels in the 1930s in his Usonian houses was limited to non-structural infill, and they did not incorporate significant thermal properties. Most interesting was the concept of modularization of the building process, in that the panels had fully completed interior and exterior finishes in order to reduce onsite labor requirements. It was not until the 1950s that one of Wright's students extended the concept by developing a structural panel with an insulating core, which is the configuration most commonly in use today. These structural insulated panels (SIPs) are produced by many different manufacturers, most following the substantially similar format of a layer of encapsulated polystyrene (EPS) board sandwiched between two layers of plywood or oriented strand board (OSB). Other variants based on materials other than wood include structural outer skins of steel, cement, or gypsum, and alternate insulation board layers include poyurethane or polyisocyanurate rigid foam and recycled agricultural fiber. Manufacturers have developed many different proprietary systems for structurally connecting their panels, which continues to be one of the biggest challenges for the SIPs industry. The panels provide very high levels of insulation for their thickness, are relatively inexpensive, can utilize primarily recycled content, and are relatively free of unhealthy gas emissions. Another unique quality of SIPs is their avoidance of trapped air spaces where moisture can build up and condense. In conventional construction, with batt-type insulation, attics and other spaces in the structural roof and floor systems must be properly ventilated to avoid mold and rot from moisture, but when using panels that are solidly filled with polystyrene, there are no trapped spaces within the structure. This makes them ideal for use in flat-roof or low-pitched-roof structures, which are otherwise very difficult to properly ventilate.

In Japan, it is common to ventilate wall cavities as well, but this is almost unheard of in North America, surprisingly enough. With the increasing interest in reducing potential sources of mold and mildew in residential construction, it is likely that more attention will be placed in the future on avoidance of all unventilated spaces in building envelopes, and SIPs can provide a reliable solution. A weakness in the panels, with regard to moisture, on the other hand, is in the material used for their outer faces. OSB is much more susceptible to moisture damage than plywood, and it becomes particularly important with SIPs construction to protect the panels from the weather before, during, and after the construction process.

In their simplest applications, SIPs can be considered structural in and of themselves, able to support spanning loads when used in horizontal applications such as floors or roofs, and compressive loads when used vertically as walls. From a structural engineering perspective, the two structural faces of the panels provide the same compressive and tensile properties as the top and bottom flanges of a steel I-beam, and the foam center serves to separate the flanges and keep them in place, in much the same way as the central web in an I-beam or truss. Under relatively light loading, the panels can operate independently, with only light connectors aligning them to each other to form walls, floors, and ceilings. More typically, they are used in conjunction with some kind of spline that connects the panels and expands their load-carrying capacities. The splines can be a proprietary part of the manufacturer's building system or engineered specifically for the particular conditions of each project, using dimension lumber, engineered wood products, or steel.

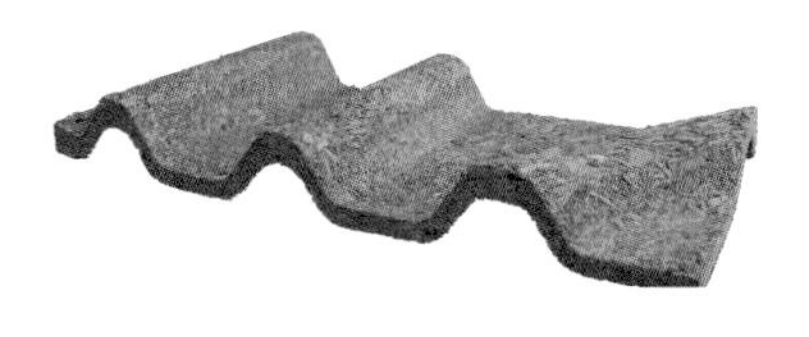

Experimental formed-skin OSB panel studies based on existing industrial processes

Since the most common SIPs are based on the dimensions of the plywood or OSB sheets that make up their faces, they typically are manufactured in four-foot (1.2 m) widths, with variable lengths up to a standard maximum of twenty-four feet (7.3 m). Some manufacturers make wider panels, in six- or eight-foot (1.8 or 2.4 m) widths. The thickness of the panels varies in response to the desired amount of insulation and other structural requirements, typically ranging from four to twelve inches (10.16 to 30.48 cm). In some cases the manufacturers will fabricate the panels to exact specifications, including cutting window and door openings or making custom widths and other special cuts, but the largest manufacturers typically just fabricate their standard panel widths to two-foot length increments, which for most applications requires additional cutting or fabricating in the field by the end user.

As a manufactured system, installation drawings, engineering, detailing, custom prefabrication, and site delivery are all provided by the manufacturer, allowing for great flexibility in design, ease of construction, and low cost compared to more conventional systems of structure and enclosure. The ability to use the same system for roof, walls, and floor provides for simplification in ordering materials as well as in construction process, providing economies of scale and reduced coordination requirements for even relatively small buildings. Although the assembly of SIPs structures is relatively straightforward, there is a learning curve and cause for hesitation for carpenters who have not worked with them before. The modules are large, which is efficient for assembly, but relatively unforgiving if there are, for instance, irregularities or dimensional errors in other portions of the structure with which they must interface. Nevertheless, SIPs are a versatile material, able to be used for a wide variety of applications as the complete system on their own, or in conjunction with other purely structural systems, such as timber frame or structural steel.

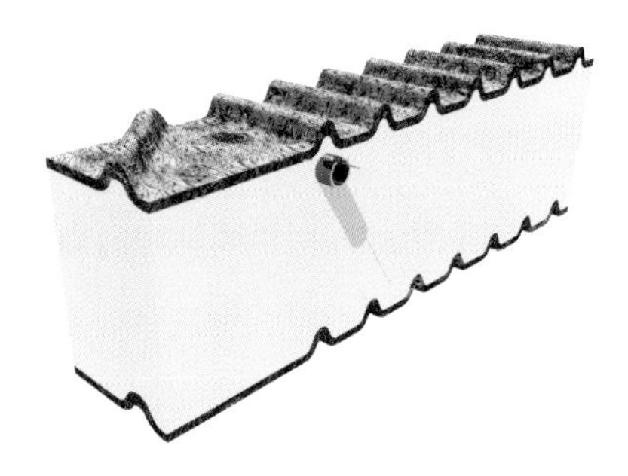

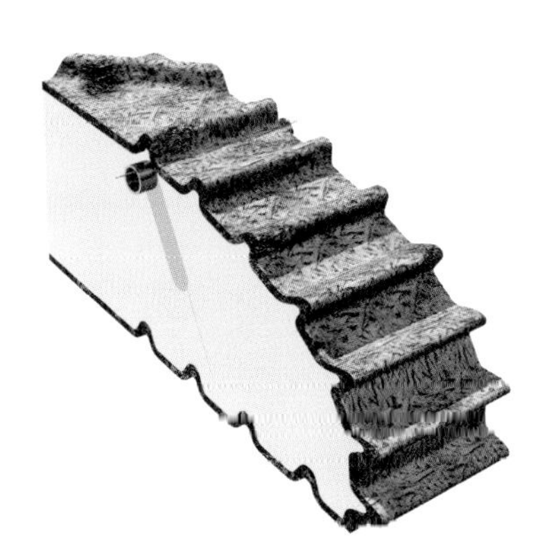

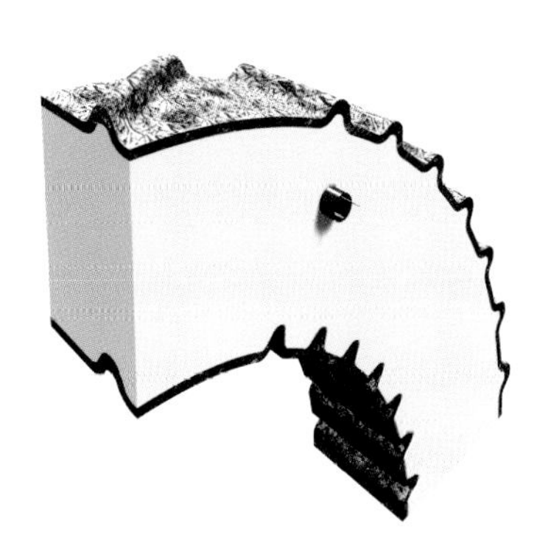

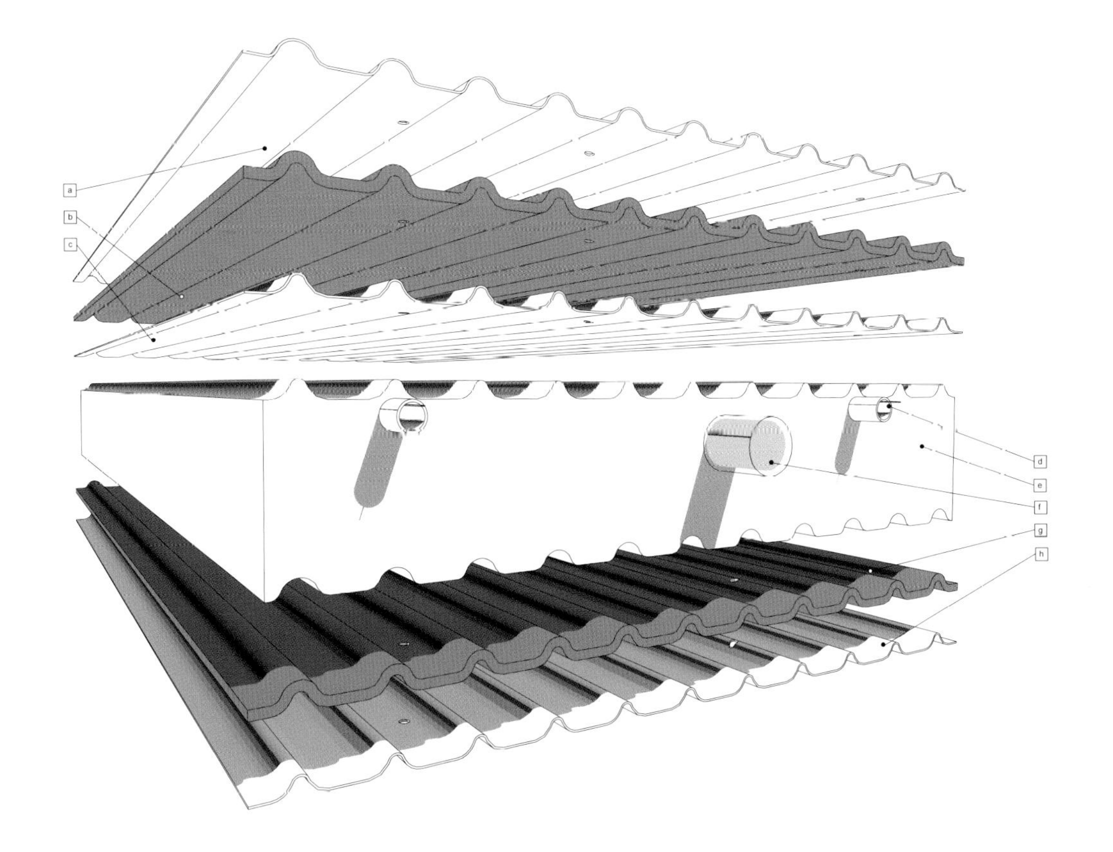

Parts List:

a. Metal finish surface
b. Corrugated cast OSB
c. Reflective metallic barrier film
d. Electrical conduit
e. Polystyrene
f. 4 in. diameter plumbing chase
g. OSB
h. Interior cast gypsum board

Chameleon House

Northport, Michigan / 2002

This prototype has been built near Northport, Michigan, two hundred miles north of Lansing, on a small peninsula of Leelanau County that juts into Lake Michigan. The house is part of a series of projects that explore the opportunities for using SIPs panels in very cost-effective ways to build structures from standardized components that can accommodate a variety of complex site conditions with minimal disturbance to the natural topography, water flow, and vegetation.

The Chameleon House sits above a cherry orchard on a hill with a spectacular westward view of Lake Michigan and the surrounding agricultural landscape. With no other houses in sight, agricultural buildings make up most of the built environment along the rural road that runs through the orchards to the site. Although few restrictive or unusual zoning constraints exist at this location, the challenging topography and geotechnical conditions play a strong role in defining the overall design strategy. The small ground-floor building footprint/foundation reduces the cost of this expensive area of the house and allows the foundation to step up the site with the slope of the hill. From our earliest meetings with the owners, it was agreed that getting up as high as possible would best take advantage of the views out and across the orchards to Lake Michigan. The owners wanted airy, open spaces where they and their young children could live and play together and small bedrooms that would encourage everyone to gather in the main living areas and outdoors. The result is a small, 1,650-square-foot (153 sq. m) house with nine different living levels, including an occupiable roof deck.

The concept of a prefabricated house was well received from the beginning, perhaps with special interest as one of the owners works for Steelcase Manufacturing, which is itself moving to expand its primarily office furniture product line to include entire prefabricated office building systems. In order to keep costs and onsite labor to a minimum, SIPs panels compose the exterior walls and roof structure, which also brings a high level of insulation that will be welcome in the cold winters. Although the SIPs are used as structural elements throughout, the addition of a two-story prefabricated steel moment frame on the lake-view side allows for the double-height window wall and the open loftlike spaces within the main living area. With the use of commonly available materials and industrial detailing such as prefabricated steel stairs and railings, the cost was kept low and the period of construction was less than half that expected of a site-built home.

Despite its development from off-the-shelf components, the house is carefully integrated into the rolling topography of its site, peering out to the westward views of Lake Michigan and the surrounding agricultural landscape. The site is minimally disturbed, other than the mounding of two earthen enclosures adjacent to the tower, created from the excavated earth of the foundation and offering a ground to contrast the tower experience above the treescape. Due to the slope of the site, the family enters at the third level, descending down to the kids' bedrooms and bath or moving up to the main living spaces that look out over the orchards to Lake Michigan.

A more conventional house would appear as an unsympathetic intrusion in this pure landscape, and with its singular vertical presence rising above the orchard, the tower is intended to reflect the austere, scaleless nonparticularity of the occasional farm buildings dotted elsewhere on the hills. To help mask the scale, the building is wrapped in a skirting wall of recycled translucent acrylic slats, standing two feet out from the galvanized sheet metal cladding of the wall surface on aluminum frames that serve also as window washing platforms and emergency exit ladders. The translucent polyethylene material set out over the dully reflective wall cladding was chosen for its ability to gather the light and color of its landscape, dissolving the finely shadowed and haloed structure into the seasonal color cycle of white snow and ice and black twig tracery; pale pink blossom clouds; pollen, green leaf, and grass; and golden straw and vivid foliage. The ever-changing appearance of the house and ability to mirror its surroundings led to its being called the "Chameleon House." The double skin creates a microclimate and thermal differential around the structure, creating a rippling mirage updraft of steaming condensation in the summer and dripping icicles in the winter.

Cantilevered steel stair to roof as seen from ground

OPPOSITE
Northwest corner of house as seen from driveway approach

Chameleon House

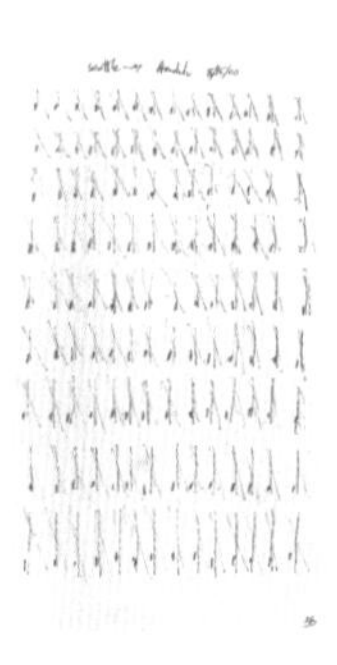
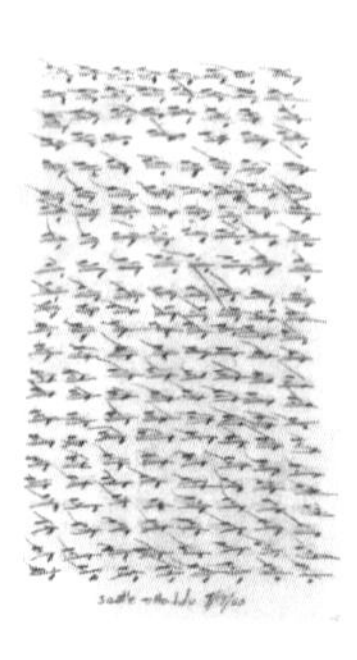
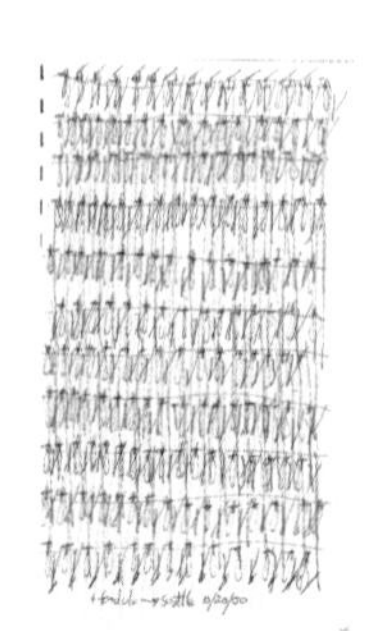
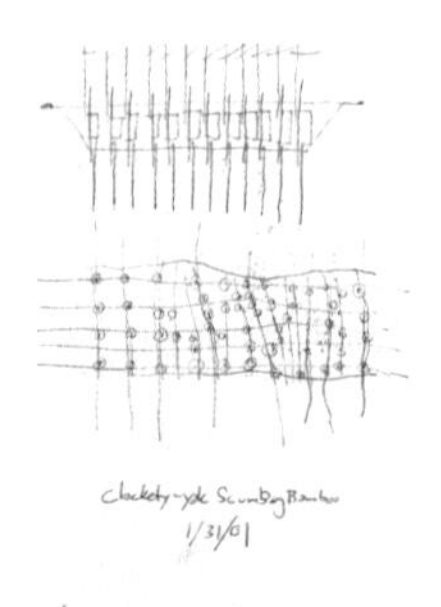
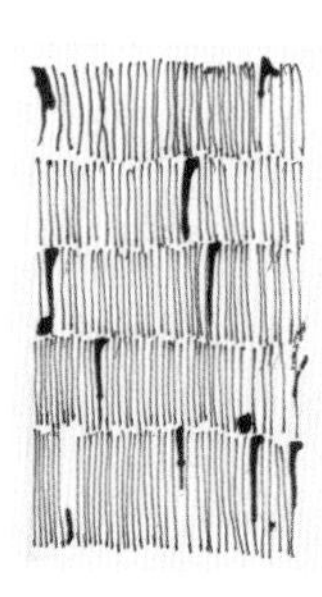
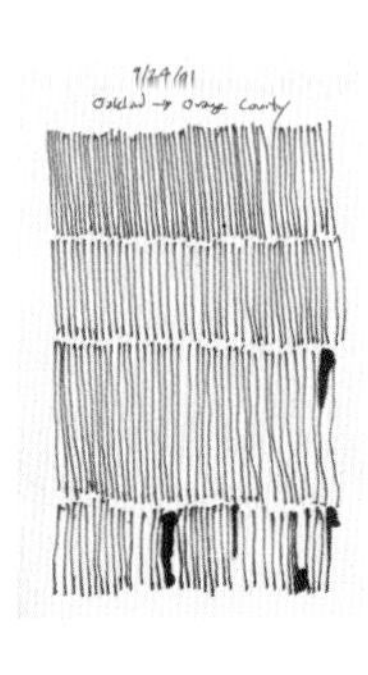
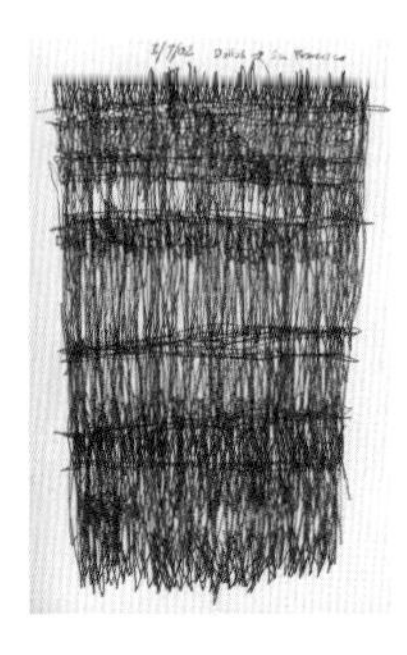
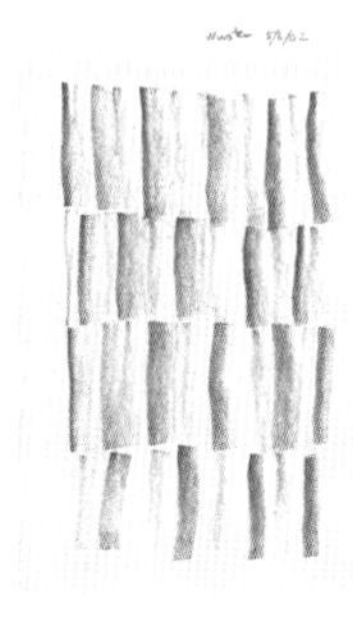
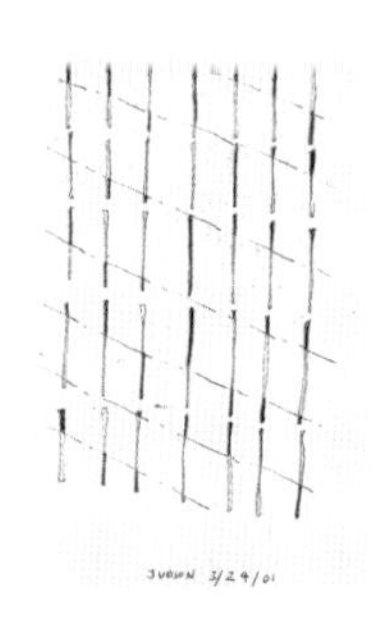
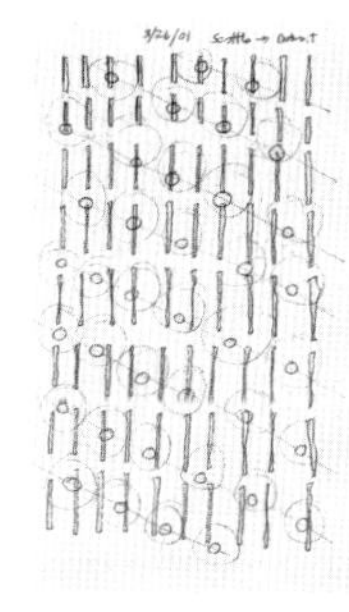
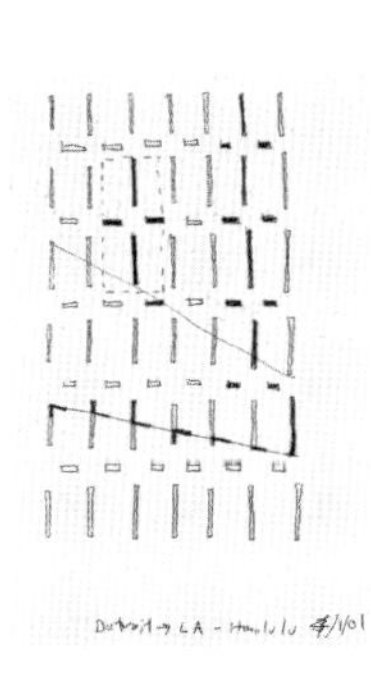
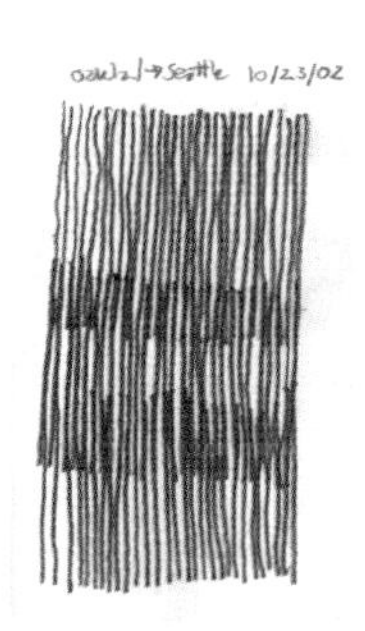

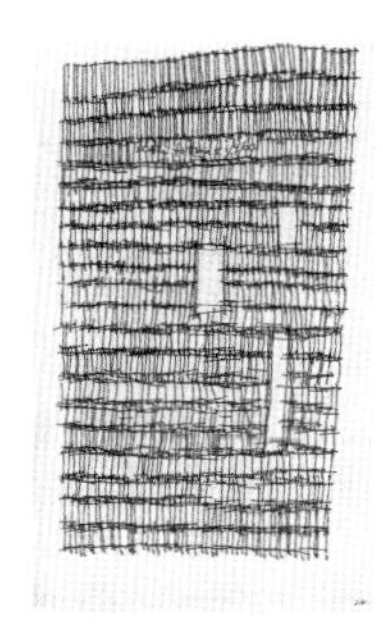
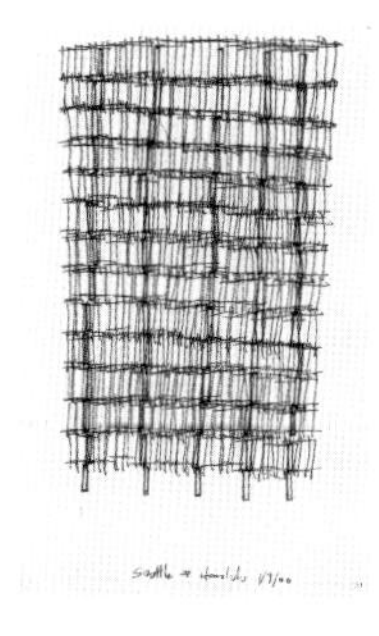

Top: Drawing study series of screen system patterns
Bottom: Early study model folded and sewn from single polycarbonate sheet, with redlined control drawing its laser fabrication (top) and study sketches (bottom)

OPPOSITE
Study models at increasing levels of refinement

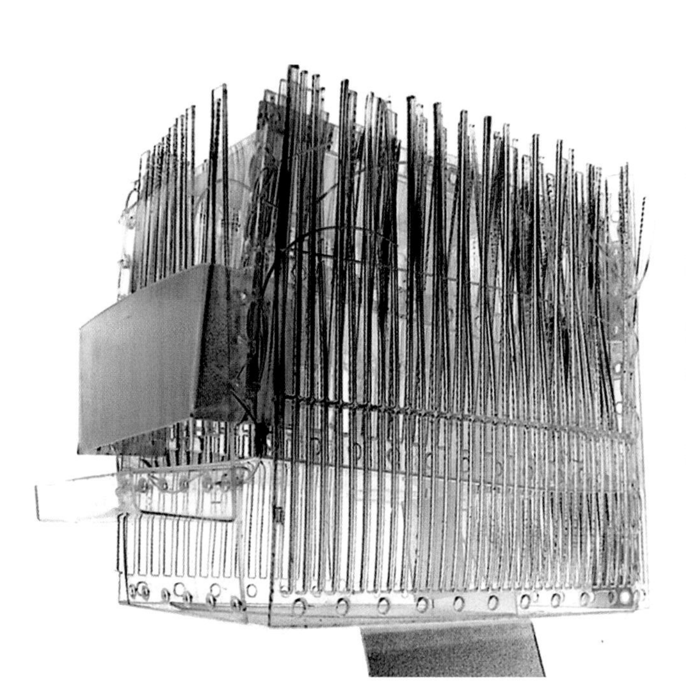
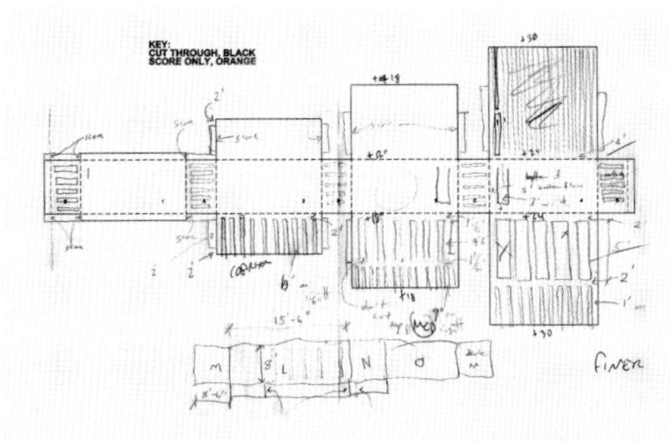

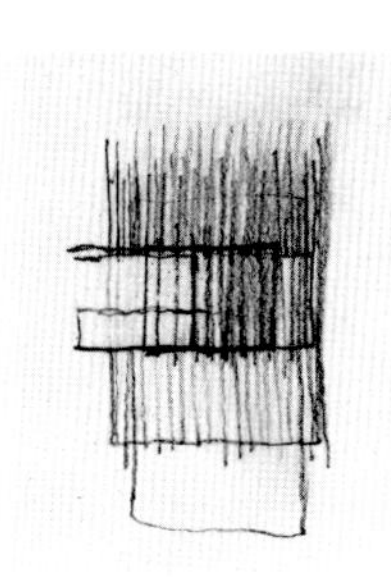
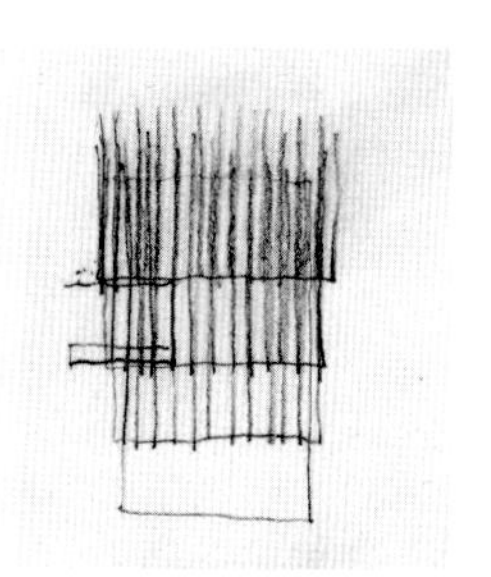

Sandwich Panels

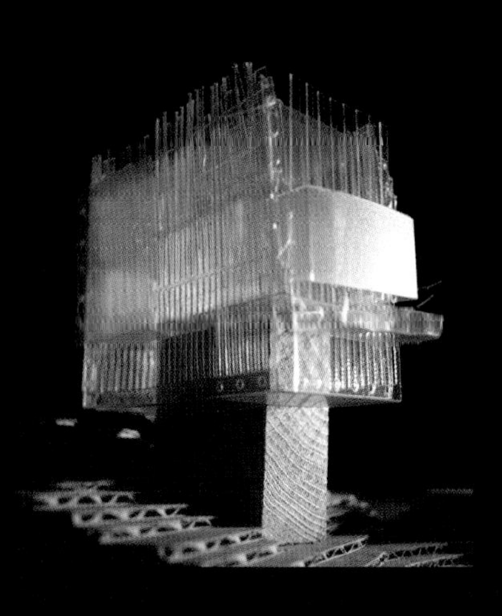

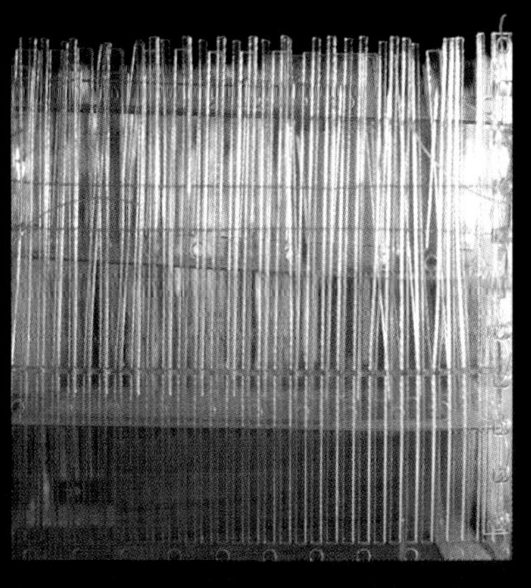

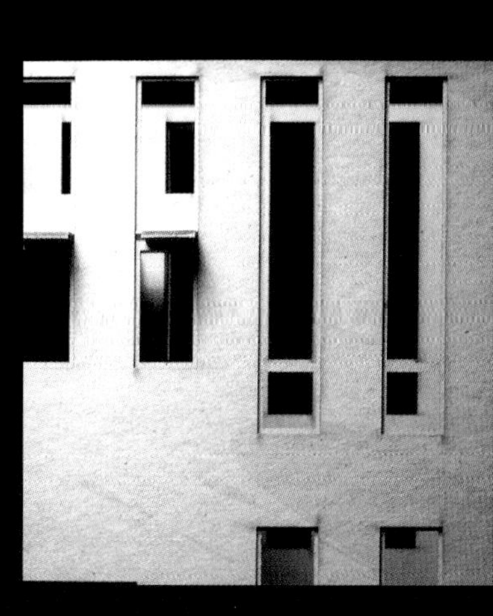

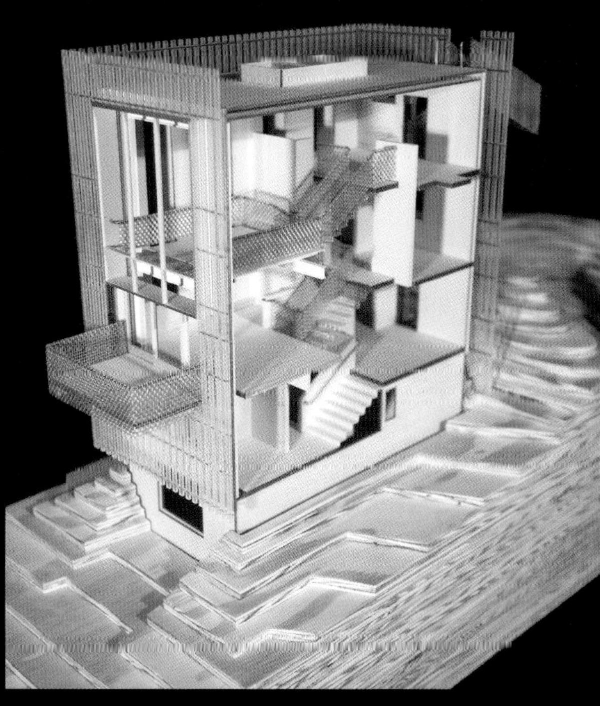

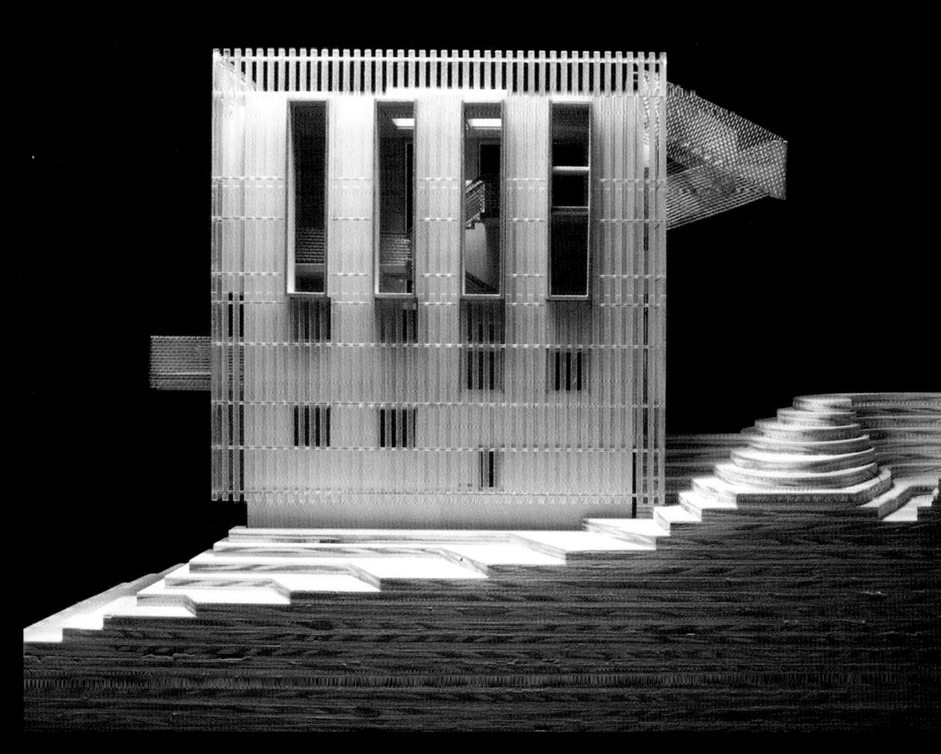

Chameleon House

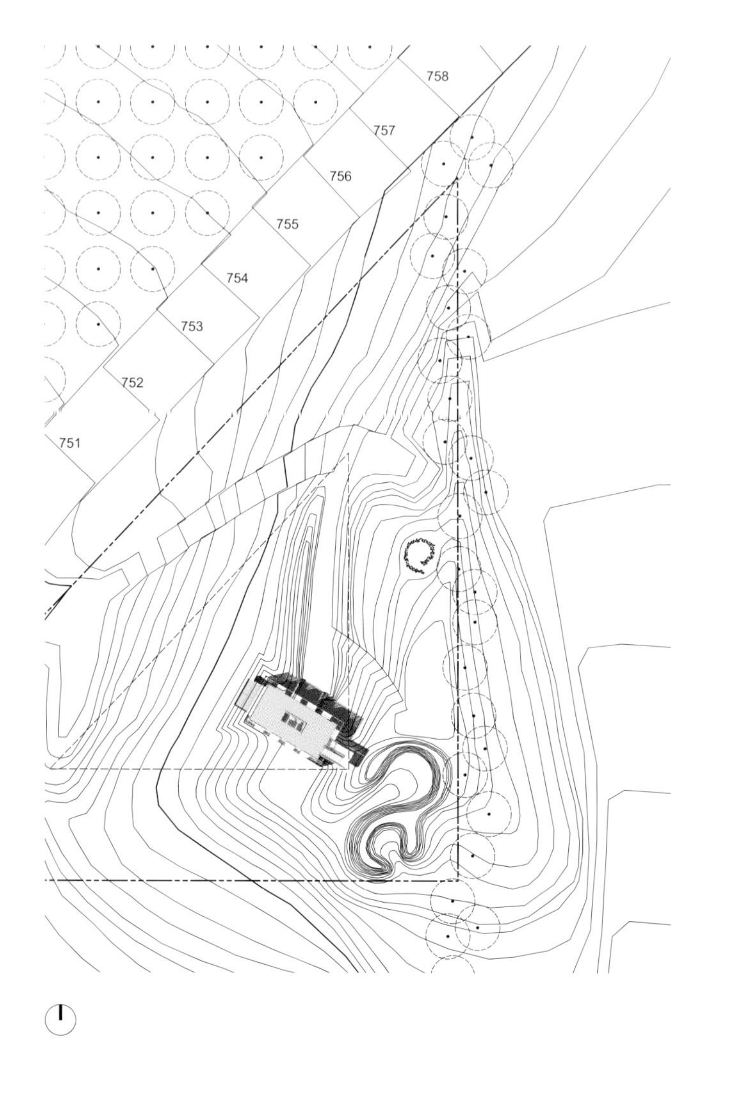

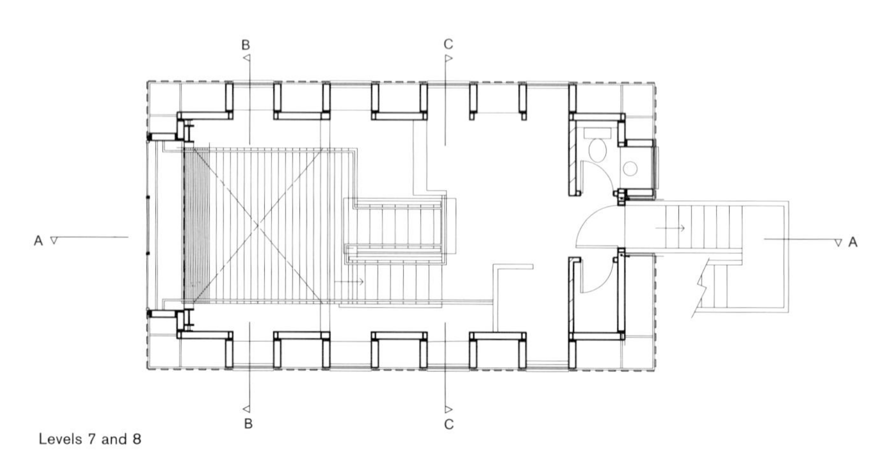

Levels 7 and 8

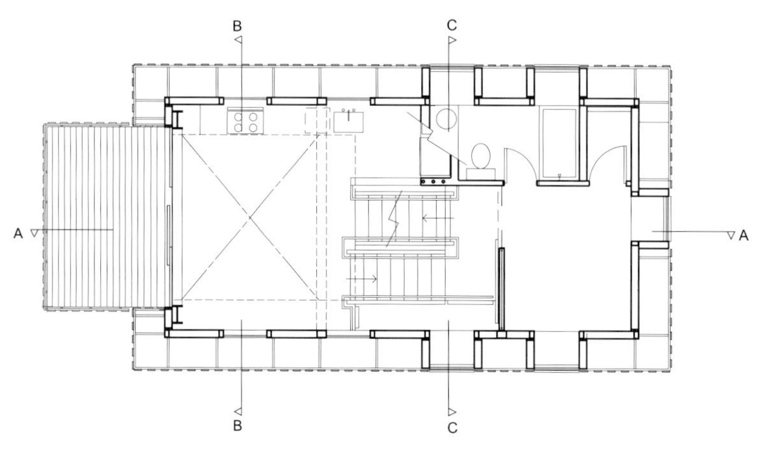

Levels 5 and 6

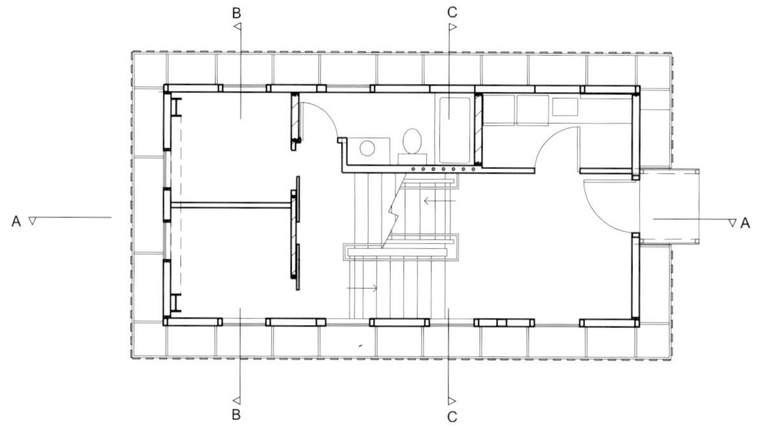

Levels 3 and 4

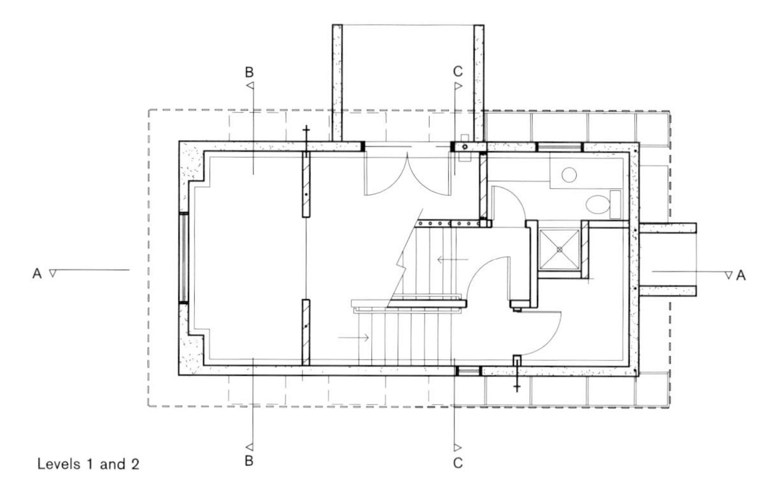

Levels 1 and 2

Sandwich Panels

OPPOSITE
Left: Site plan with house and landscape berm utilizing earth excavated for basement
Right: Building plans of half-levels 1 through 8. Half-levels 9 and 10 occur at exterior stair landing and rooftop terrace.

Right: Rendered image of building with exploded view of primary prefabricated building components. Numbers indicate order of assembly.
Bottom: Section view looking south showing open central stair with skylight above and half-level floor arrangement

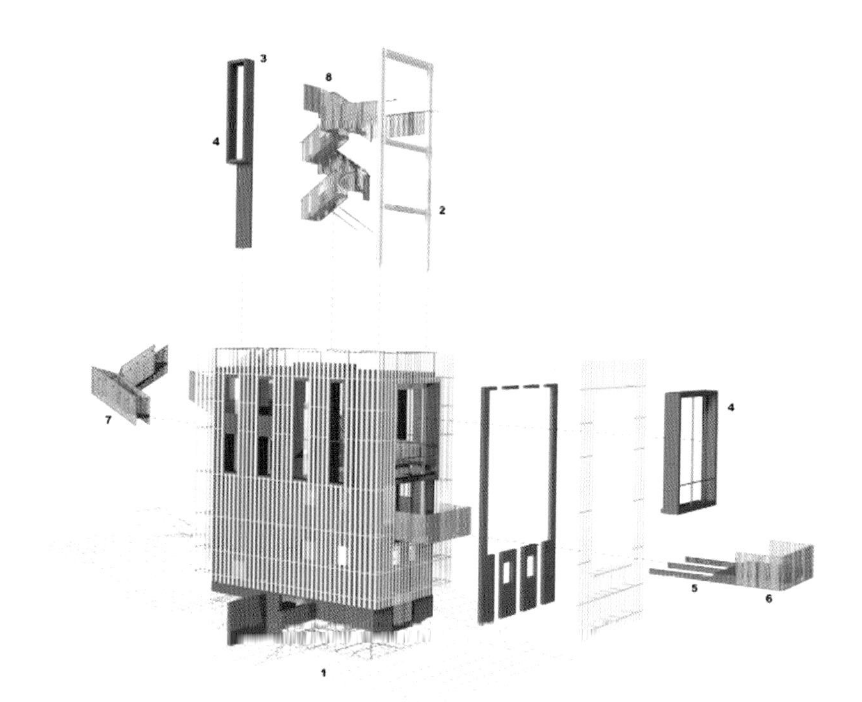

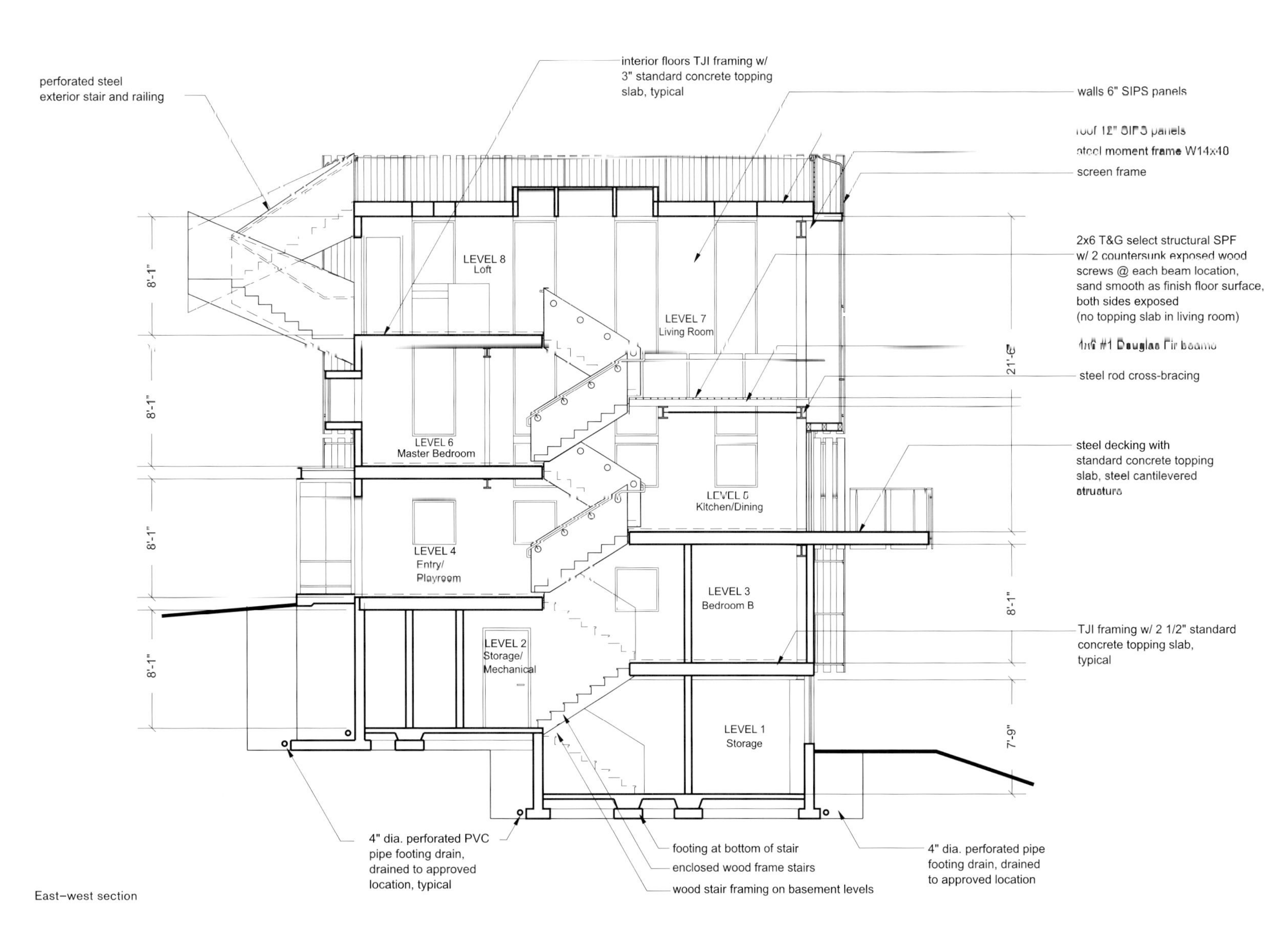

East–west section

Chameleon House

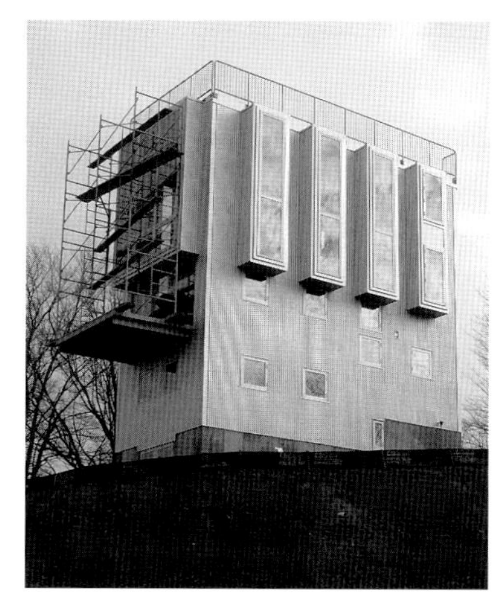

Sequence of construction images showing footings, foundation, steel moment frame, installation of sandwich panels, cladding, and exterior frames for acrylic slats

Sandwich Panels

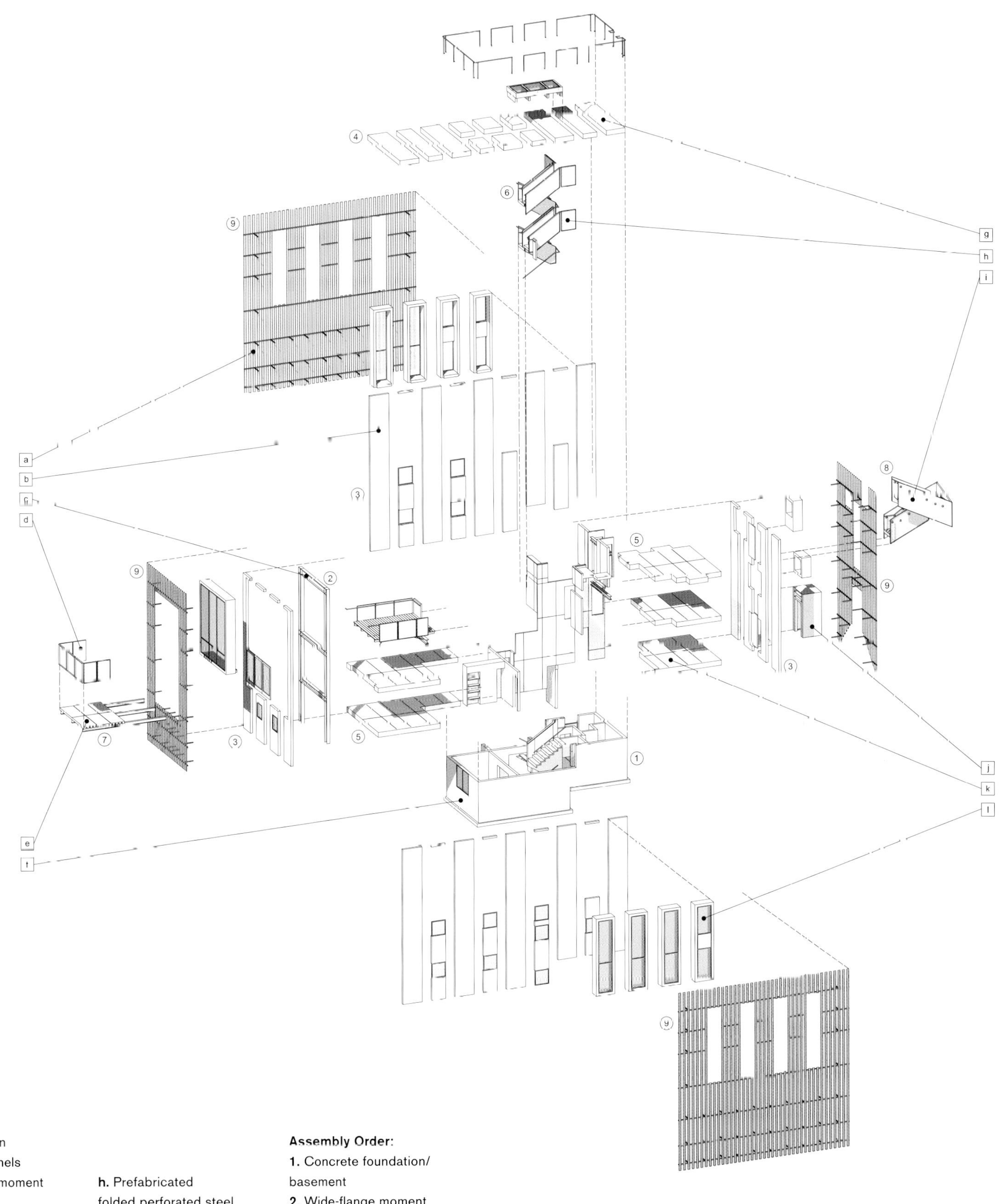

Parts List:

a. Acrylic screen
b. SIPs wall panels
c. Wide-flange moment frame
d. Prefabricated folded perforated steel railings
e. Cantilevered main balcony
f. Concrete foundation/ basement
g. SIPs roof panels
h. Prefabricated folded perforated steel stairs/railings
i. Prefabricated folded perforated steel roof access stair
j. Acrylic entry box
k. Panelized floors
l. SIPs window boxes

Assembly Order:

1. Concrete foundation/ basement
2. Wide-flange moment frame
3. SIPs wall panels
4. SIPs roof panels
5. Panelized floors
6. Prefabricated folded perforated steel interior stairs
7. Cantilevered main balcony
8. Prefabricated folded perforated steel roof access stair
9. Acrylic screen with tube steel frame

Chameleon House

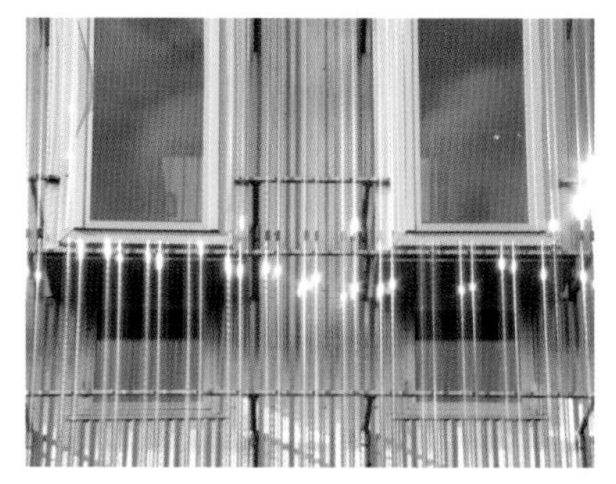

Exterior detail views showing varied reflective surfaces of acrylic slats, glazing, aluminum window frames, and galvanized steel skin, structure, stairs, and railings

Interior detail views showing contrast of natural plywood panel interior building liner with galvanized and painted steel structure, stairs, and railings

OPPOSITE
Morning sun reflecting on northeast face

Above: Southwest corner at dusk

Slice House

Los Angeles, California / 2003

The Slice House is representative of a flexible SIPs-based building system that uses panels and other prefabricated components in a modular approach that is readily adapted by varying building length by adding or deleting units of a predesigned building section. By fully employing the structural capabilities of the panels and using them in their most raw, as manufactured form, the costs and complexities of supplemental structural systems and onsite fabrication are dramatically reduced. Within this modular balloon-framed section, where two-story panels stretch from foundation to roof, the interior spaces can be as open or divided as the program requires since only the exterior walls are structurally load-bearing. In refining this approach, we have worked to balance an interest in developing a kit of rational building elements with a focus on demonstrating their case study implementation, while at the same time satisfying the needs and aspirations of a typical program and demonstrating a simple means of carefully deploying a relatively universal prefabricated building system into an example of a unique landscape and setting. The goal has been to design a cost-, time-, and resource-efficient means of constructing a spatially rich living experience on a particular site.

In this project we have used only readily available, off-the-shelf materials and production systems that we know can be financed through standard banking channels in a timely manner and permitted by the applicable building department without extensive review of unfamiliar construction methods. To meet the strict time and budget conditions common to many building projects, the design employs well-proven and readily available building components and systems that already have building code approvals and can be easily permitted and financed. The panelized SIPs wall system can be purchased from numerous sources throughout the country and is utilized in full-panel factory-direct module sizes that require virtually no site fabrication and produce no waste. The single-piece vertical panel arrangement allows for the offsite installation of exterior and interior finish surfaces so that panels will arrive fully completed, ready for assembly. At present, few fabrication facilities offer this kind of secondary manufacturing of the panel products, and it is more cost-effective to install these finishes on site, but this is a promising area for future development of SIPs-based building systems.

The alternating application of solid panels and vertical window strips highlights the panelized construction approach and illustrates the ability to place solids and voids to take advantage of the particularities of site and program. The organization of windows into preassembled components accommodates many different manufacturers' standard window units without the need for custom curtain wall assemblies. This approach can adapt to alternate window dimensions by altering the spacing between the SIPs. To speed site assembly, the vertical window strips can be mulled together in the factory in full-story or building-height modules, using temporary braces and protective surfaces to keep them rigid and squared up until fixed in place on site.

The modular approach to the design allows for many opportunities to use products chosen for environmental sustainability and resource efficiency. The SIPs building panels have excellent insulation properties, with R-23 ratings for the six-inch panels specified and R-30 for an eight-inch option. Recycled products are specified in many applications throughout the house, including the use of byproduct agricultural fiber panels for interior and exterior finishes.

Particular attention has been given to configuring the overall house to reduce energy consumption for heating and cooling needs. Windows and sun-screening devices are arranged to admit more light in the winter to warm the interior and to screen out direct light in the summer to keep the interior cool. Concrete slabs are used on the floors as part of the high-efficiency radiant heating system, providing the thermal mass to store heat from the solar energy that strikes them during the winter. These slabs, shaded from summer sun, help maintain cool temperatures during the hottest months. Additional cooling is provided by the extensive flow-through ventilation from opening windows and doors on all sides of the house. The large number of operable windows allows the option of selecting low openings on windward sides and high windows in the lee direction to optimize the ventilation capabilities, with the opportunity to reconfigure according to changing season and weather patterns.

As conceived in this example of the Slice House system, Phase One of the project includes the basic building program and utilizes a base building shell with the addition of exterior wood slatted screens, a perforated metal screen entry component, and exterior deck as bolt-on adaptations to this particular site. The Phase Two structure is proposed as a freestanding building to house a garage and workspaces for home offices, a guest apartment, or a music studio.

Model views showing relationship of modular building segments to sloping topography of site. Excavated earth is recycled as sculpted landscape berms.

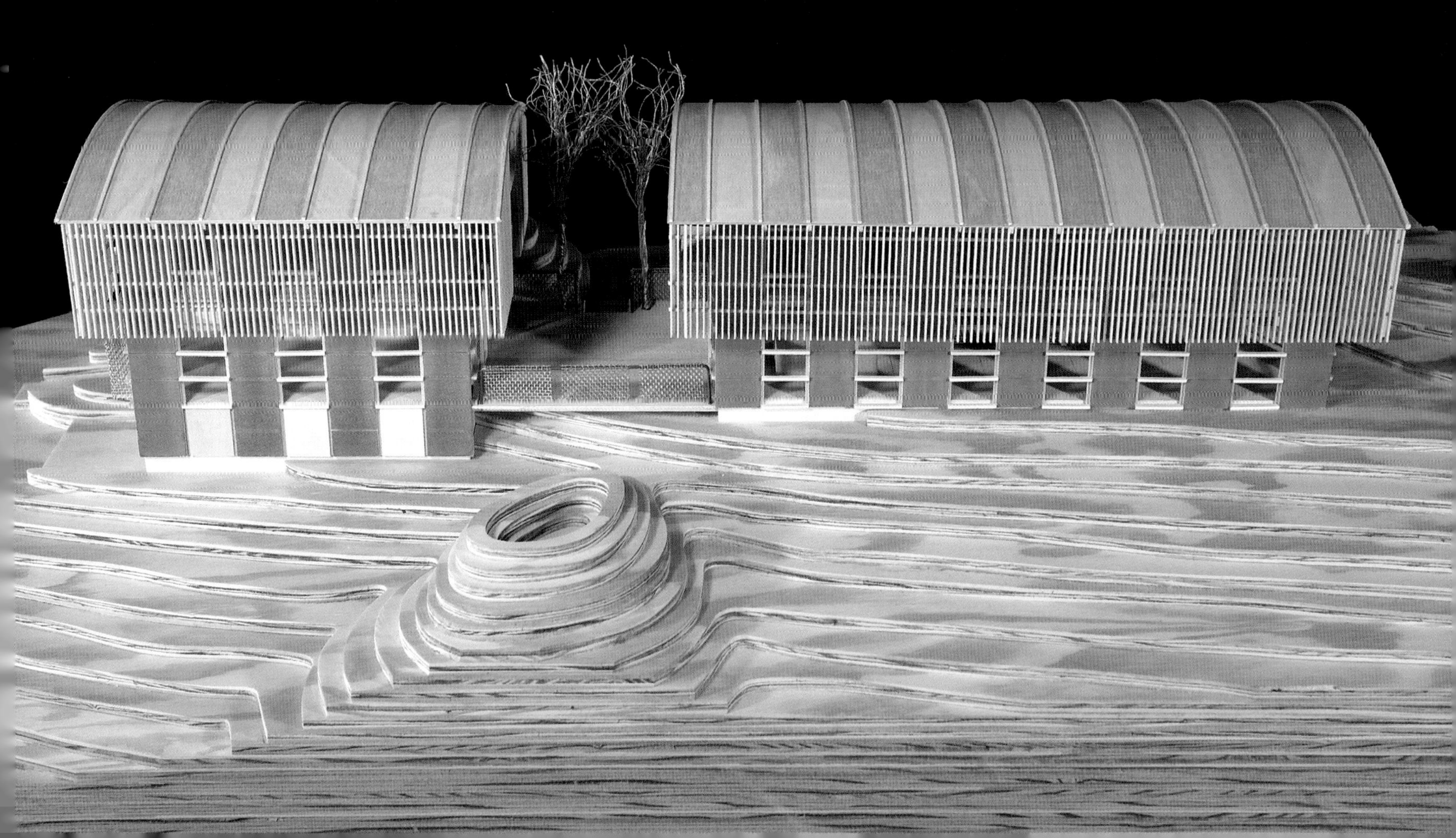

Slice House

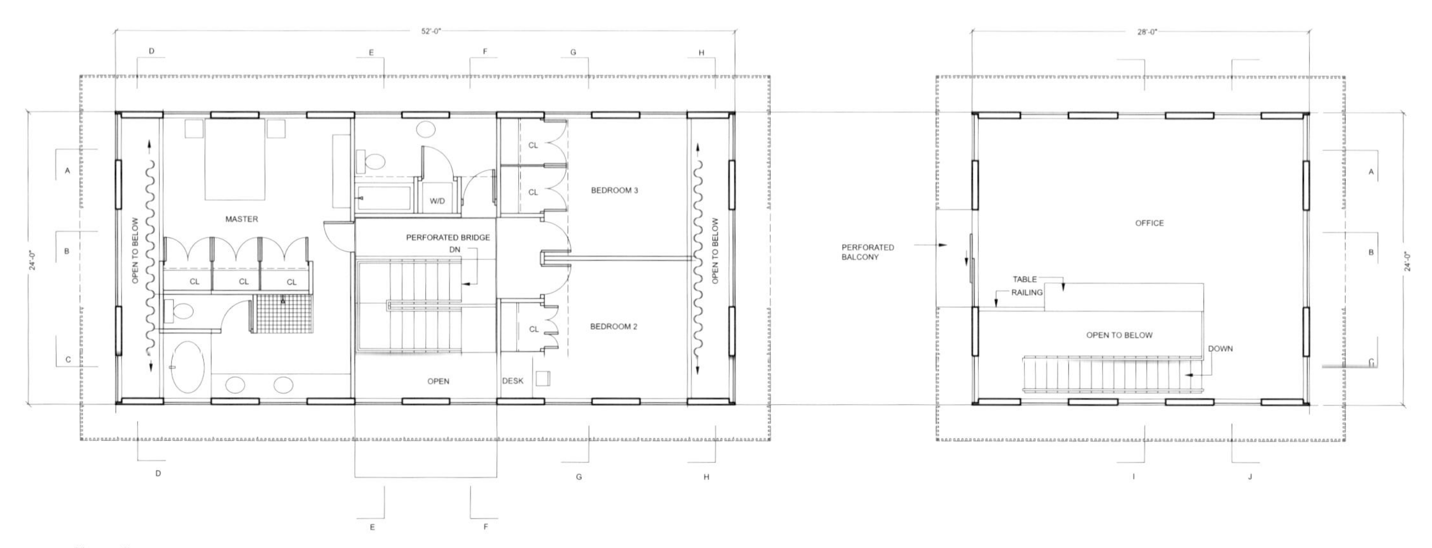

Upper floor

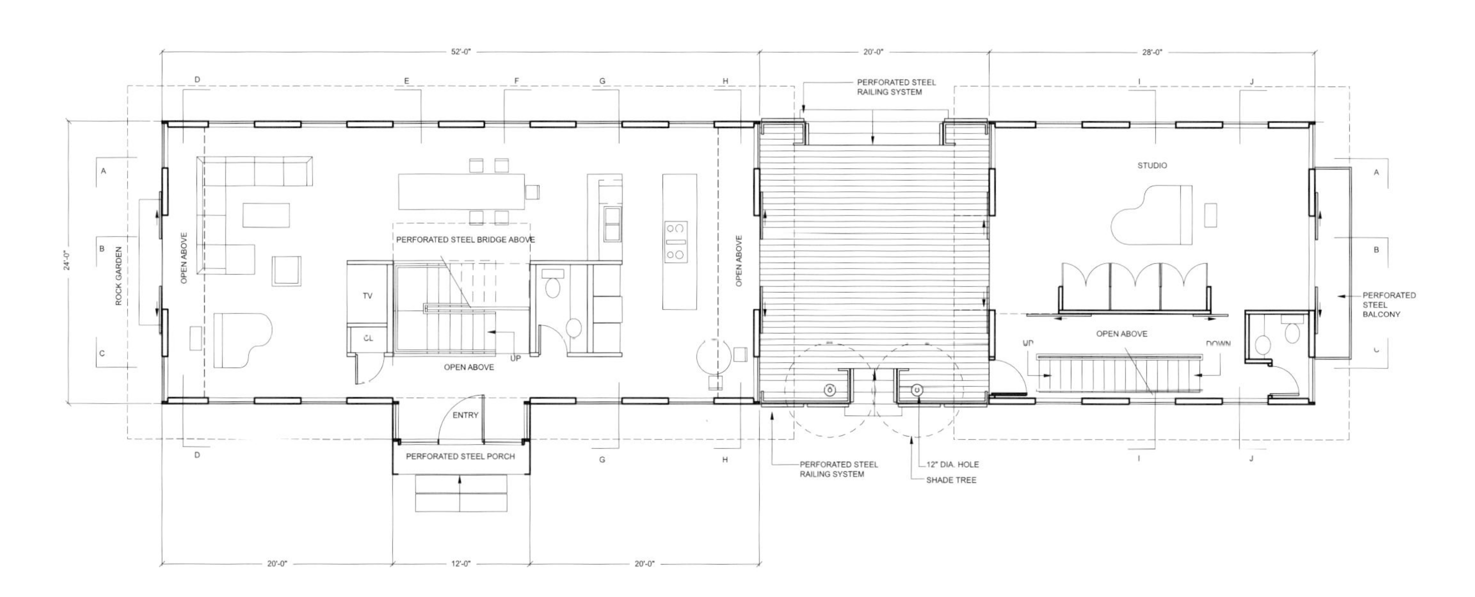

Main floor

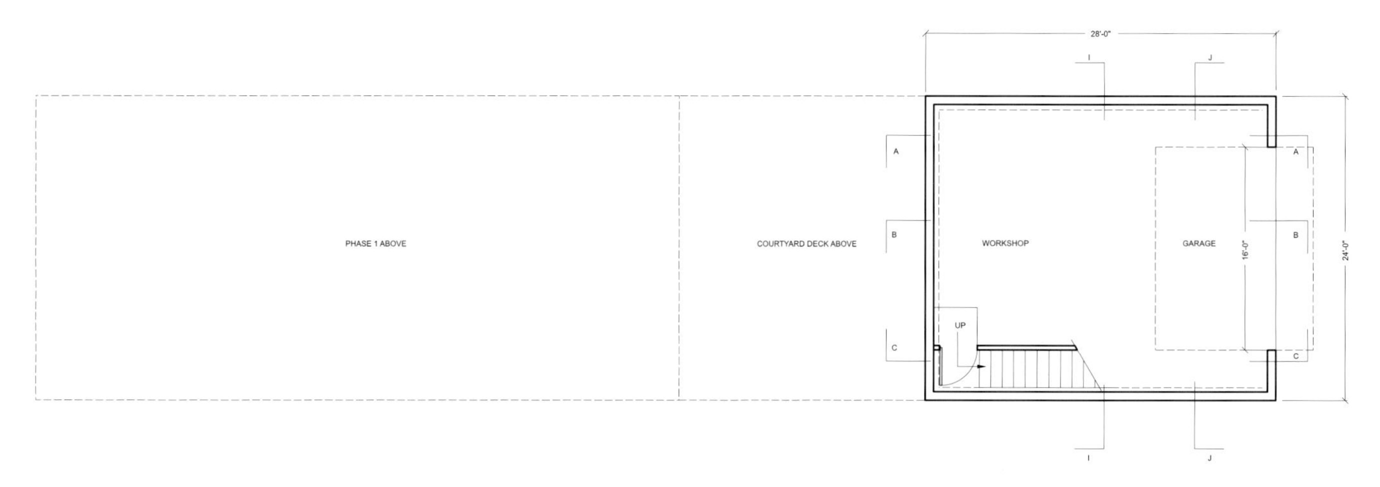

Basement garage

OPPOSITE
Plans

Right: Elevations and section showing potential for spatial variation in slice concept

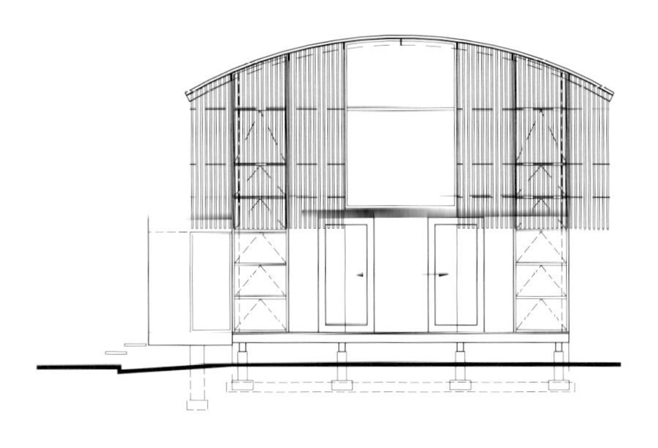

East elevation phase 1

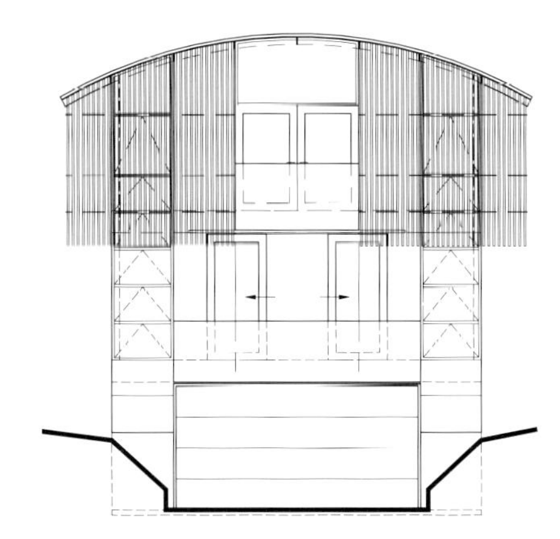

East elevation phase 2

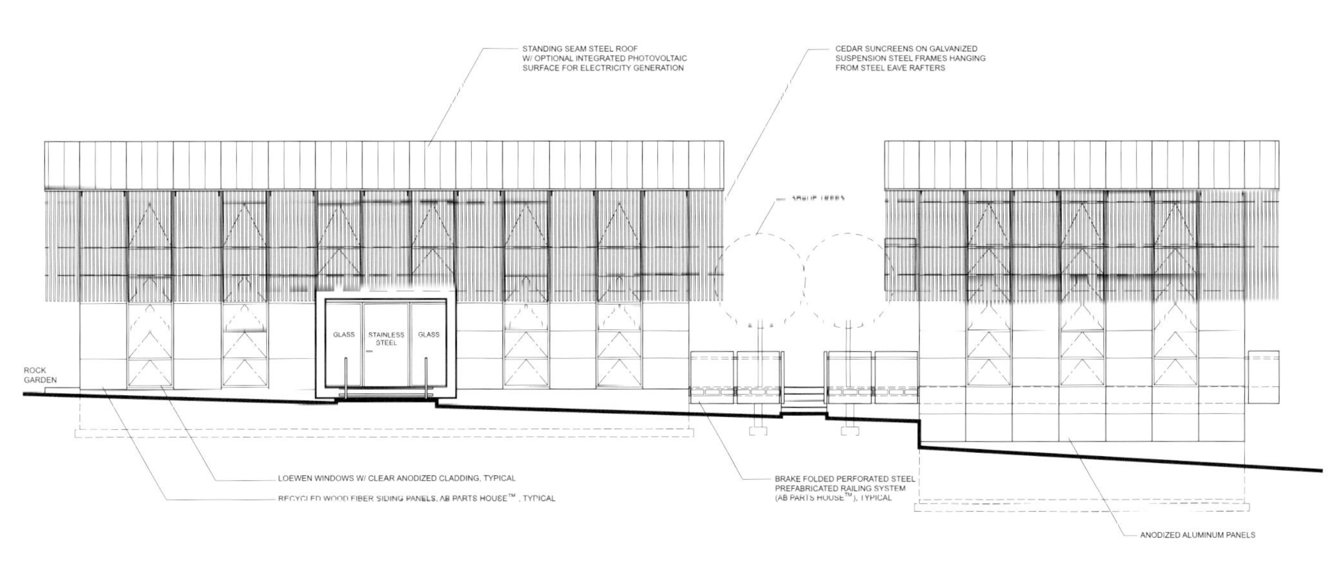

South elevation phase 1

South elevation phase 2

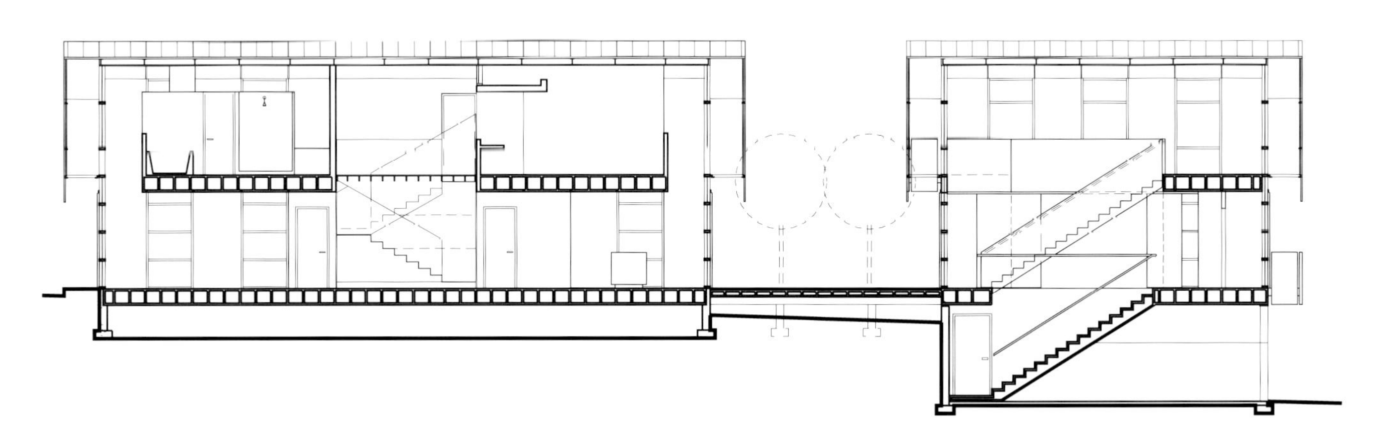

Section C phase 1

Section C phase 2

Slice House

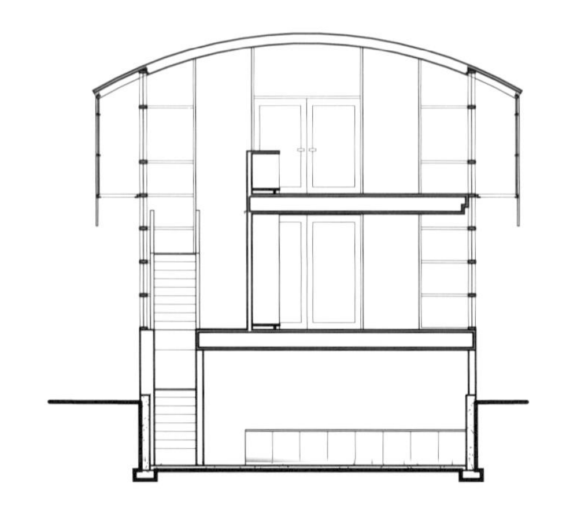

Section I

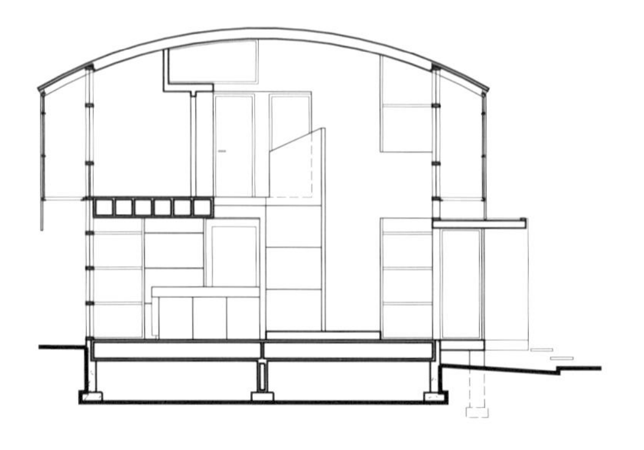

Section F

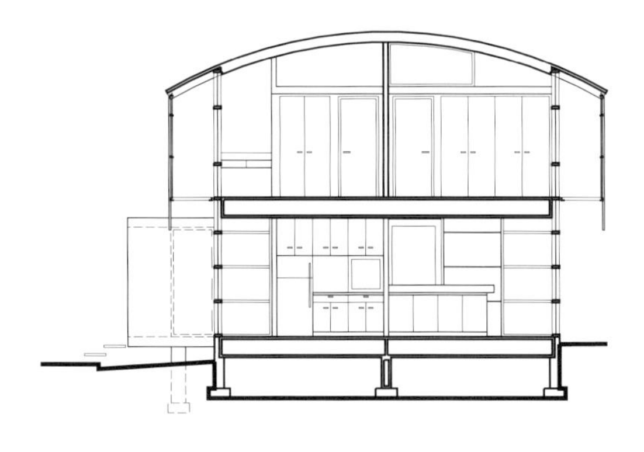

Section G

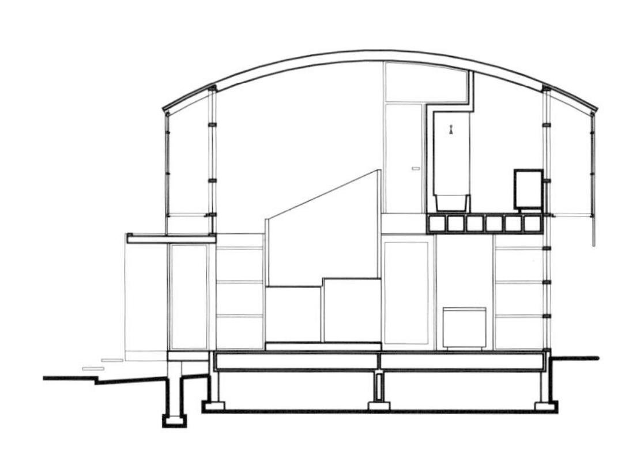

Section E

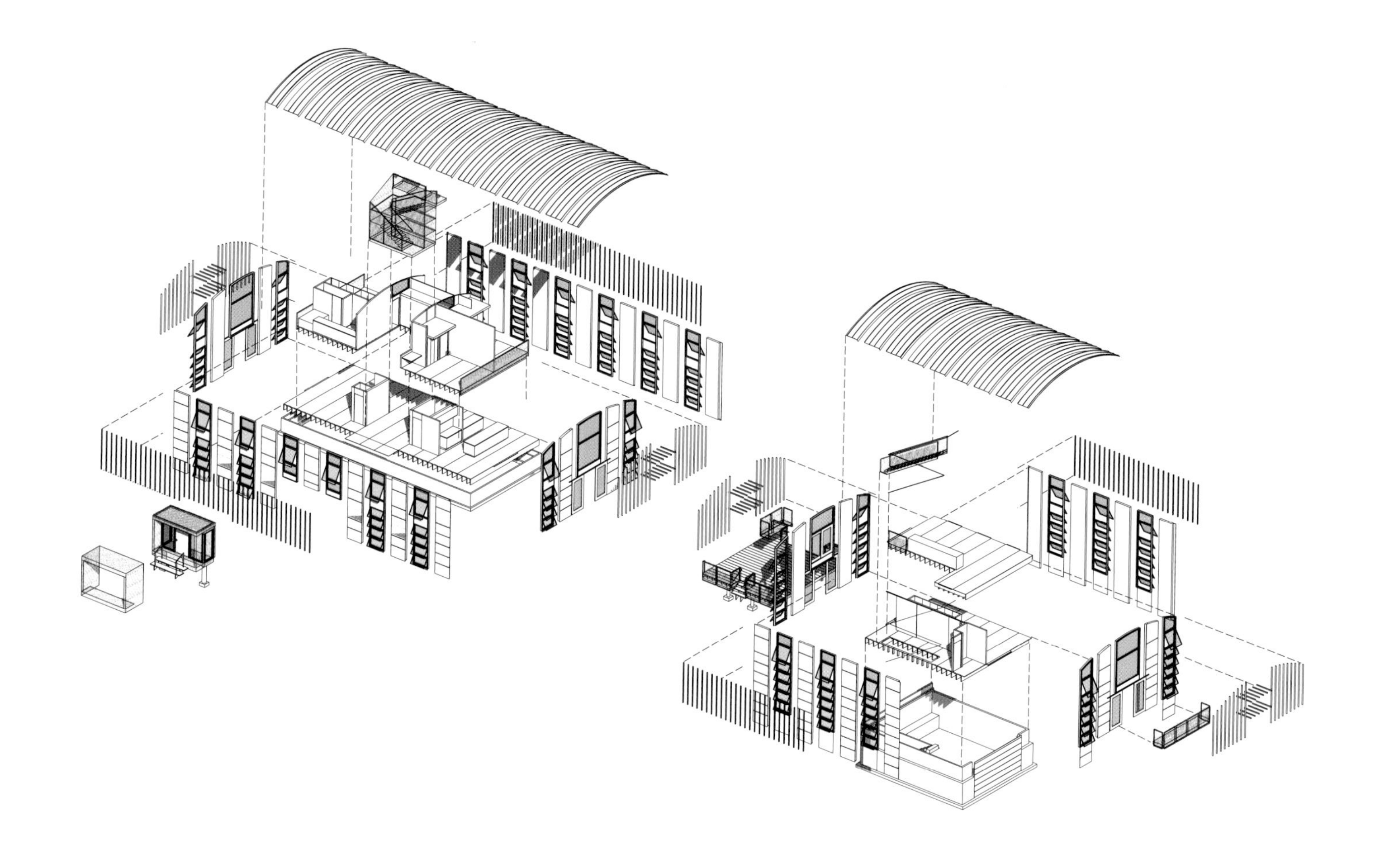

Left: Section drawings through house show interlocking single- and double-height spaces
Below: exploded axonometric view of primary building components

OPPOSITE
Drawings and rendered views of prefabricated perforated sheet steel stair module

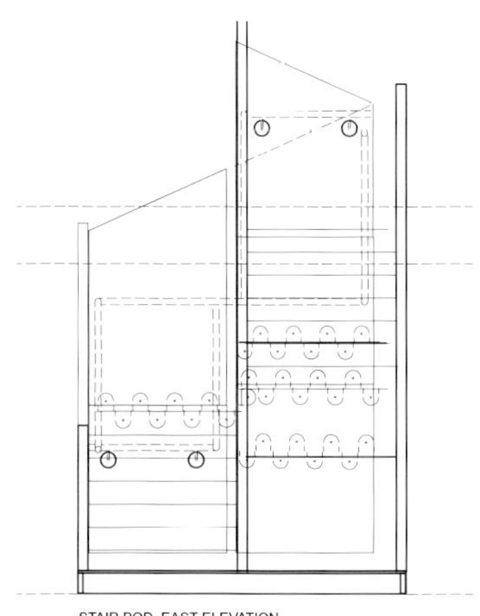

STAIR POD, EAST ELEVATION

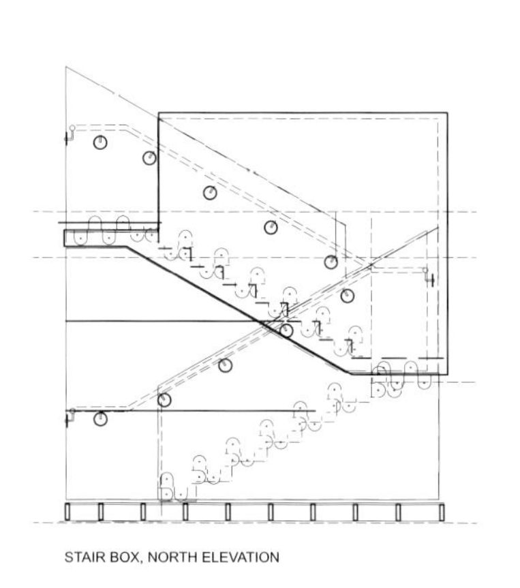

STAIR BOX, NORTH ELEVATION

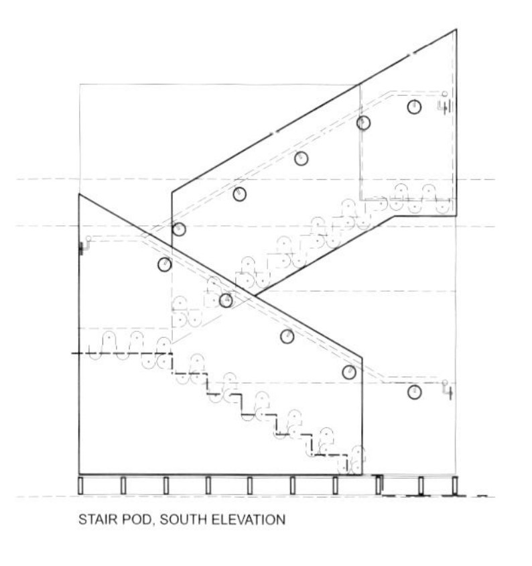

STAIR POD, SOUTH ELEVATION

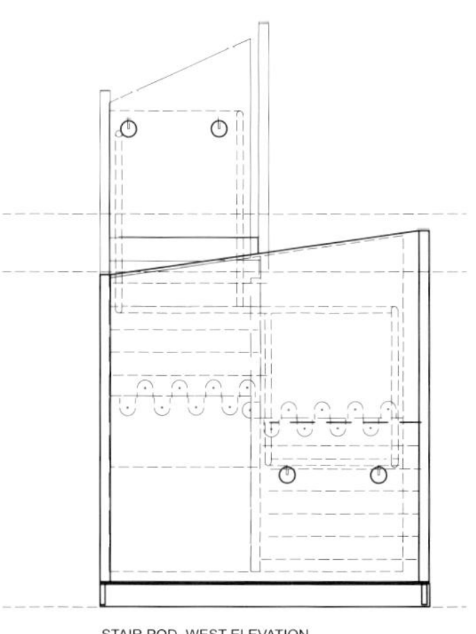

STAIR POD, WEST ELEVATION

Abiquiu House

Abiquiu, New Mexico / 2003

Designed for an anthropologist and a concert pianist, retiring from Phoenix, Arizona, to this small New Mexico town on a desert site fronting the Rio Chama—not far from Georgia O'Keefe's famous home on the bluff above—this house uses several relatively standard prefabrication systems. SIPs are used for the wall panels only, while the roof and floors are constructed of prefabricated 2x4 long-span trusses. Although it was originally intended to use panels as the roof and floor structure as well, the house was switched shortly before construction to a truss system to simplify the assembly and to reduce the structural lumber splines required in the long spans of the panels.

The owners have a number of animals—dogs and cats and occasional injured strays—that they were concerned with protecting from the prevalent local hawks, eagles, coyotes, and rattlesnakes. Rather than compromise the design with the addition of a retrofitted chain link dog run, we developed a thoroughly integrated animal house. For budget reasons, local contextualism, and appropriately barnyard practicality, we settled on chain link as a major material system for the house, protecting domestic animals and people from other animals or from accidental falls from the upper terraces.

Chain link is an ingenious prefabricated system that can be rolled out and hung from above like curtains, stretched and bolted to the walls and frames with large, round, specially cut steel washers that can be inexpensively manufactured in quantity and made available as modular parts in the system. In some places the chain link stands away from the house, providing enclosure to exterior living spaces, and in other areas it hugs tight to the steel-siding-clad wall surfaces, providing visual continuity and textural relief to the large flat planes while at the same time providing a trellis for creeping plants that will grow up from the ground to further soften the profile of the house.

Left: Preliminary design sketches
Right: Site plan

OPPOSITE
Model views of east (top) and south (bottom) sides showing chain link mesh used as skin, enclosure for animals, shading device, and privacy veil

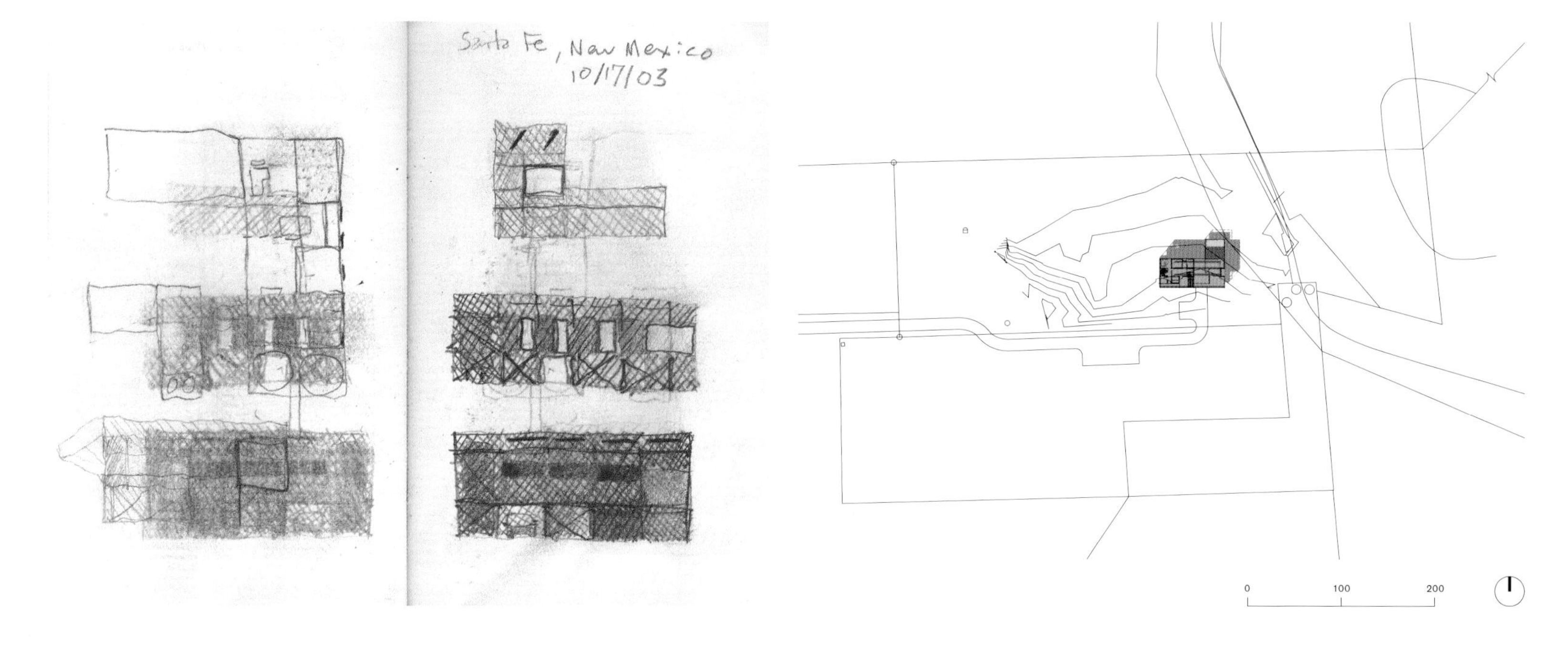

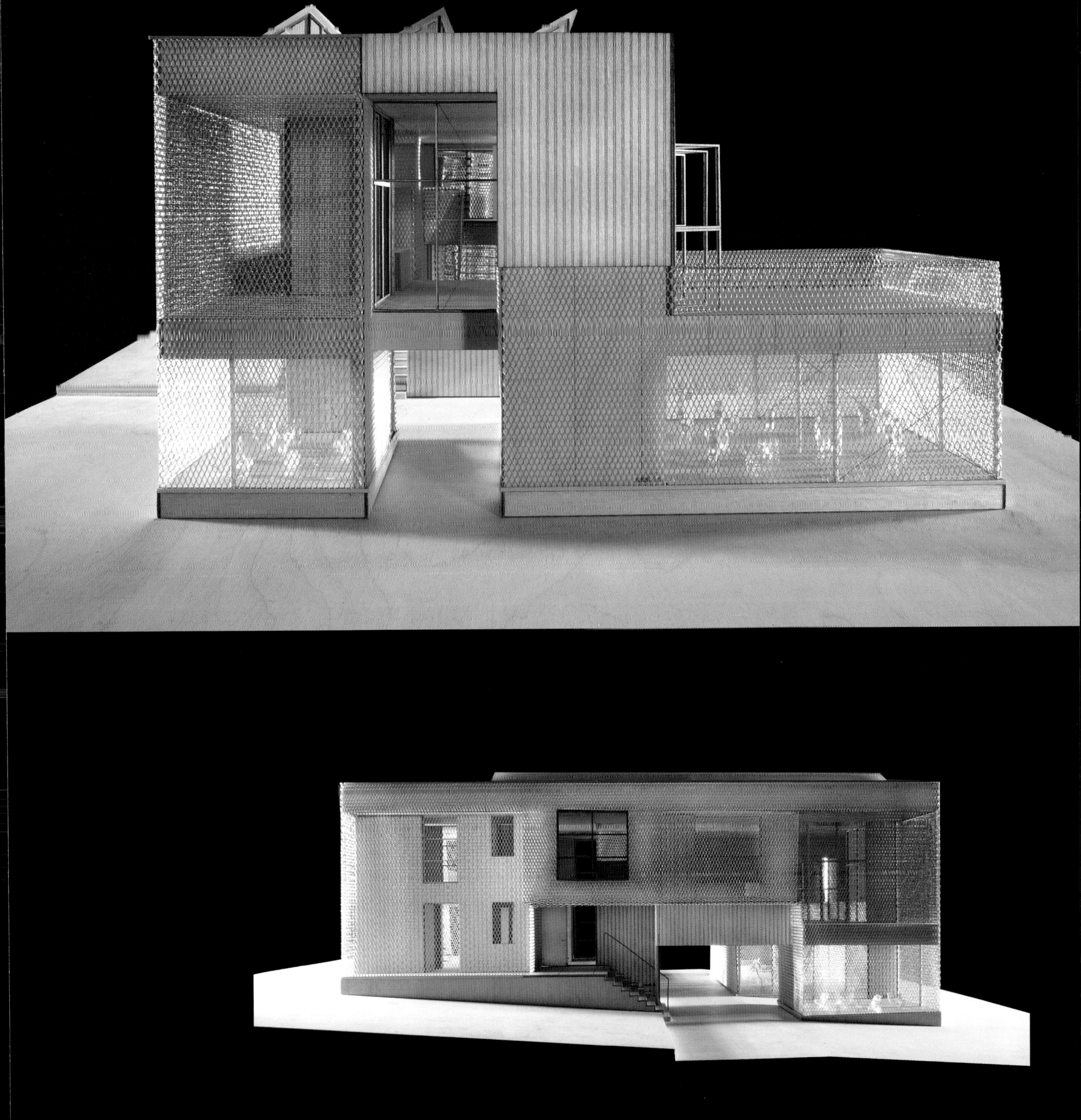

Abiquiu House

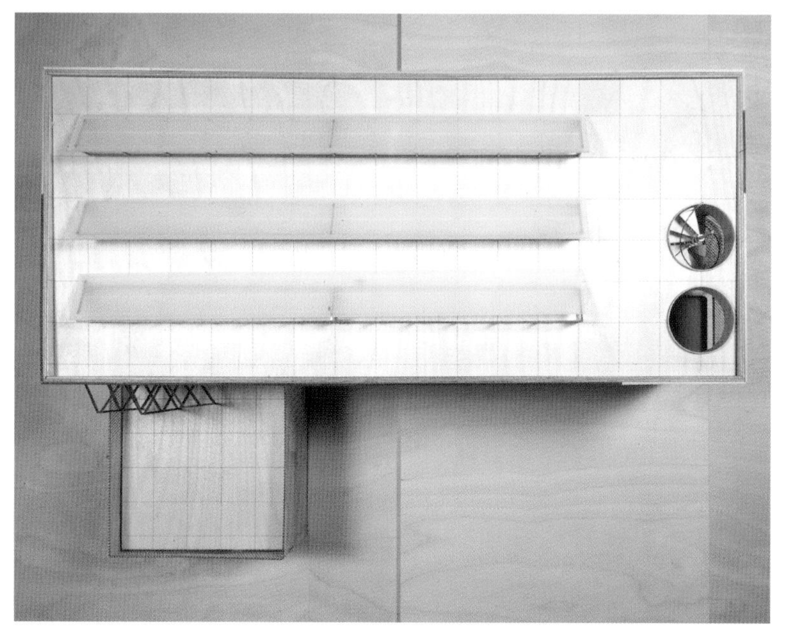

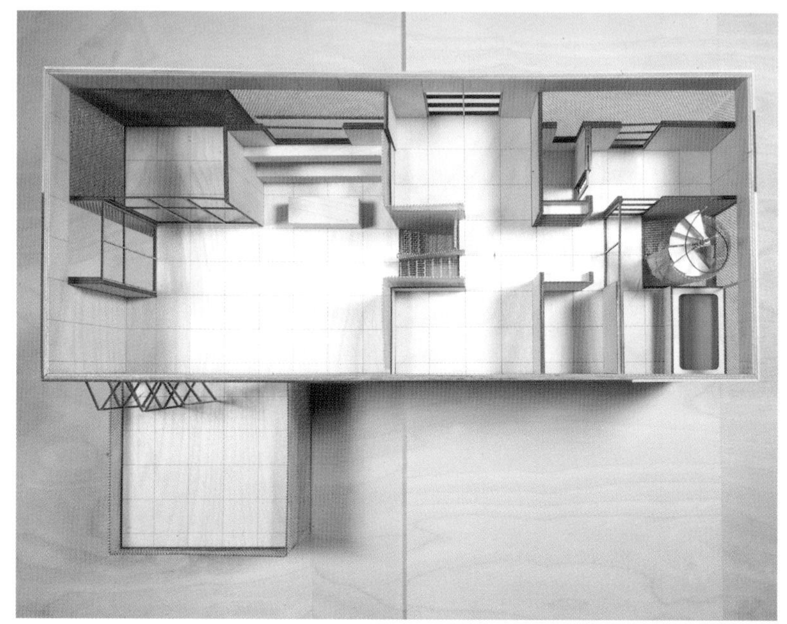

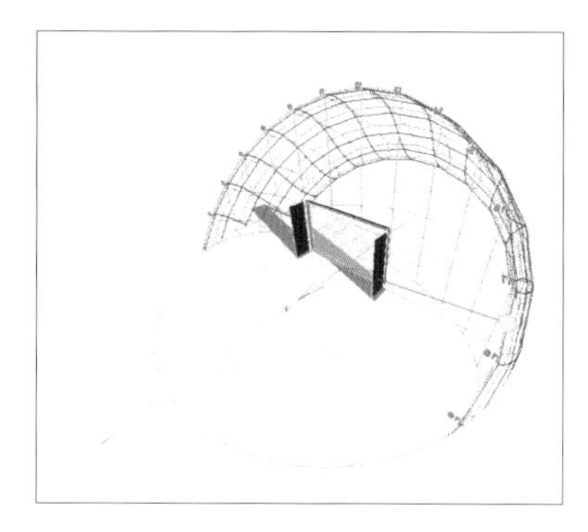

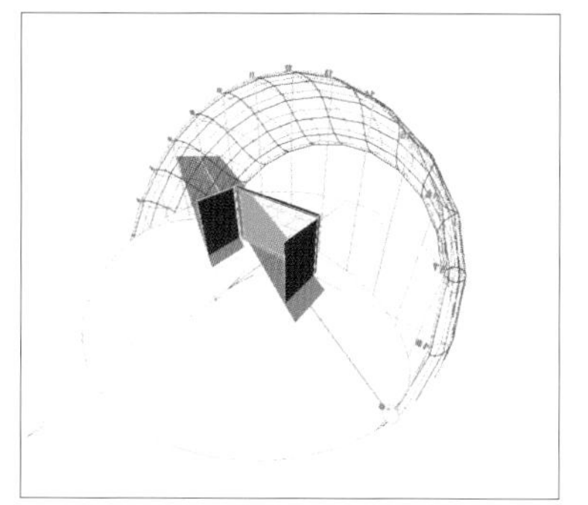

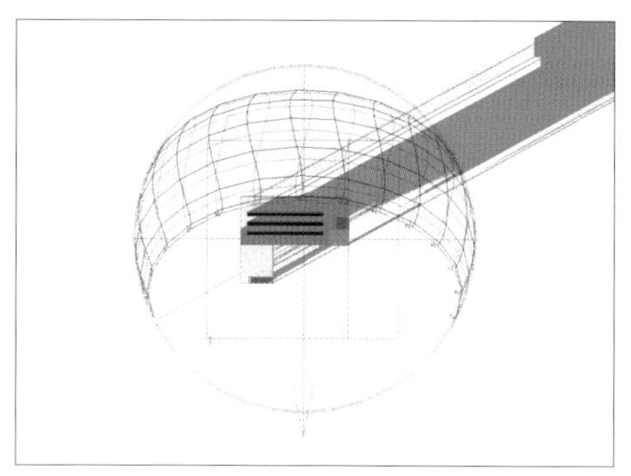

Above: Model views of northeast and southeast corners, rooftop deck with solar panels, and interior with roof removed
Left: Sun study diagrams used to track shading patterns. Prepared in collaboration with solar consultant Olivier Pennetier

OPPOSITE
Extensive computer model daylighting studies of apertures and interior spaces were used to create strategy for maximum solar gain in winter and passive cooling in summer

Sandwich Panels

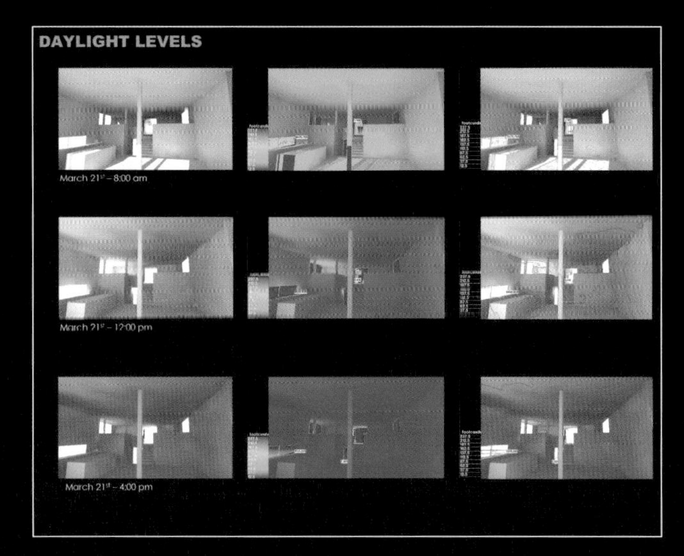

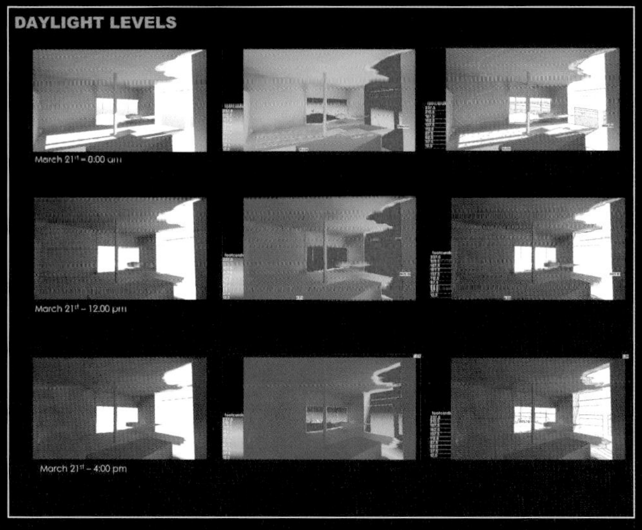

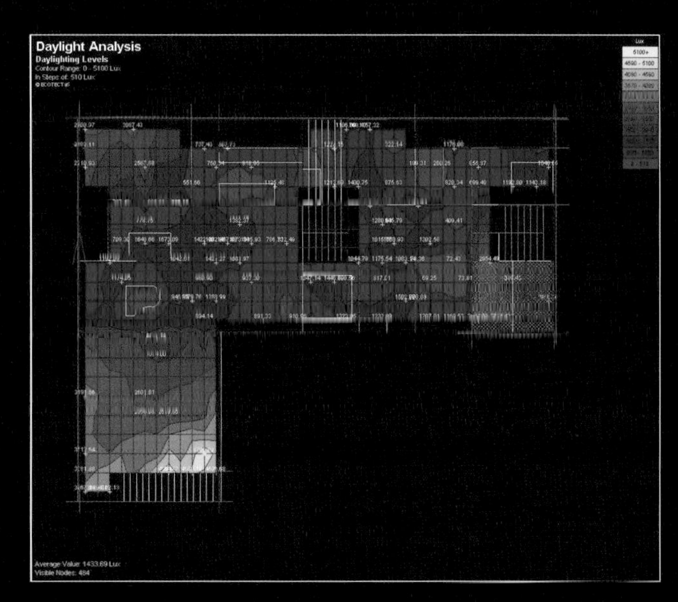

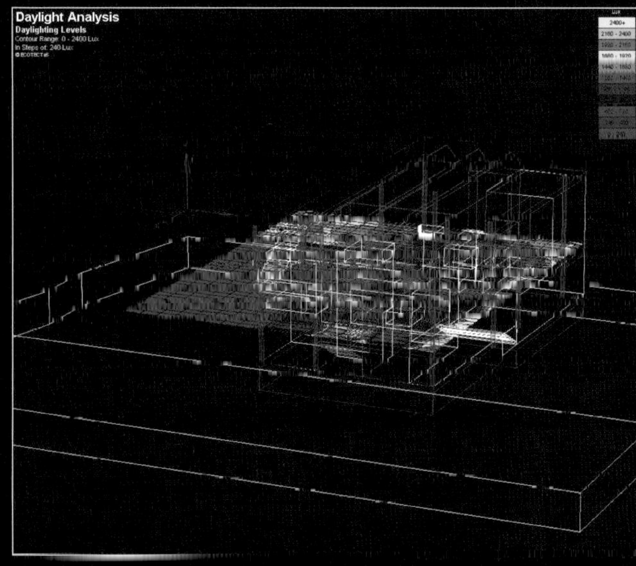

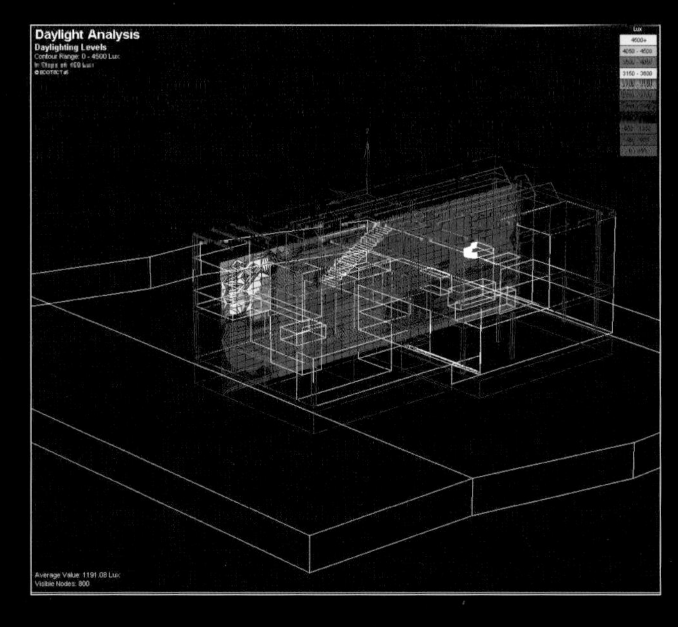

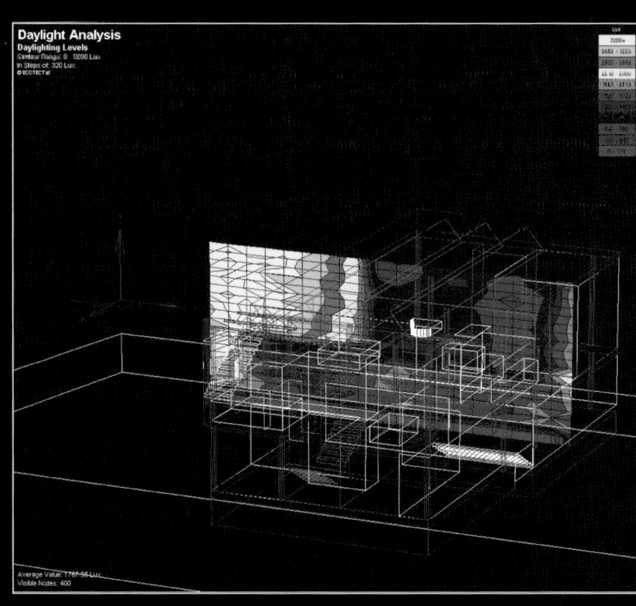

Abiquiu House

Top: Upper-level plan
Bottom: Ground-level plan

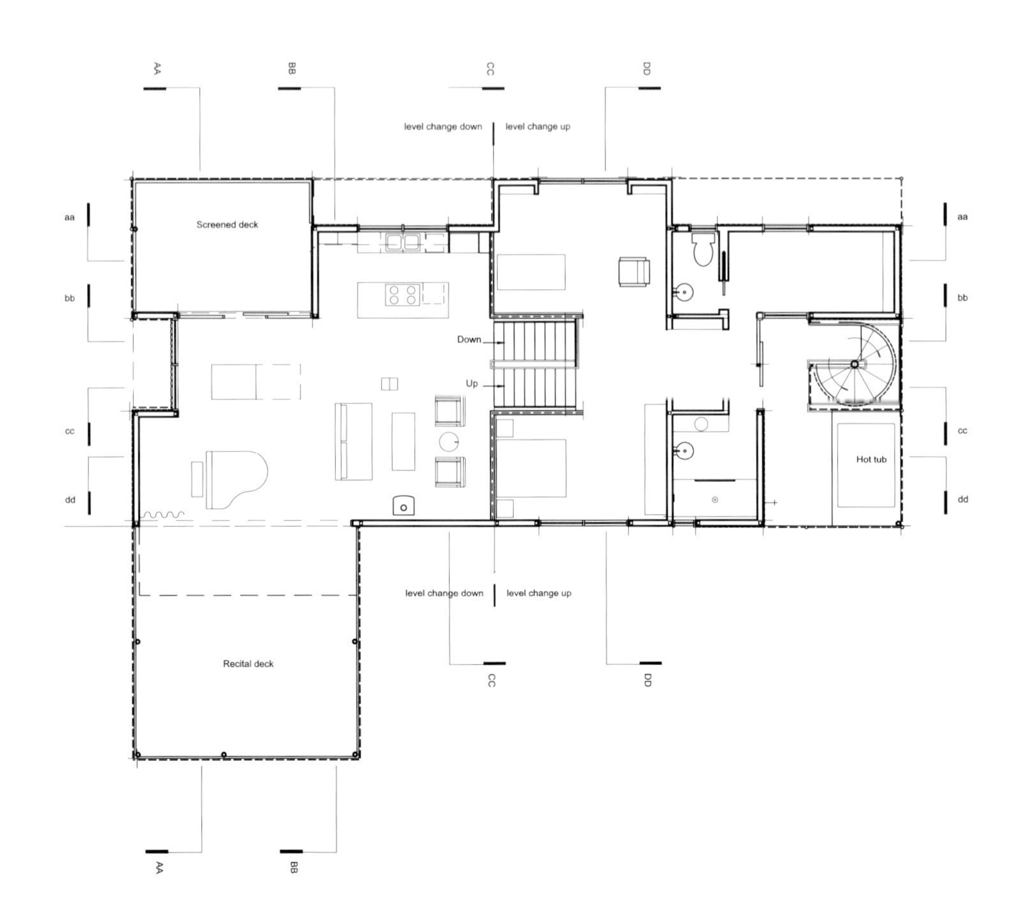

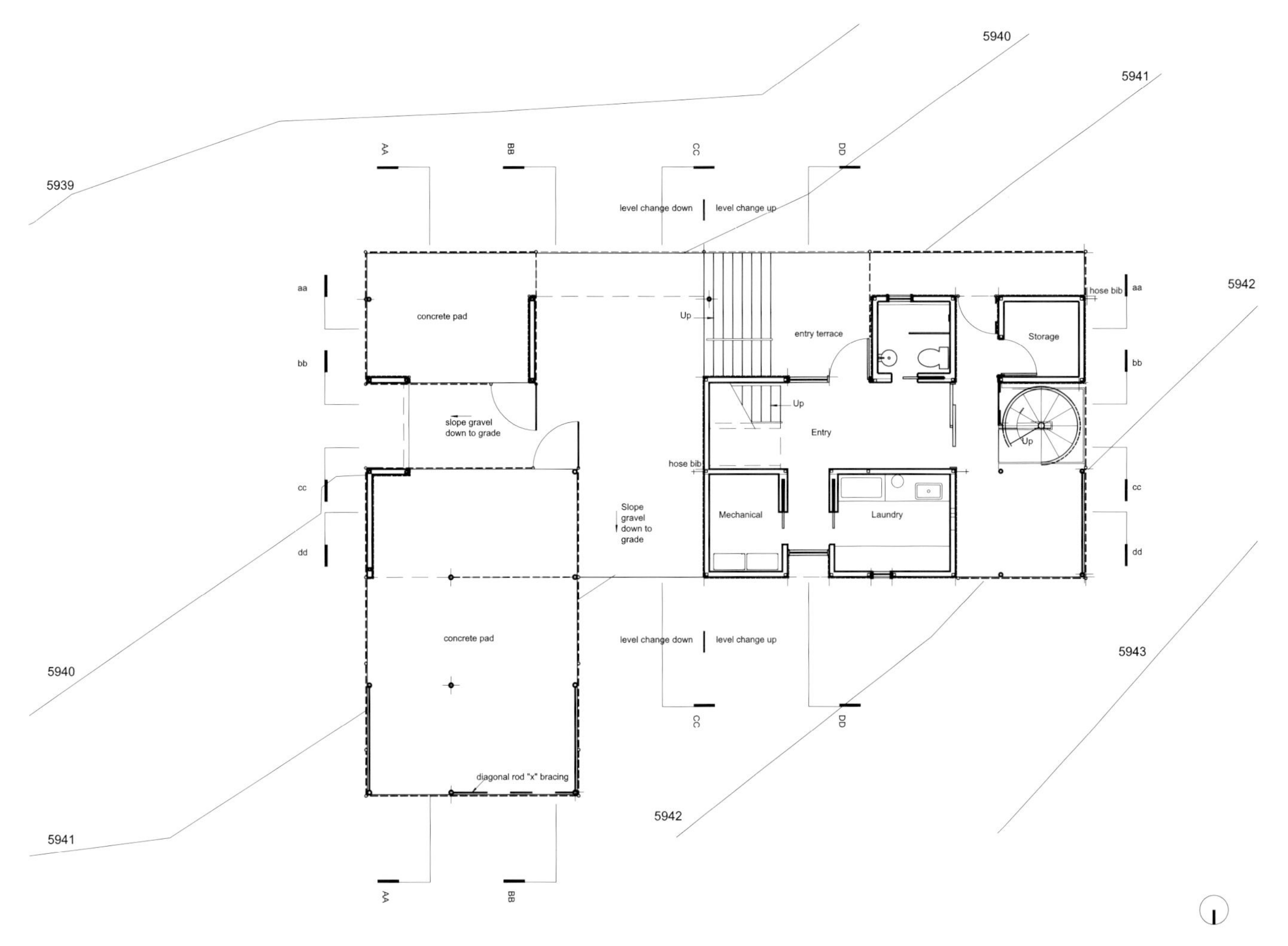

Sandwich Panels

Section and detail drawings of stair assembly with instructions to steel fabricator and detailor

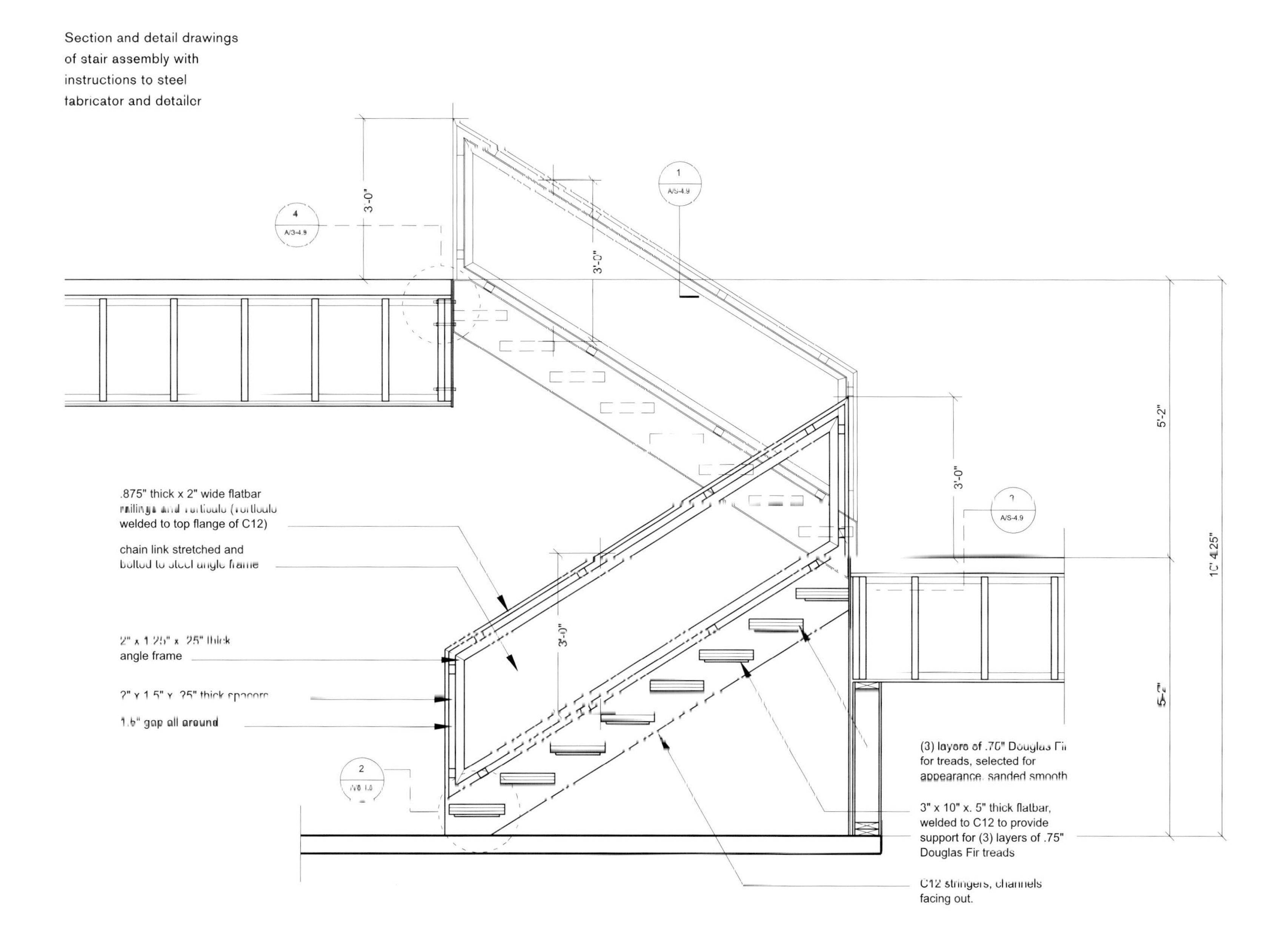

4
A/S-4.9
1
A/S-4.9
3'-0"
3'-0"
3'-0"
3'-0"
5'-2"
5'-2"
10' 4.25"
2
A/S-4.9
.875" thick x 2" wide flatbar railings and verticals (verticals welded to top flange of C12)
chain link stretched and bolted to steel angle frame
2" x 1.25" x .25" thick angle frame
2" x 1.5" x .25" thick spacers
1.5" gap all around
(3) layers of .75" Douglas Fir for treads, selected for appearance, sanded smooth
3" x 10" x. 5" thick flatbar, welded to C12 to provide support for (3) layers of .75" Douglas Fir treads
C12 stringers, channels facing out.

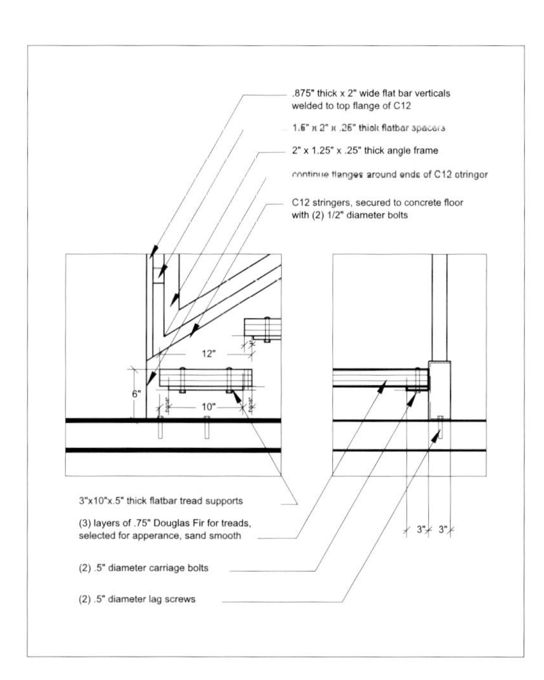

.875" thick x 2" wide flat bar verticals welded to top flange of C12
1.6" x 2" x .26" thick flatbar spacers
2" x 1.25" x .25" thick angle frame
continue flanges around ends of C12 stringer
C12 stringers, secured to concrete floor with (2) 1/2" diameter bolts
12"
6"
10"
3"x10"x.5" thick flatbar tread supports
(3) layers of .75" Douglas Fir for treads, selected for apperance, sand smooth
(2) .5" diameter carriage bolts
(2) .5" diameter lag screws
3"
3"

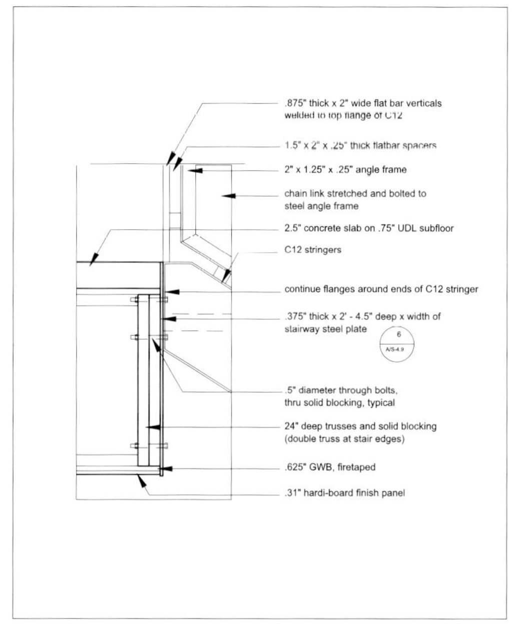

.875" thick x 2" wide flat bar verticals welded to top flange of C12
1.5" x 2" x .25" thick flatbar spacers
2" x 1.25" x .25" angle frame
chain link stretched and bolted to steel angle frame
2.5" concrete slab on .75" UDL subfloor
C12 stringers
continue flanges around ends of C12 stringer
.375" thick x 2' - 4.5" deep x width of stairway steel plate
6
A/S-4.9
.5" diameter through bolts, thru solid blocking, typical
24" deep trusses and solid blocking (double truss at stair edges)
.625" GWB, firetaped
.31" hardi-board finish panel

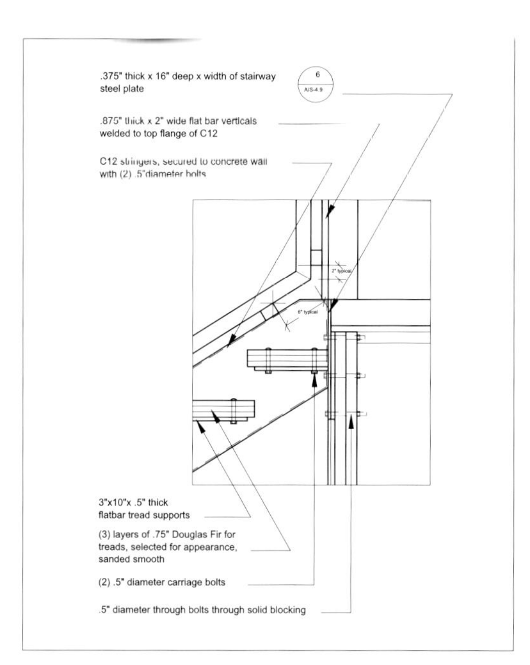

.375" thick x 16" deep x width of stairway steel plate
6
A/S-4.9
.875" thick x 2" wide flat bar verticals welded to top flange of C12
C12 stringers, secured to concrete wall with (2) .5"diameter bolts
2" typical
6" typical
3"x10"x .5" thick flatbar tread supports
(3) layers of .75" Douglas Fir for treads, selected for appearance, sanded smooth
(2) .5" diameter carriage bolts
.5" diameter through bolts through solid blocking

Abiquiu House

Top: Section drawing showing relationship of stepped interior and exterior spaces as they relate to grade of natural slope toward river

Bottom: Views from site including Pedernal Mountain to the west made famous in the paintings of Georgia O'Keefe, cloud formations to the north over the open desert, and sequence of construction views

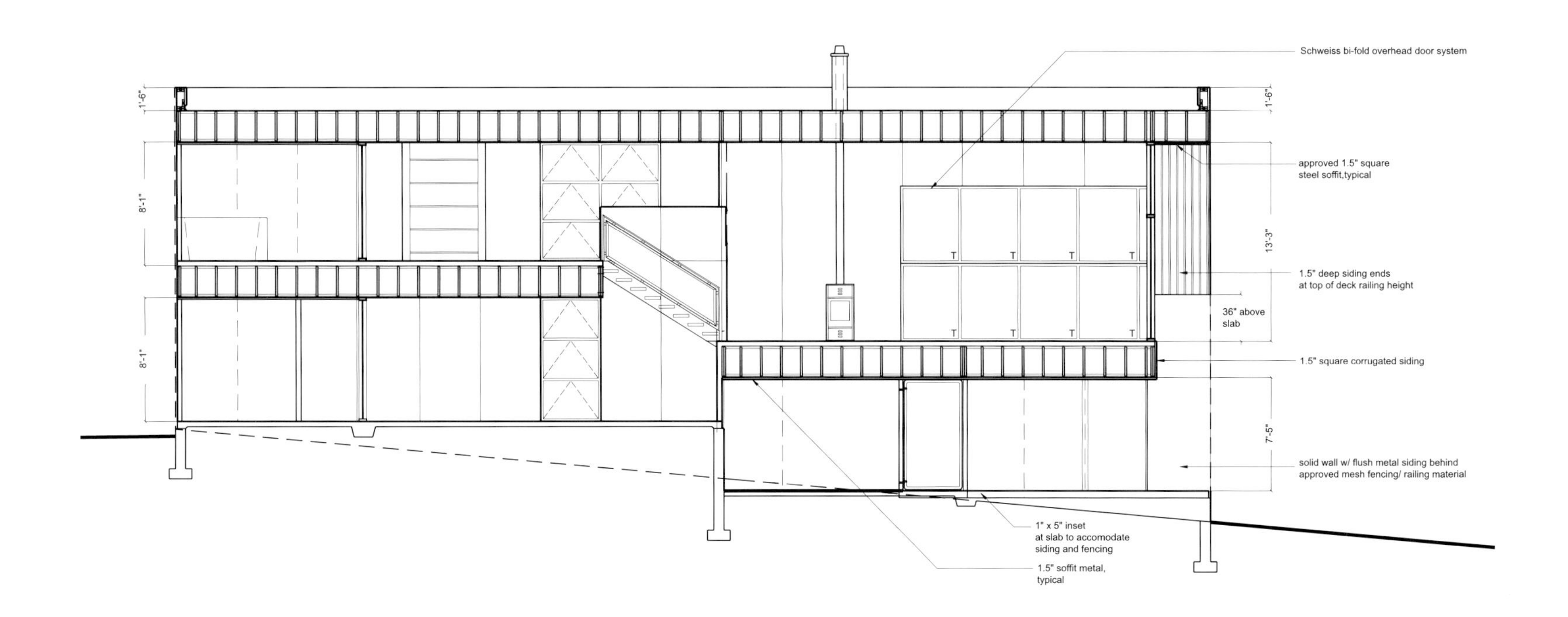

Sandwich Panels

Exploded perspective view of southeast corner

Parts List:

a. Roof trusses

b. Floor trusses

c. SIPs panels used as structure and enclosure

d. Photovoltaic and solar water heating panels

e. Chain link mesh screen walls

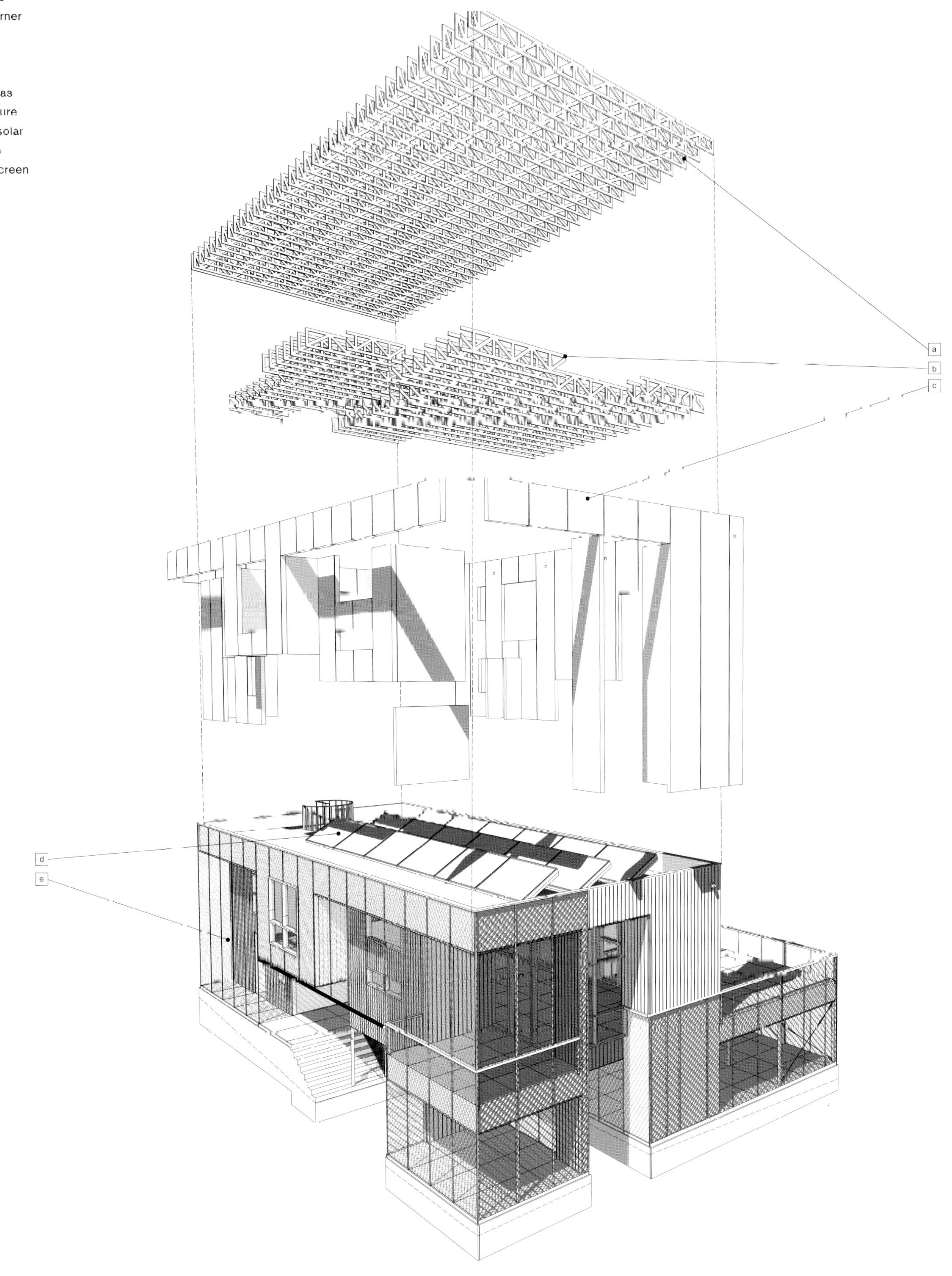

Abiquiu House

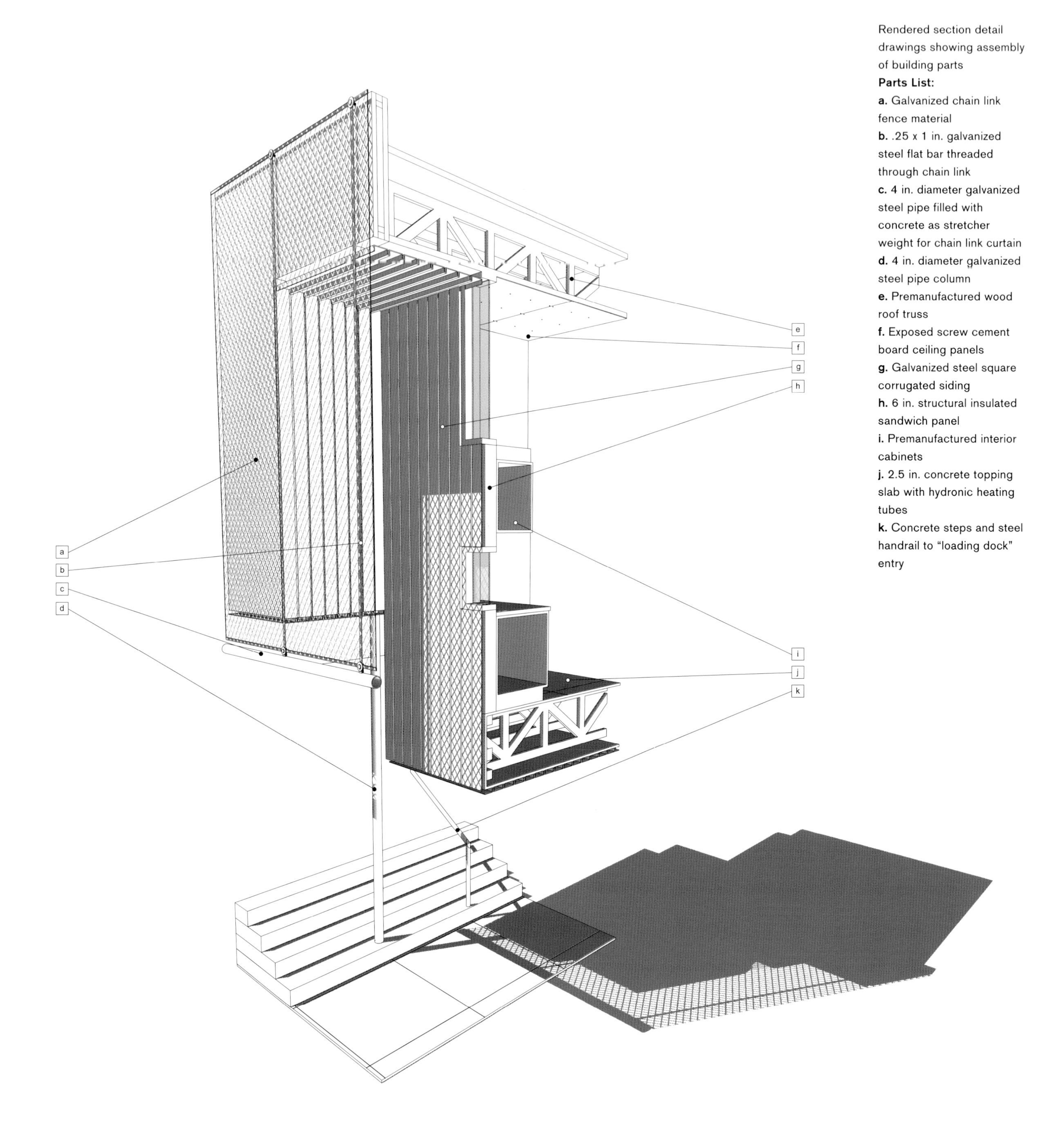

Rendered section detail drawings showing assembly of building parts

Parts List:

a. Galvanized chain link fence material

b. .25 x 1 in. galvanized steel flat bar threaded through chain link

c. 4 in. diameter galvanized steel pipe filled with concrete as stretcher weight for chain link curtain

d. 4 in. diameter galvanized steel pipe column

e. Premanufactured wood roof truss

f. Exposed screw cement board ceiling panels

g. Galvanized steel square corrugated siding

h. 6 in. structural insulated sandwich panel

i. Premanufactured interior cabinets

j. 2.5 in. concrete topping slab with hydronic heating tubes

k. Concrete steps and steel handrail to "loading dock" entry

Sandwich Panels

Parts List:

a. .5 x 6 in. galvanized steel lag bolt with 2 in. diameter galvanized steel washer

b. Galvanized chain link fence material set out 1 in. in front of wall plane

c. Flat 24 gauge galvanized sheet steel siding surface screwed to wall panels at locations of overlapping chain link

d. Galvanized 4 x 4 x .25 in. hollow steel tube canale scupper with 6 in. diameter x .25 in. galvanized steel plate washer

e. 24 gauge galvanized sheet steel square corrugated, 1.5 in. deep siding panels with outer surface flush with chain link

f. Steel frame electric bi-fold, overhead hangar door with aluminum sash and tempered glass

g. 2.5 x 2.5 x .25 in. galvanized steel railing cap continues as transition flashing between deep corrugated siding and chain link screen panels

h. Exposed, structural concrete deck slab

i. W12 x 22 galvanized steel deck structure

j. .25 in. x 1 ft. galvanized steel flat bar tensioner laced through chain link at 4 ft. panel points

k. Chain link screened animal enclosure for protection from coyotes and bald eagles

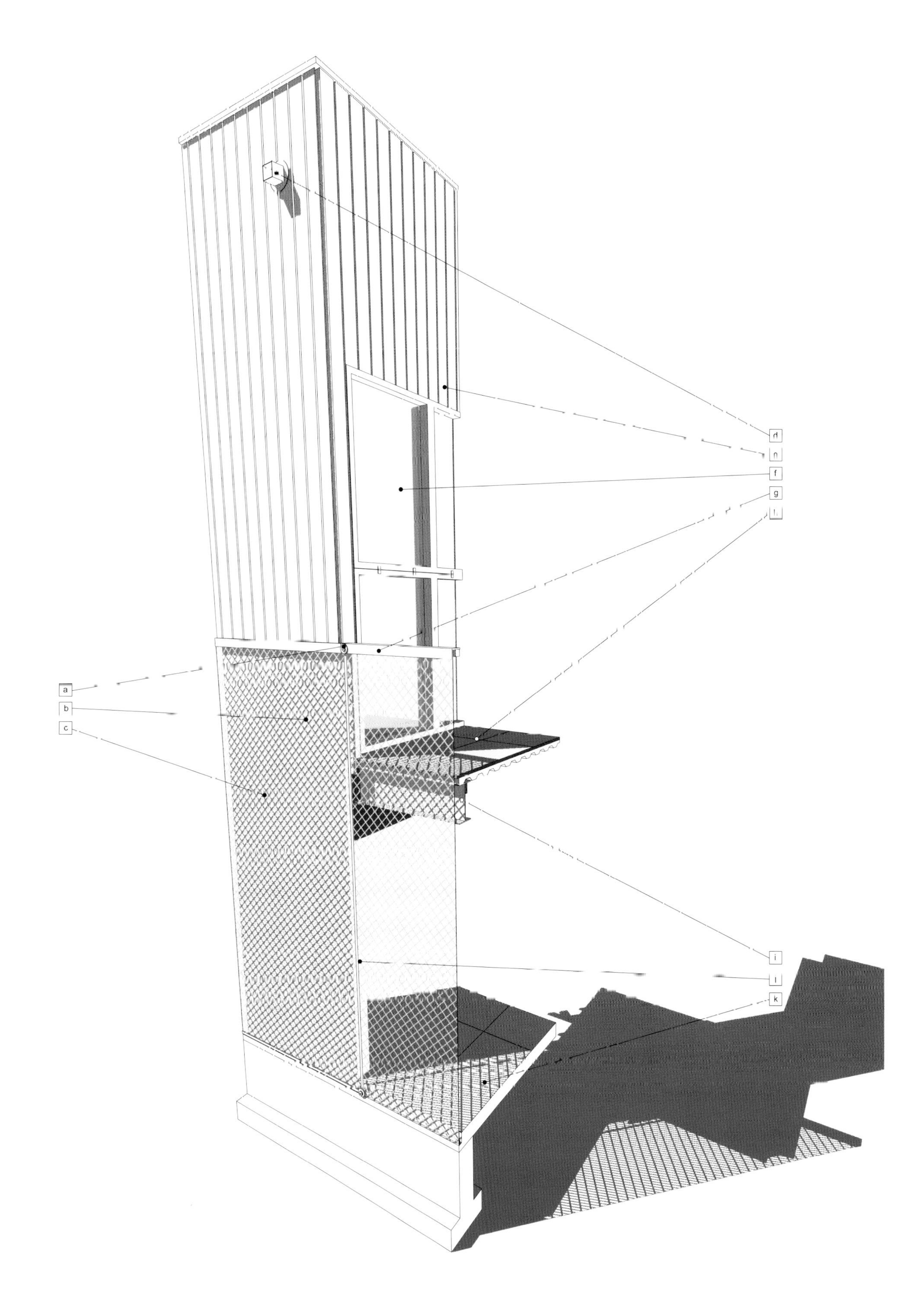

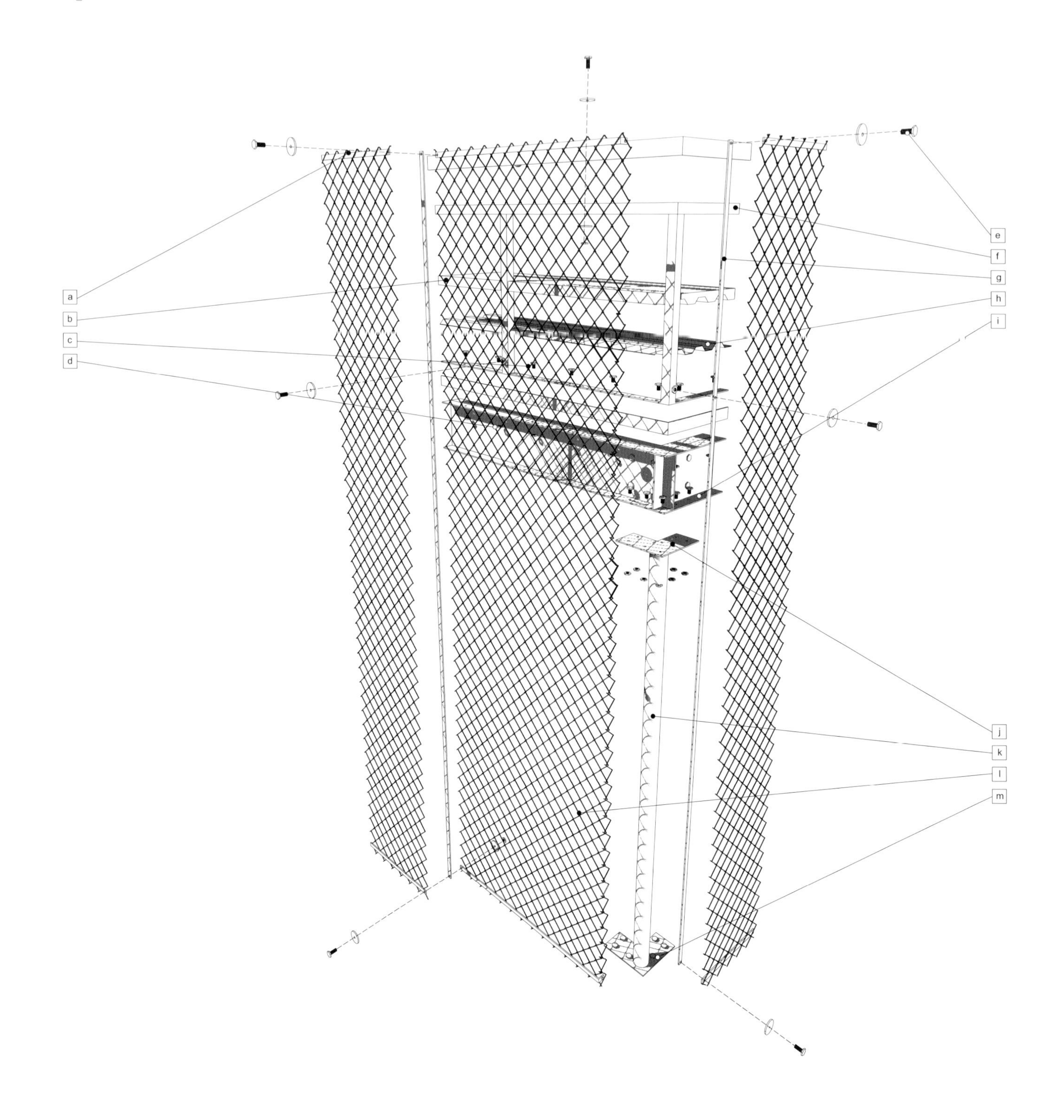

Perspective section details exploded to show assembly of chain link screen and railing systems

Parts List:
a. .25 x 1 in. galvanized flat tensioning bar laced through chain link
b. Exposed 4 in. concrete structural deck slab
c. .5 in. diameter galvanized steel through-bolts attaching railing structure to deck beams
d. Galvanized chain link railing and animal enclosure
e. .5 in. diameter through-bolt and 2 in. diameter washer
f. 2 x 2 x .25 in. galvanized hollow steel tube railing structure
g. .25 x 1 in. galvanized flat bar laced through chain link at 4 ft. o.c. and at corners and material seams
h. Galvanized sheet steel structural pan for concrete slab
i. W12 x 22 galvanized steel wide flange beams
j. .25 in. galvanized steel bolting plate welded to top of vertical column
k. 4 in. diameter x .25 in. wall galvanized steel pipe column
l. Galvanized chain link railing and animal enclosure
m. .5 in. concrete anchor bolt connecting chain link to foundation inset

Sandwich Panels

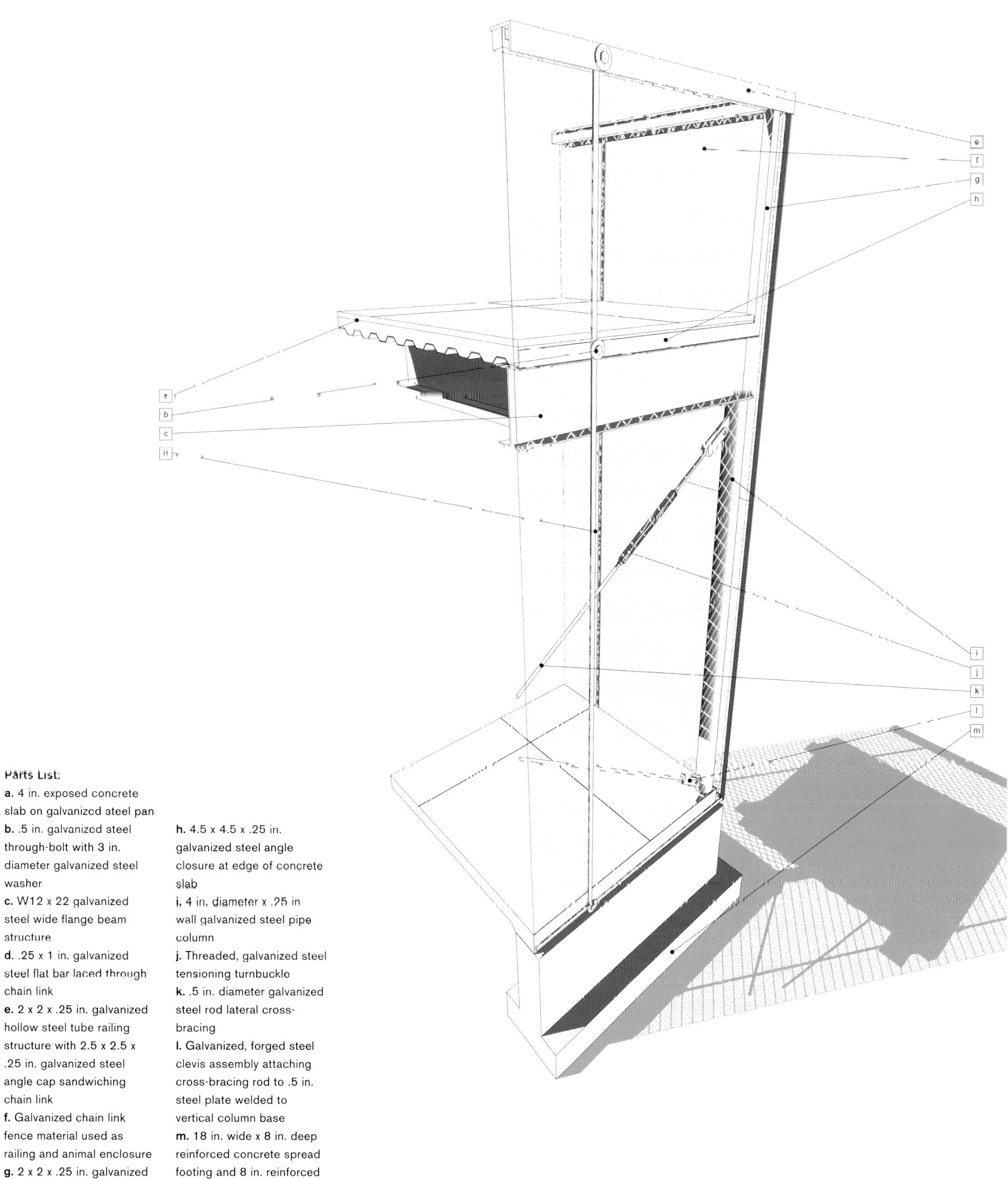

Parts List:

a. 4 in. exposed concrete slab on galvanized steel pan

b. .5 in. galvanized steel through-bolt with 3 in. diameter galvanized steel washer

c. W12 x 22 galvanized steel wide flange beam structure

d. .25 x 1 in. galvanized steel flat bar laced through chain link

e. 2 x 2 x .25 in. galvanized hollow steel tube railing structure with 2.5 x 2.5 x .25 in. galvanized steel angle cap sandwiching chain link

f. Galvanized chain link fence material used as railing and animal enclosure

g. 2 x 2 x .25 in. galvanized hollow steel tube railing vertical

h. 4.5 x 4.5 x .25 in. galvanized steel angle closure at edge of concrete slab

i. 4 in. diameter x .25 in wall galvanized steel pipe column

j. Threaded, galvanized steel tensioning turnbuckle

k. .5 in. diameter galvanized steel rod lateral cross-bracing

l. Galvanized, forged steel clevis assembly attaching cross-bracing rod to .5 in. steel plate welded to vertical column base

m. 18 in. wide x 8 in. deep reinforced concrete spread footing and 8 in. reinforced concrete stem wall foundation

Chapter 6

Modular Systems

The term *modular*, in the world of architecture and construction, has often been used to refer to largely completed or whole sections of buildings built at a factory and trucked to a site for quick deployment. The most fully finished examples come with completed exteriors, mechanical systems, interior finishes, and fixtures, ready to be used for housing, classrooms, offices, and other purposes. The greatest promise of most modular approaches to construction is that the more standardized the system can be, the easier it is to produce each unit in greater volume, and the more efficient and cost effective it becomes. The paradox is that the more standardized the units become, the less flexible they are, and the narrower the range of possible applications. There have been a few "sweet spots" in modular buildings, mostly found in low-end, cost-driven housing or short-lifespan institutional or commercial buildings that need to meet short-term needs at lowest possible costs, such as in construction site offices or temporary classrooms that are intended to be replaced as soon as "real" buildings can be completed. The fact is that many of these structures remain in place long after their intended lifespan, becoming poorly planned and placed buildings clinging tenuously and awkwardly to their sites. The standardization of their rigid chassis-based structural systems, which allows them to be transported in large pieces over roads and highways, leaves limited options for integration and adaptation once they come to rest.

Architects have dreamed about and developed experimental modular systems for more than a century, but few have achieved design distinction or financial viability, in part because of the large capital investment needed to create systems that are competitive in the marketplace. When proprietary modules are built in factories in small quantities, they achieve few of the efficiencies afforded by large-scale production and retain all of the challenges posed by transporting large, heavy pieces and then installing them in place.

It is unfortunate that the terms "modular" and "prefabricated" have become interchangeable in many people's vocabularies as it greatly confuses the viability and applicability of different available prefabrication systems, many of which are outlined in this book. A far more useful definition—straight from the *American Heritage Dictionary*—describes a module as a standardized, often interchangeable component of a system or construction that is designed for easy assembly or flexible use. Following this direction, a modular approach to design and building construction need not imply any particular scale for the modules themselves, and it in fact points out that a greater number of smaller modules can lead to a greater level of flexibility.

At one end of the spectrum, one could consider a single brick a very small modular component to be used in quantity to form any number of flexible solutions, while at the other extreme, one could consider a single-wide mobile home a very large component, offering just one inflexible result varied only by its placement on a site.

The most promising applications for modular design and construction fall somewhere between these extremes, where the idea of modularization can be applied not only to the building components themselves but also to the process of design and implementation. This idea is not new to architects who design large hotels, high-rise open-plan office buildings, or other structures where design and construction efficiencies are achieved by repeating identical guest room layouts or floor plates many times to create the larger structure. It is less common to consider pre-designed building block modules in creating single family houses or other structures that must adapt to more fine-grained programmatic or site-induced requirements, but it is not difficult to imagine a library of fully designed and engineered components at the scale of individual rooms that need only be assembled and minimally adapted in CAD files to create any number of unique solutions with great efficiency. Beyond the streamlined design time, if these modules can have associated parts lists and specifications, then other aspects of the construction process, such as materials ordering, can similarly be approached in a modular fashion, with the same efficiencies of time and cost. If the preparation of a materials list for a project could consist of a small number of modular parts groupings, instead of thousands of discrete objects, then a certain kind of prefabrication can have occurred in the building process even before the actual construction has begun.

Modular Systems

Model, rendered, and exploded structural view of steel frame modular system as aggregated for a six-story apartment building prototype designed for an urban block in Los Angeles. Two levels of modularity exist in this approach, since the steel wide flange beams and columns are standardized and repeated within each dwelling-unit-sized module, enabling the smallest possible number of unique fabricated steel elements and connection types to be utilized to create the largest possible component of shippable dimensions. Multiples of these units are then assembled on site to create larger buildings.

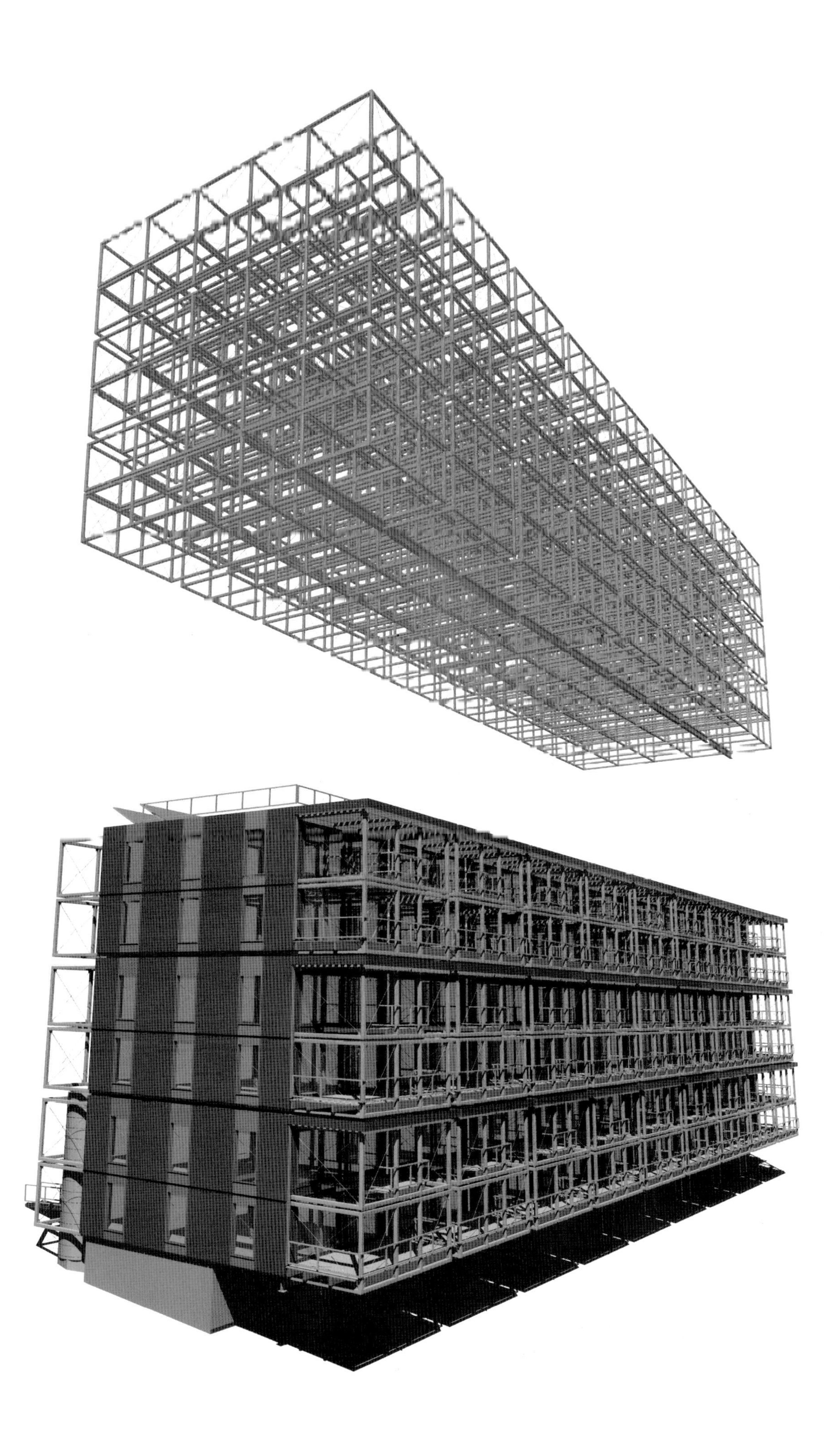

Great Hanshin Earthquake Community Shelter

Kobe, Japan / 1995

Following the Great Hanshin earthquake of January, 1995, centered in Kobe, Japan, a multinational task force was assembled to propose immediate reconstruction aid strategies and temporary housing and community shelter construction for the hard-hit communities. As this quick project developed, the Hyogo prefectural government reported that they were well-prepared with emergency housing services, but as much of this temporary housing was constructed on community open space and playgrounds and had filled most schools and community centers, the more immediate need was for slightly larger and more complicated community centers for group meetings, classes, and recreation.

Invited to participate in an emergency housing reconstruction team appointed by Washington State and Hyogo Prefecture, we designed this temporary community center in conjunction with a program of assistance being provided by the Washington State and U.S. Federal governments to the people of Kobe. Using readily available building materials donated by Washington's construction products industry, the design makes repetitive use of a small number of interchangeable wood-framed panel modules for floors, ceilings, and walls, together with other standardized components with simple connections, to facilitate quick erection of the buildings by volunteer labor without special skills or tools. Intended as prototypes for customization by each community, the buildings are flexible and open so they can be used by different groups at different times of the day, and for different activities during the week. They have movable interior elements such as a kitchen, storage cabinets, bookshelves, and other furnishings, all on wheels so that they can be quickly reconfigured as needed. The rooftop garden terrace is designed as a sign of life above the rubble and replaces precious open space consumed at the ground level by the temporary housing that fills the city's parks.

Model of one configuration of the assembled components

OPPOSITE
Exploded axonometric drawing of principal building components showing modular approach to structure, enclosure, fittings, and furnishings.

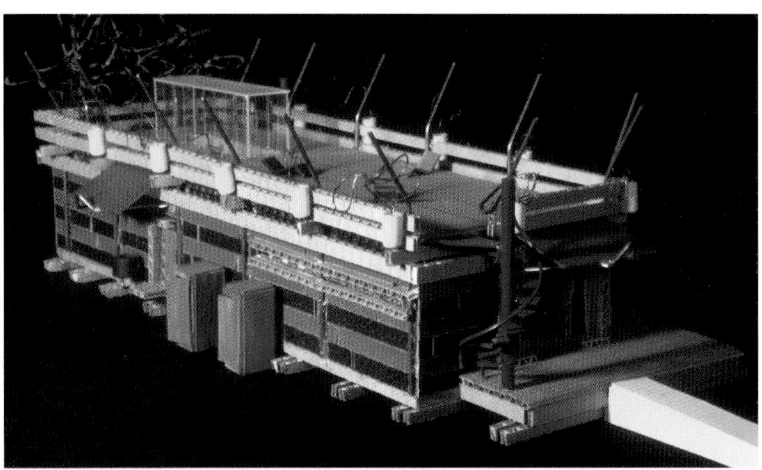

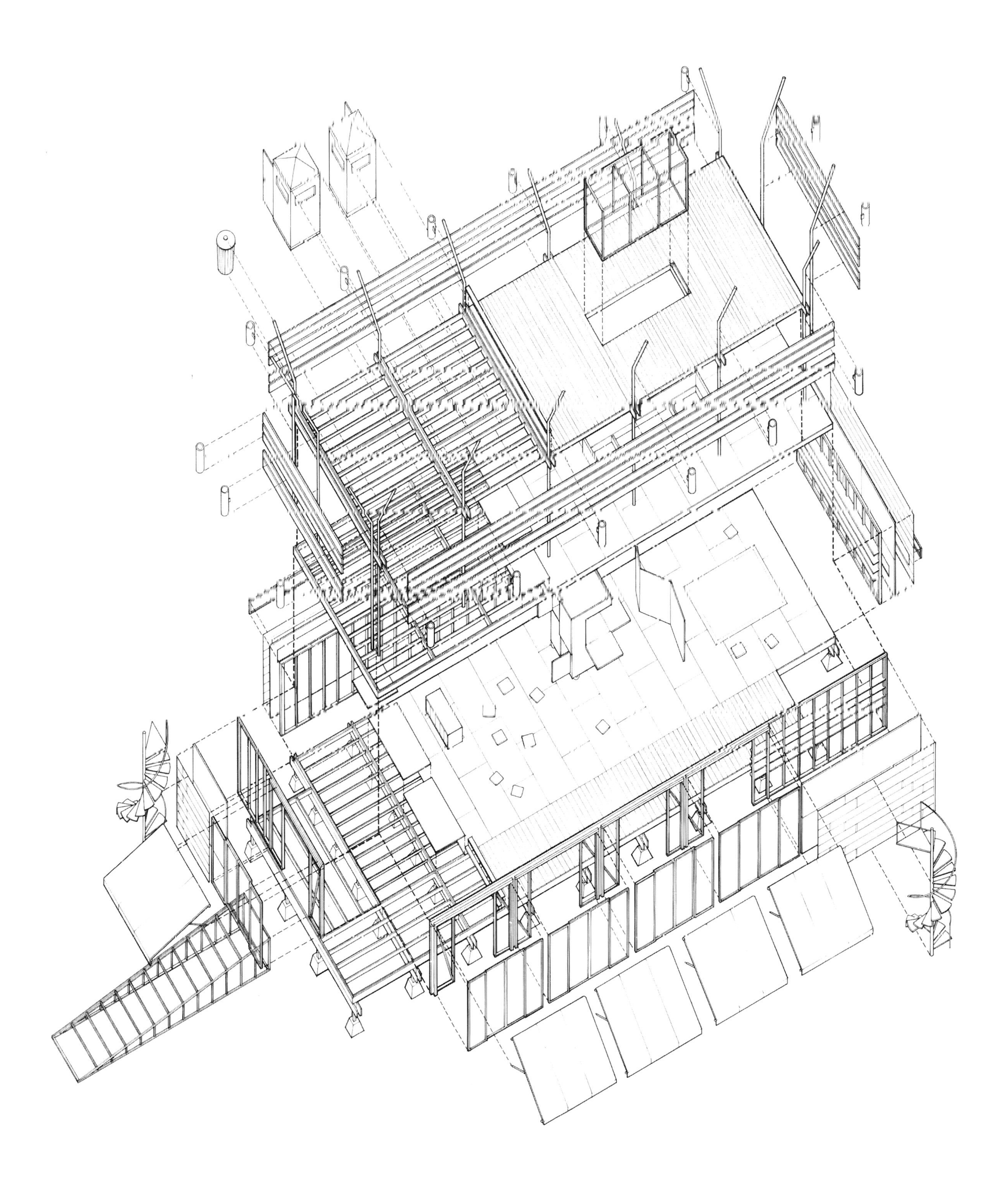

Sophia Three-Generation Modular Houses

Tokyo, Japan / 1996

This is a series of prototype panelized houses designed for a Tokyo developer of alternative elderly and handicap-accessible housing. The houses are designed in expandable modules to accommodate changing life patterns with options for traditional three-generation homes with separate family living quarters on the two levels. The homes are designed for maximum accessibility and focus on providing outdoor living areas and optimal sunlight, ventilation, and views for housebound owners. The design concept is based on the establishment of a catalog of predesigned modules that can be configured for many different site and program configurations with great efficiency of design, planning, and construction management resources. All of the materials lists and assembly instructions for each module are preconfigured, making it simpler to put together the kit of needed parts in the design studio, for the materials supplier and consolidating warehouse, and at the jobsite when the components are unloaded from shipping containers.

Top: Diagrams of sequential assembly of different program element modules showing potential for pre-planned building expansion over time as different needs arise.
Bottom: Construction photos of single-story base module constructed for use as sales office and model of building system

Modular Systems

Top and bottom right: Exploded and assembled views of possible configurations
1. Base module
2. Expansion module with larger kitchen area and living space
3. Lower-floor bedroom expansion module
4. Upper-floor bedroom expansion module
5. Flat, covered rooftop for additional outdoor living space with shading in hot weather. An additional kitchen/living module could also be placed here

Bottom left: Neighborhood plan showing contrast between typical rectangular house shapes, with no private exterior space, and various configurations of L- or U-shaped modular prototypes with courtyards, rooftop decks, and other [illegible] exterior living spaces.

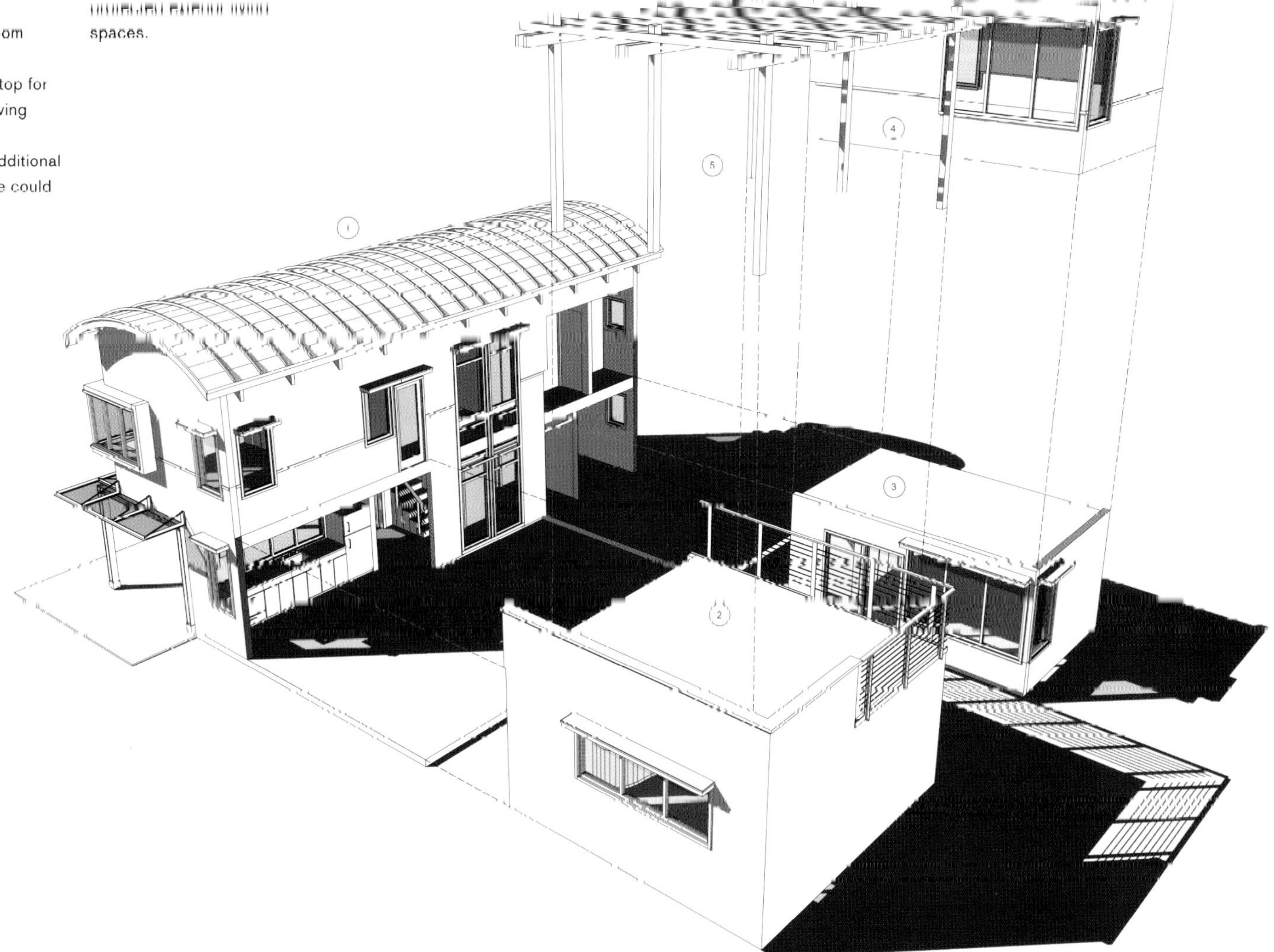

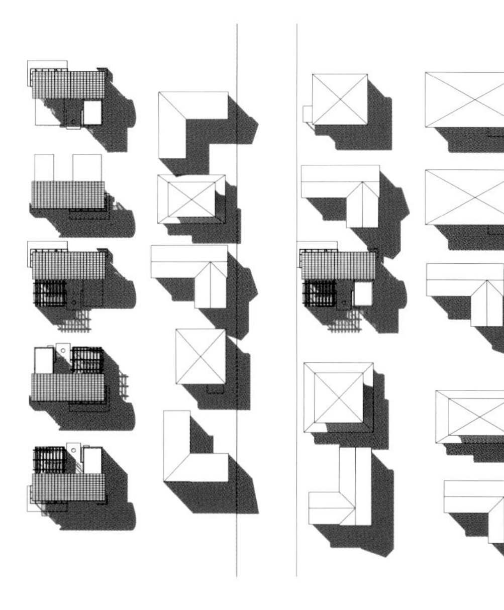

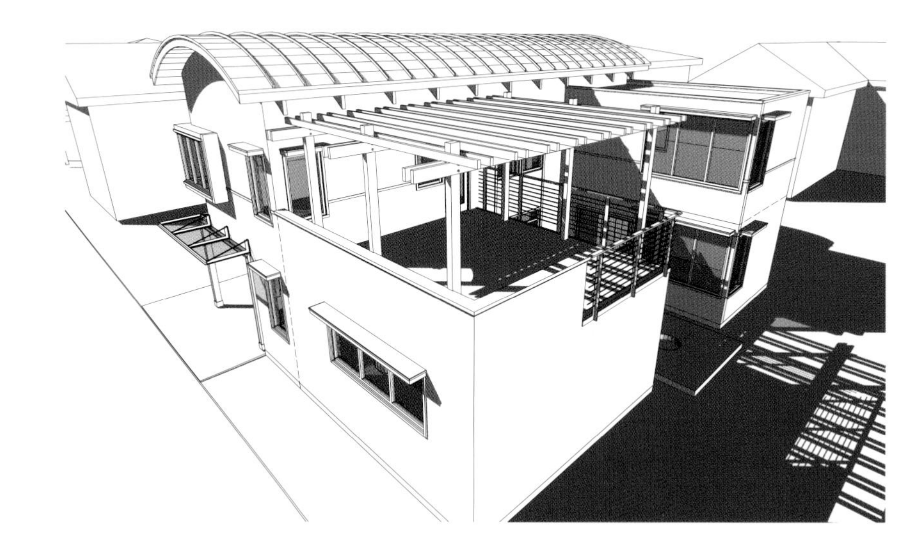

Yosemite Cabin

Madera County, California / 2003

This project represents one application of prototype development using a relatively low-cost, program-flexible steel, and SIPs panel prefabricated self-contained modular building system, designed for easy transport and lightweight, multistory lifting in remote wilderness areas and in difficult-to-access urban areas. The modules are flexible enough to provide for a wide range of applications, and the system has been adapted to multifamily apartments, high-rise penthouses, and off-grid wilderness settings.

The initial design impetus for this modular prefabricated building prototype came from a request for a vacation house to be situated on a spectacular riverside site in the middle of the national forest, high in the Sierra Madre Mountains. The site is completely isolated, far from power, phone, paved roads, or towns, and required a solution that could be built almost entirely off site. The only access road is too narrow, twisty, and rough to accommodate commercially available modular home solutions, and the steep site required a steel structural frame able to cantilever out over a creek. Shortly after this request, two additional commissions arrived, for a penthouse addition to be lifted into place on top of a thirty-two-story tower in Manhattan and a field biology research station on the California coast. Combining the criteria for these projects, all requiring offsite fabrication and rapid erection in difficult construction conditions, a common frame and enclosure system was developed to provide great program flexibility and site specificity within an affordably standardized building system.

In the two wilderness sites, the buildings are entirely off-grid, so the frame is built as an armature to accept a range of solar panel and water collection systems. For the Manhattan penthouse, the development conditions required the resiting and upgrading of HVAC systems for the entire building, and the frame in this case becomes the armature for new air handling and mechanical equipment encompassed within the overall penthouse design.

The primary structural system in all cases is composed of 8 x 12 x 8-foot (2.44 x 3.66 x 2.44 m) tube steel frames joined end-to-end or side-to-side to form the building's needed length and width and cross-braced with steel rods for lightweight lateral stability. In the Yosemite Cabin, four of these inhabitable trusses are then joined together to form the two-story building, joined laterally by 6-foot (1.83 m) bolt-on spreader beams to create the open interior hall and 22-foot (6.71 m) overall width. While still in the fabrication shop, these structural steel chassis modules are fitted with SIPs panel insulating enclosure walls with lightweight steel skins, plywood interiors, and full plumbing, cabinetry, and mechanical systems. All mechanical systems are contained within a single stacked module system, minimizing cost and the complexity of onsite connections. The small sizes of the individual modules allow them to be carried to the remote sites by four-wheel-drive flatbed trucks and lifted into place by small and maneuverable truck-mounted cranes. This unit size is substantially smaller, lighter, and more maneuverable than typical modular building units, allowing for quicker, higher lifting with smaller and more affordable cranes, causing less traffic interference in the city and less ecological upset in the country.

Once assembled into the 1,500-square-foot (139 sq. m) structure, as presented here in the Yosemite version, the complete assembly becomes rigid enough to cantilever out in all directions from a small-footprint foundation. The cross-braced steel rods that are required in all frame bays during trucking and lifting can then be removed strategically within some interior bays, allowing a freer flow of space and program. Further development of this system has expanded to include multifamily affordable housing structures on difficult-to-access urban in-fill sites.

Below: Initial design sketches
Bottom: Site photographs

OPPOSITE
Model photos of off-grid building proposal, showing implementation of four truck-sized modules with pre-applied enclosure systems on minimal cast-in-place pedestal foundation. Series below shows layers of wall enclosure panels over bare steel frame

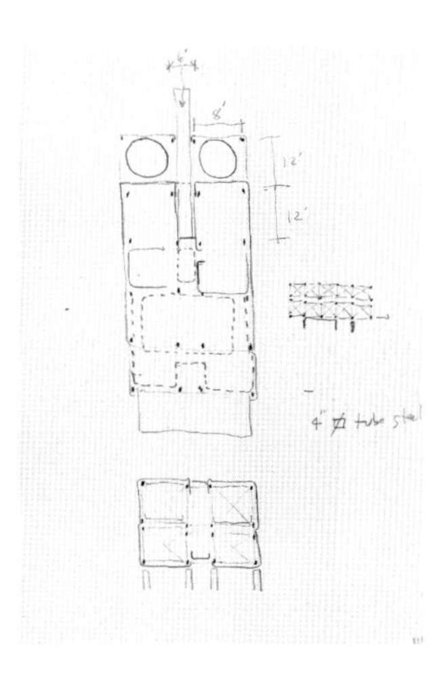

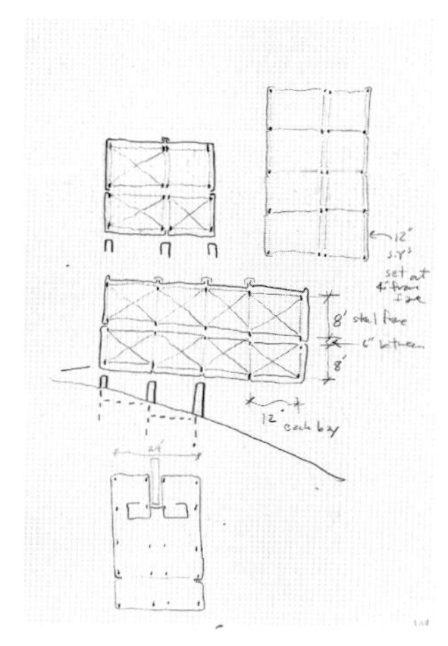

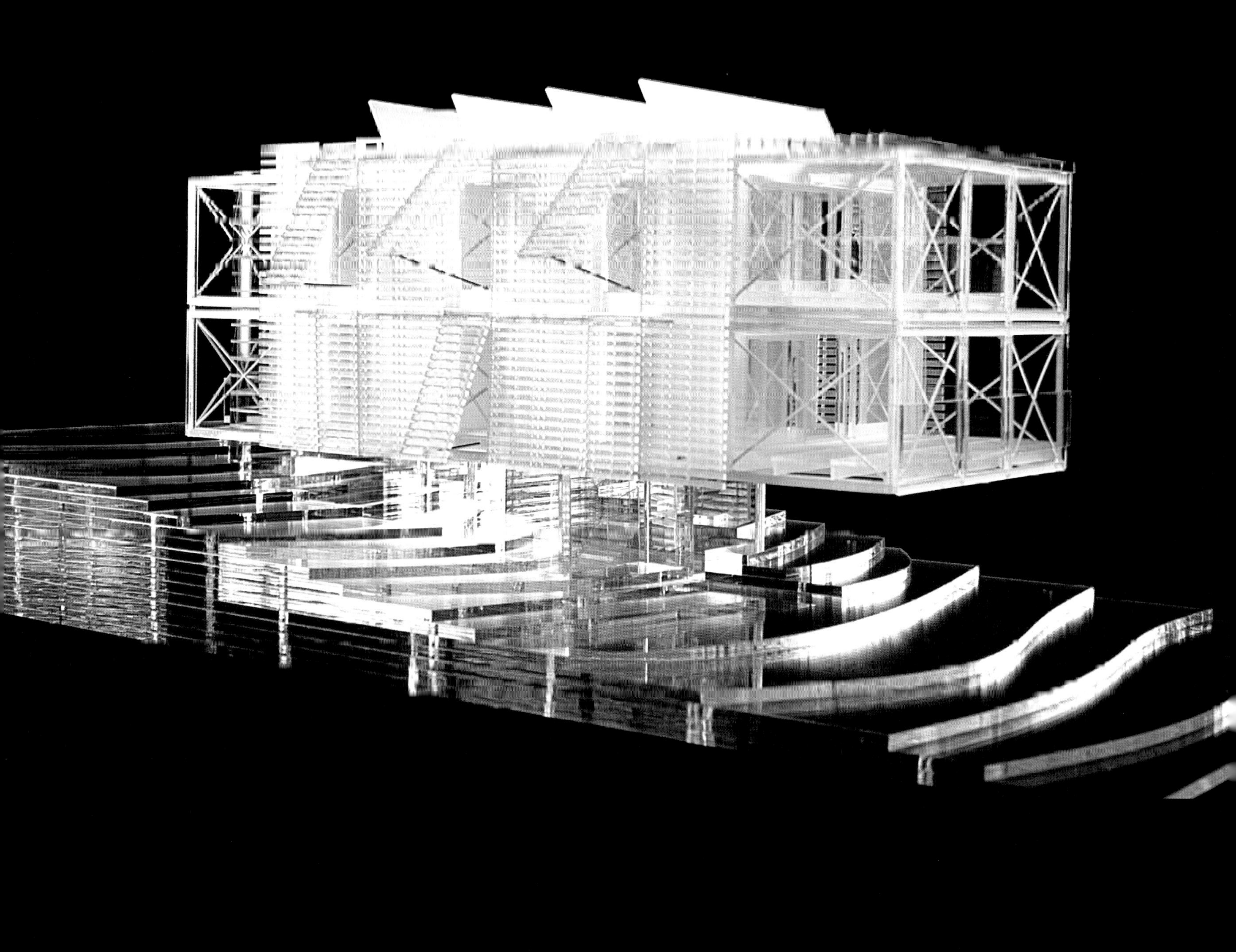

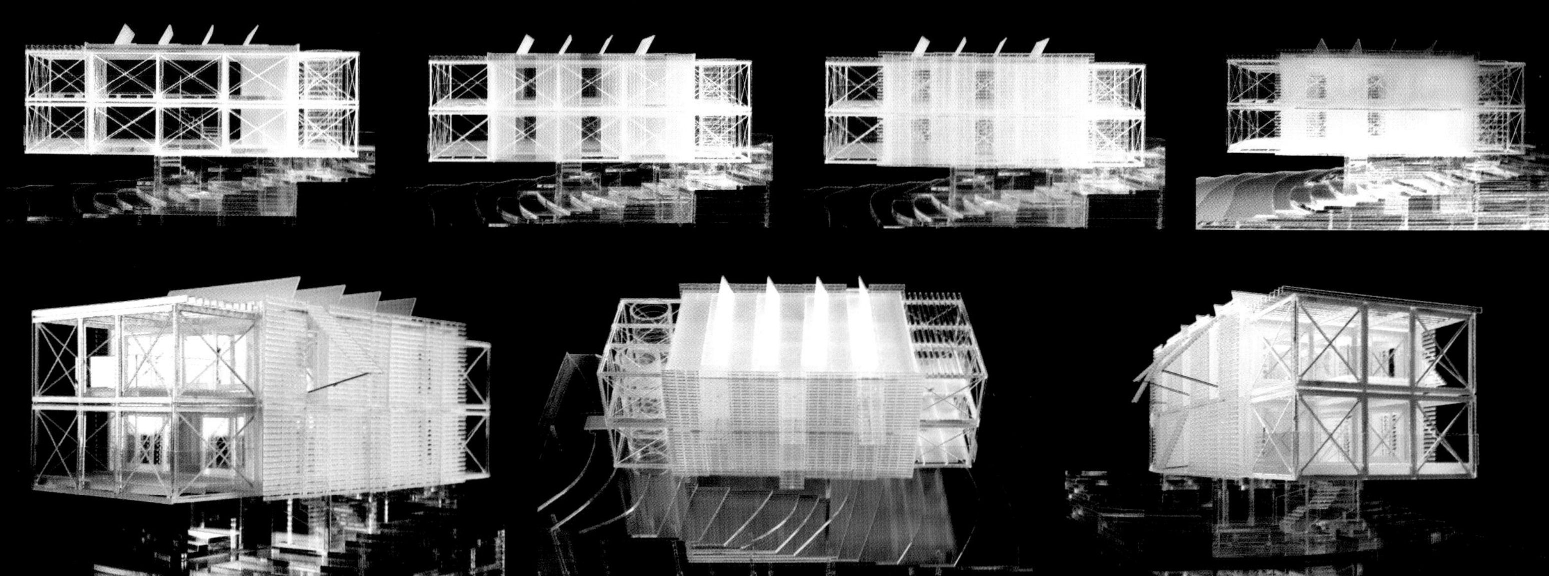

Yosemite Cabin

Top: Upper- and main-floor plans
Bottom: Structural steel frame elevations

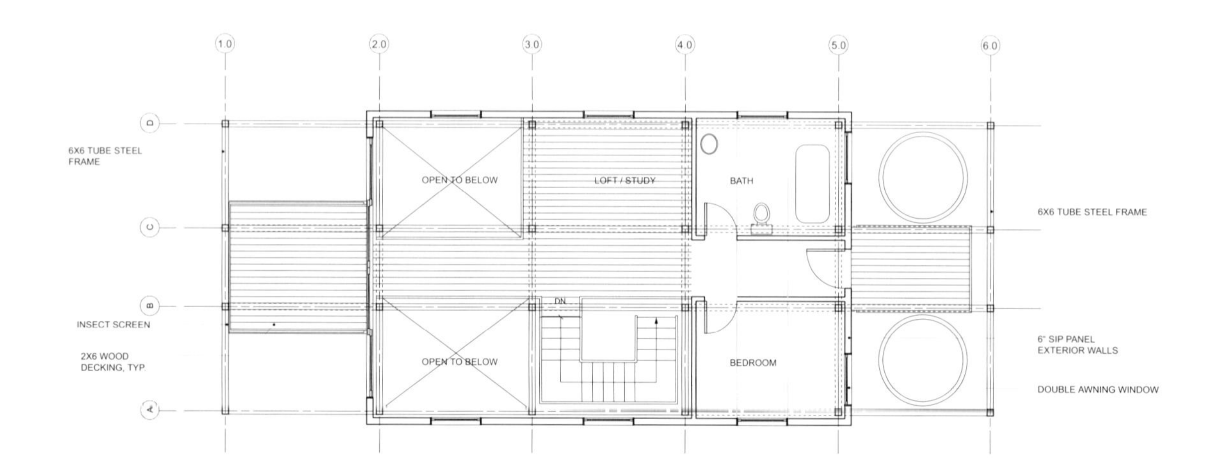

Upper floor

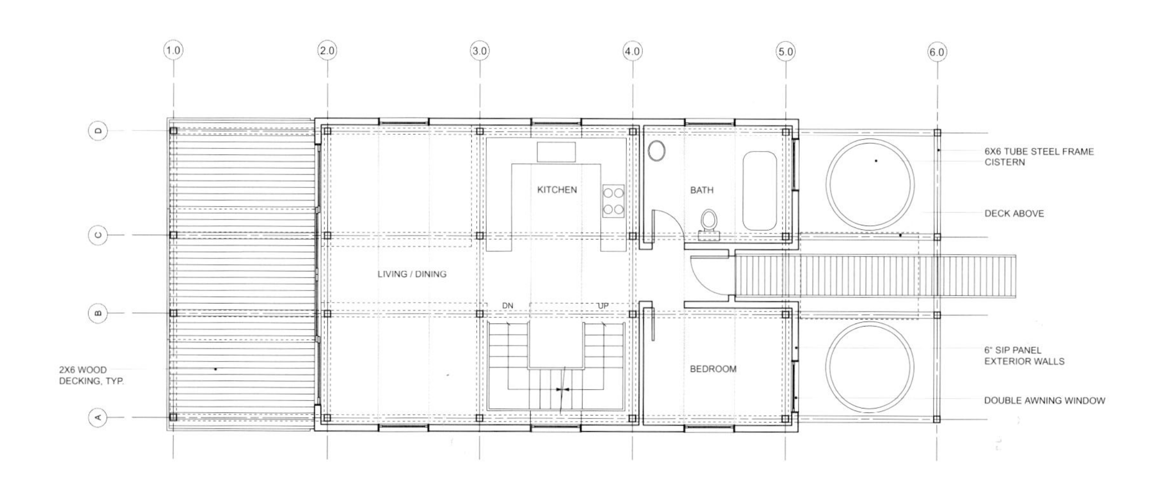

Main floor

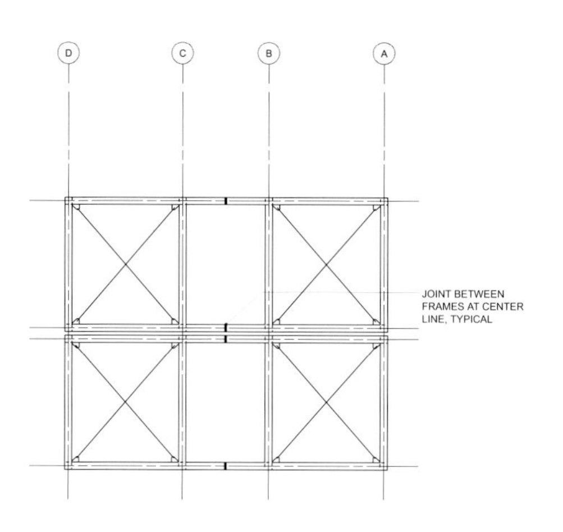

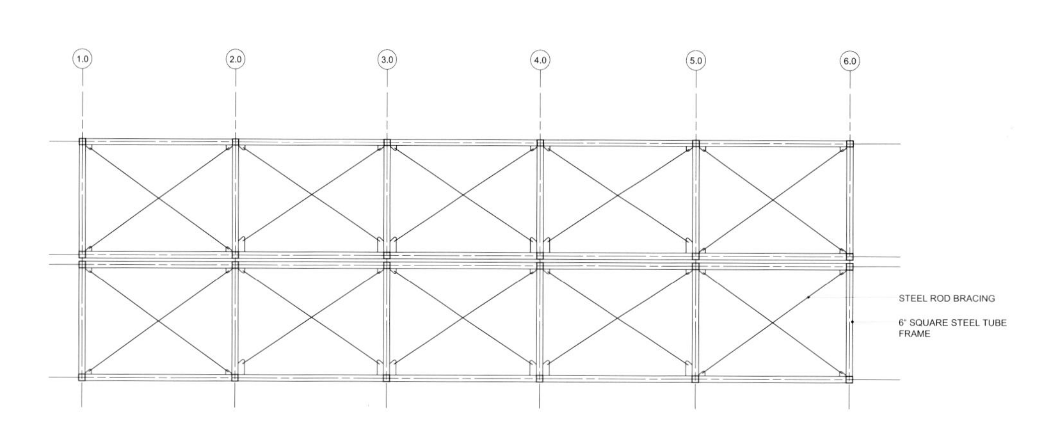

Top: Elevations
Bottom: Frame assembly details showing relationship of SIP's enclosure to structural steel

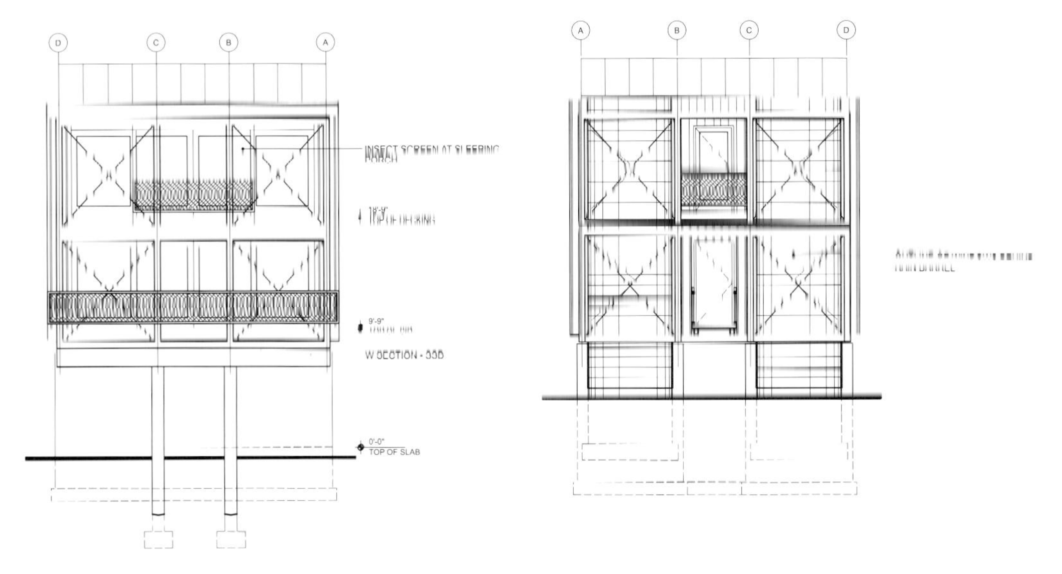

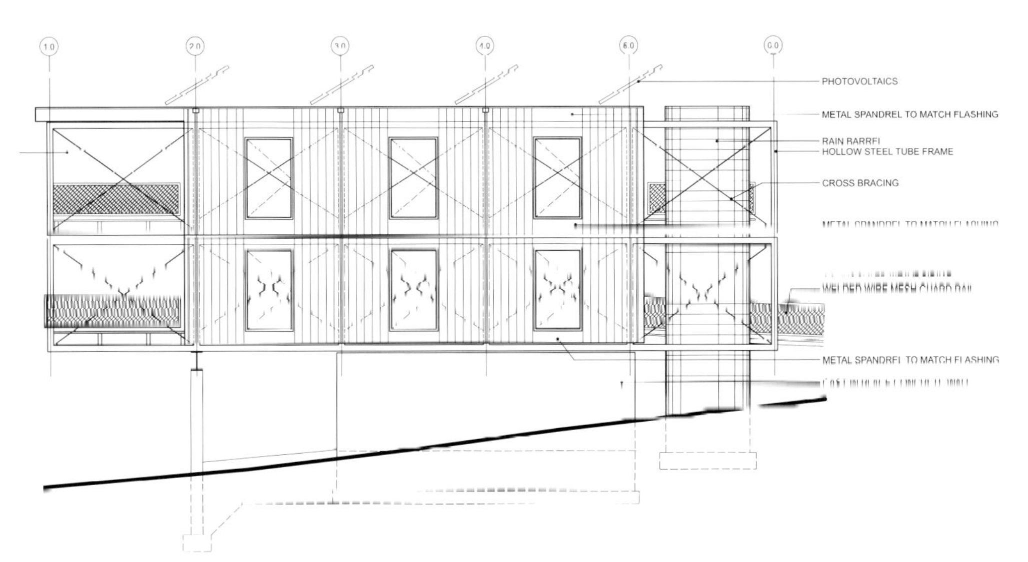

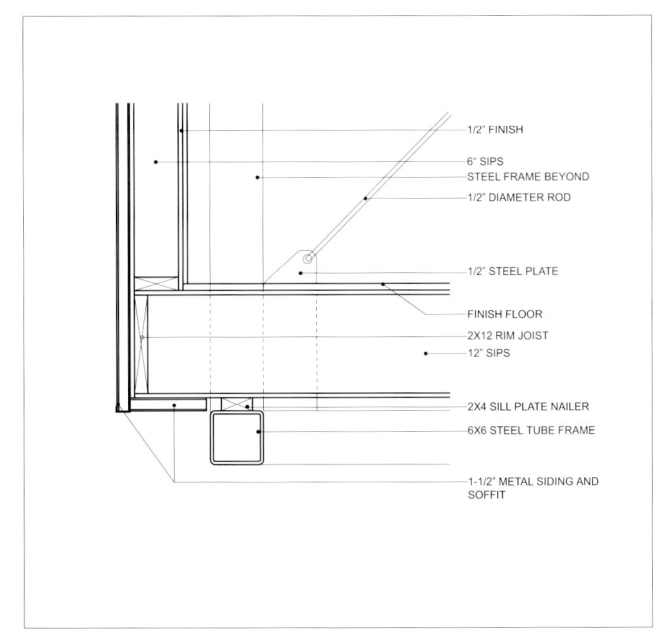

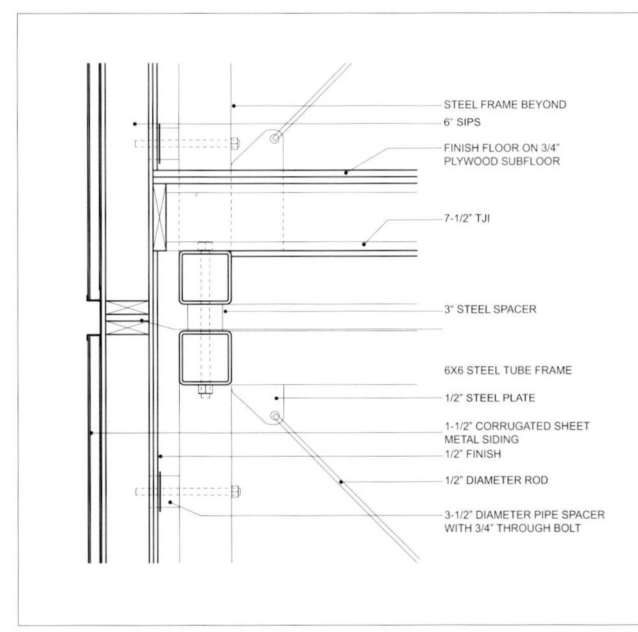

Yosemite Cabin

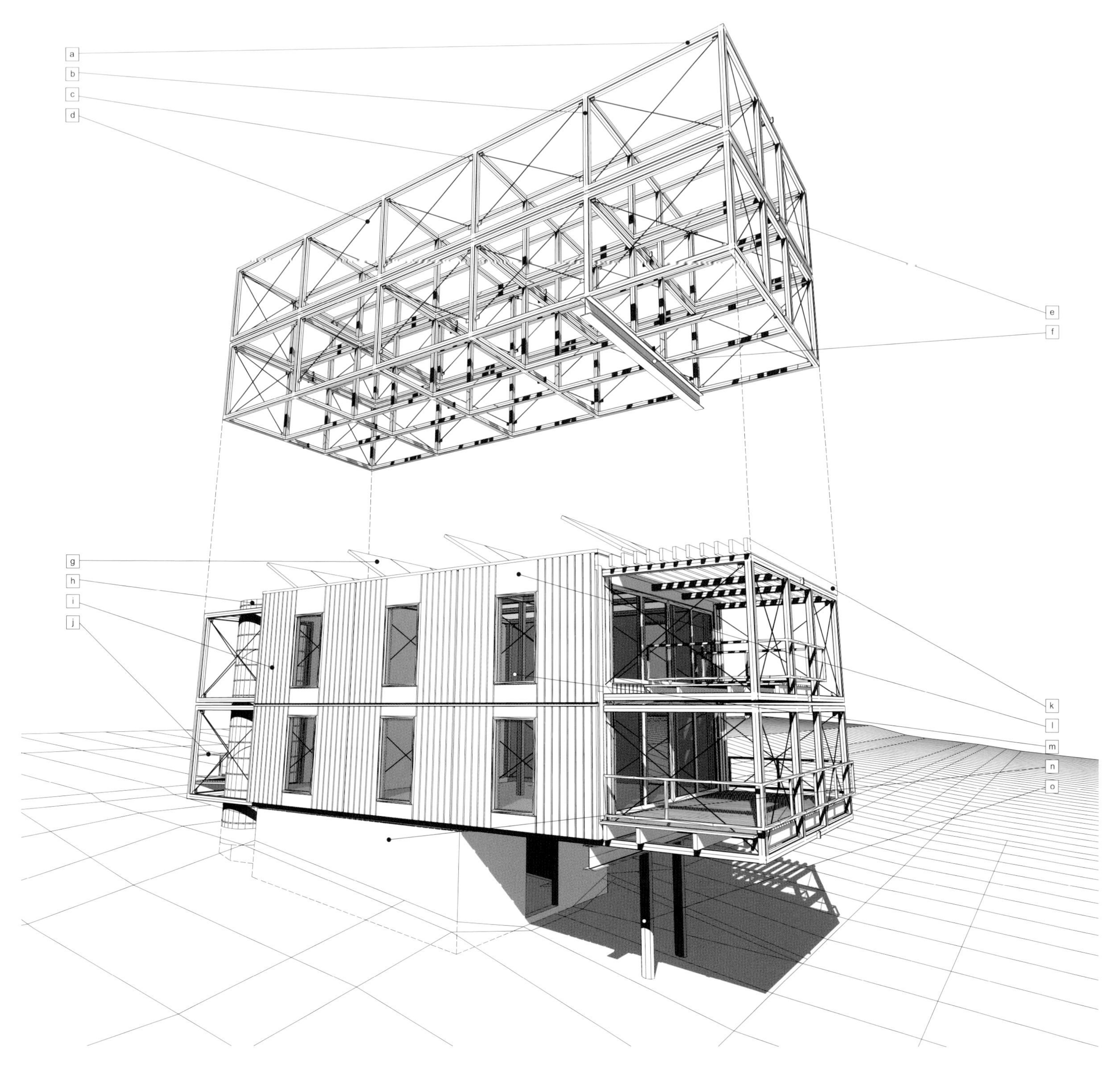

Parts List:
a. 6x6 structural steel tube frame horizontal
b. 6x6 steel vertical
c. .5 in. steel gusset
d. .5 in. diameter steel rod cross-bracing
e. .5 in. steel connector plates at module breaks
f. W12 x 101 cantilevered support beam
g. Solar panels
h. Rainwater catchment cisterns
i. 1x4 redwood siding
j. Steel entry ramp
k. Shade trellis
l. Aluminum siding panels above and below windows
m. Aluminum casement windows
n. Concrete foundation
o. 12 in. diameter concrete columns

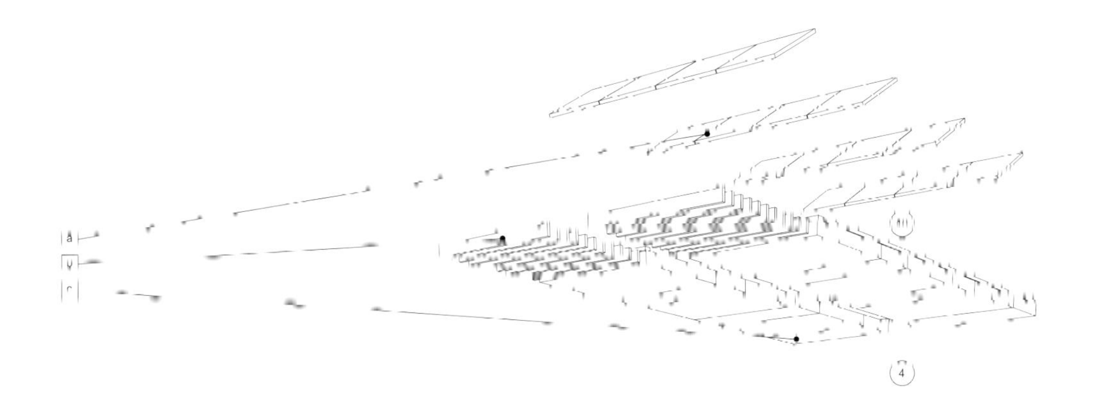

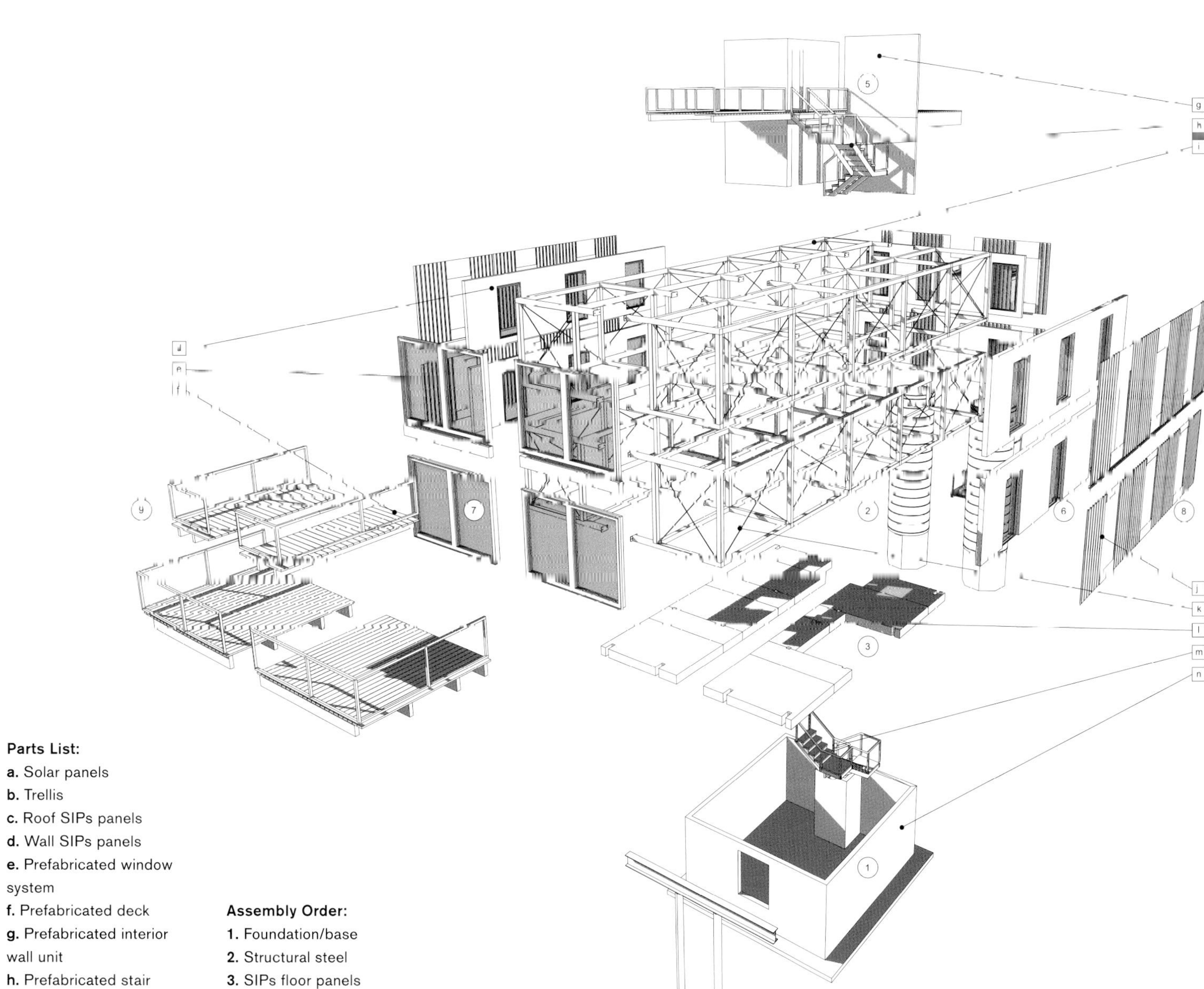

Parts List:
a. Solar panels
b. Trellis
c. Roof SIPs panels
d. Wall SIPs panels
e. Prefabricated window system
f. Prefabricated deck
g. Prefabricated interior wall unit
h. Prefabricated stair system
i. 6x6 structural steel frame
j. 1 x 4 in. wood siding
k. Cross bracing
l. Floor SIPs panels
m. Prefabricated stair system
n. Concrete foundation/ base

Assembly Order:
1. Foundation/base
2. Structural steel
3. SIPs floor panels
4. SIPs roof panels
5. Prefabricated interior wall unit
6. SIPs wall panels
7. Prefabricated window system
8. Wood siding
9. Prefabricated decks
10. Solar panels

Manhattan Penthouse

New York, New York / 2003

A developer with a penthouse apartment in a building in Uptown Manhattan asked us to investigate the possibility of adding a second story to his unit. The plan would offer the condominium association a second adjacent rooftop unit for common use in exchange for the developer's rights to construct his private penthouse addition on the north side of the roof.

At thirty-two stories, the roof is near the upper limits for feasible access by very large mobile crane. Permitting for crane access is relatively complex, and it becomes increasingly so with lengthier street-closure times. Hourly crane time is very expensive as well, and insurance and safety risks rise dramatically as the length of crane time, numbers of individual truck deliveries, and number of lifts increases. For all of these reasons, a substantially complete prefabricated building module is very sensible. A very large and heavy unit creates other problems, however, so there are still offsetting considerations for less prefabrication with smaller sub-assemblies. For this project, we settled on the complete module system, but we have simultaneously developed a system of elevator-sized, folding structural components that could be brought up to the roof in many compact parcels and then quickly deployed on the roof top.

Below: Rendered view of penthouse exterior and landscaping with vegetated shutter panels in closed position

OPPOSITE

Top: Rendered view, shutter panels in open position.

Bottom: Detail view showing components of planted screen system

Parts List:

a. Green plants

b. Perforated stainless steel

c. .5 in. stainless steel bolt

d. Large perforations on vertical leg allow filtered light

e. 1.5 x 1.5 in. steel shutter frame

f. Rainwater drip irrigation tube

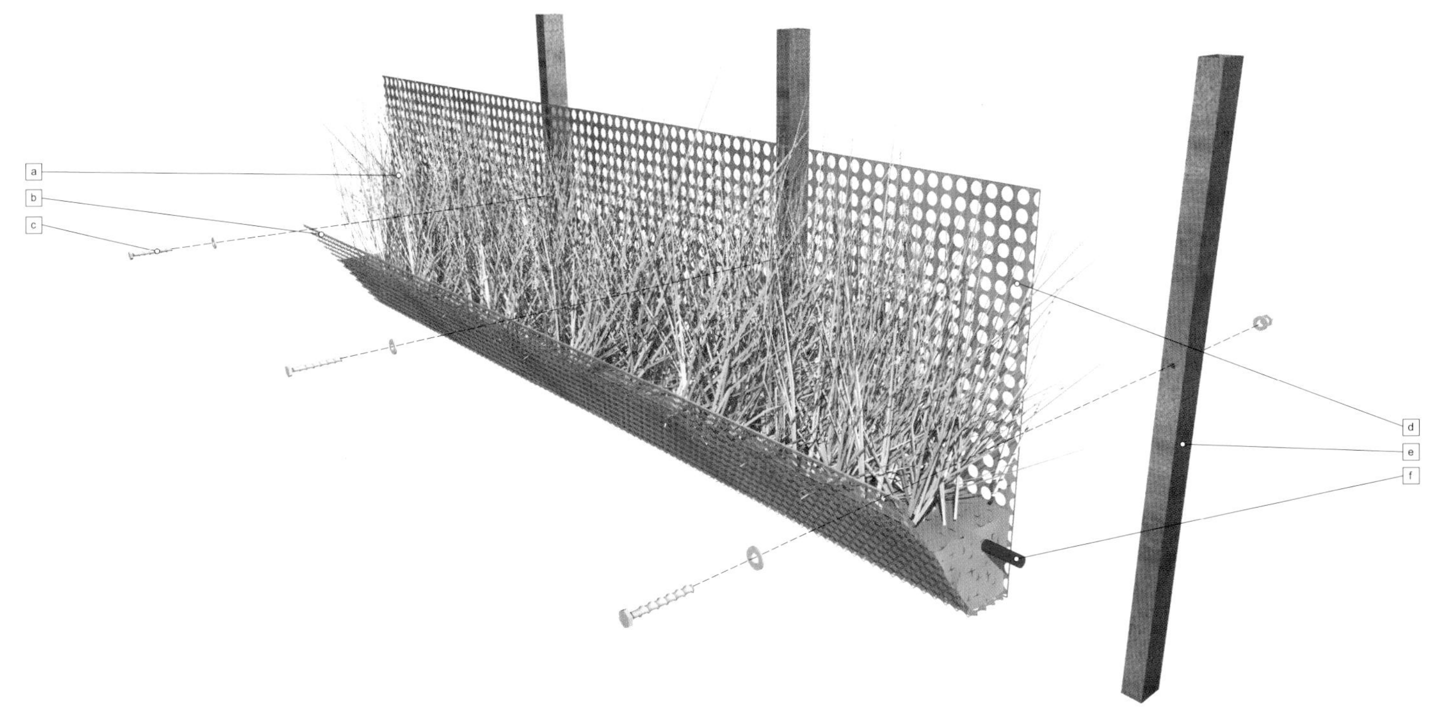
a
b
c
d
e
f

Manhattan Penthouse

Left: Plan of penthouse unit with surrounding roofscape; section through new modular units showing opening and connecting stair to existing apartment
Below: Aerial views of neighborhood and building rooftop

OPPOSITE
Diagram showing how building modules are to be lifted from ground using mobile crane

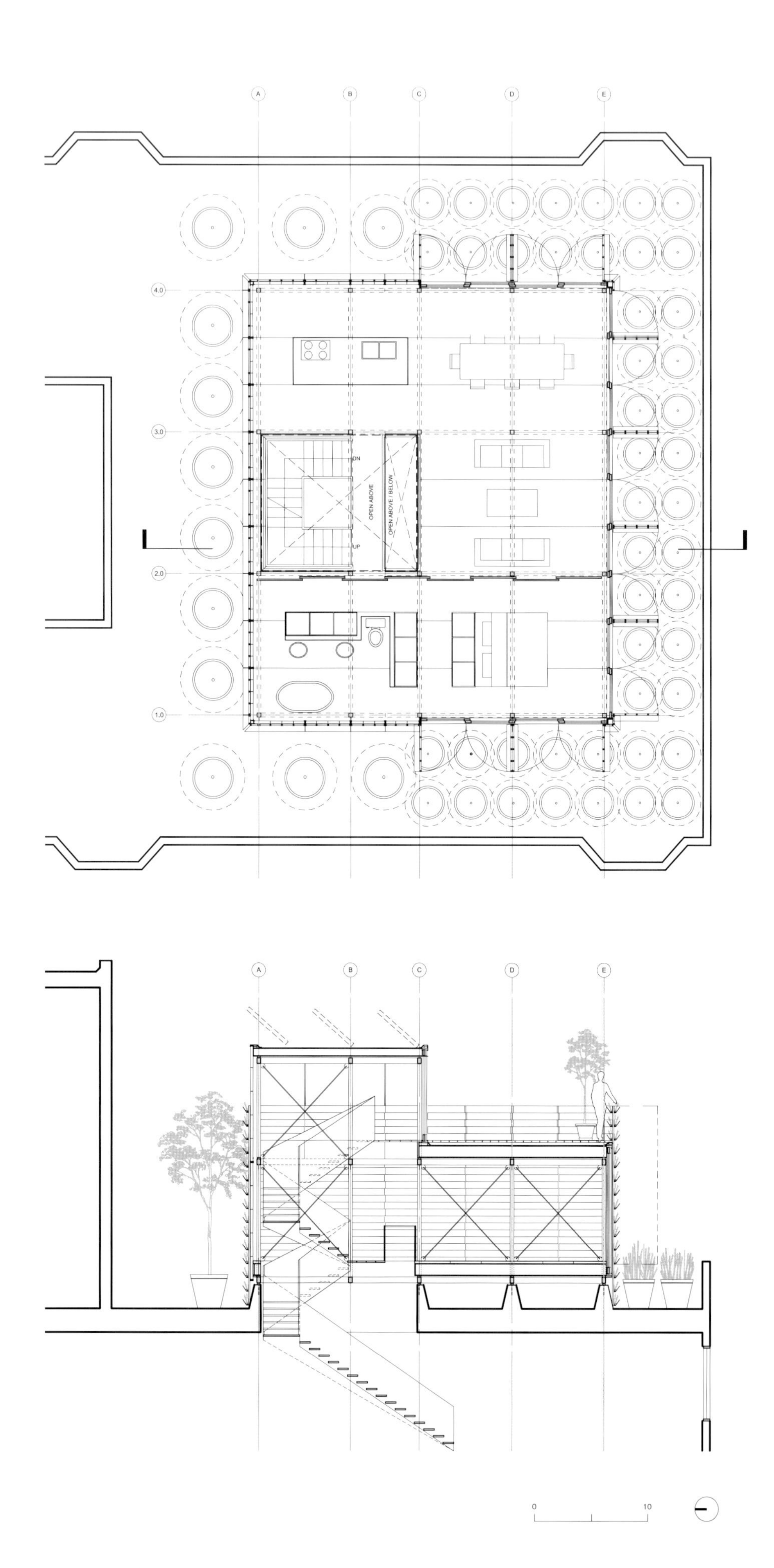

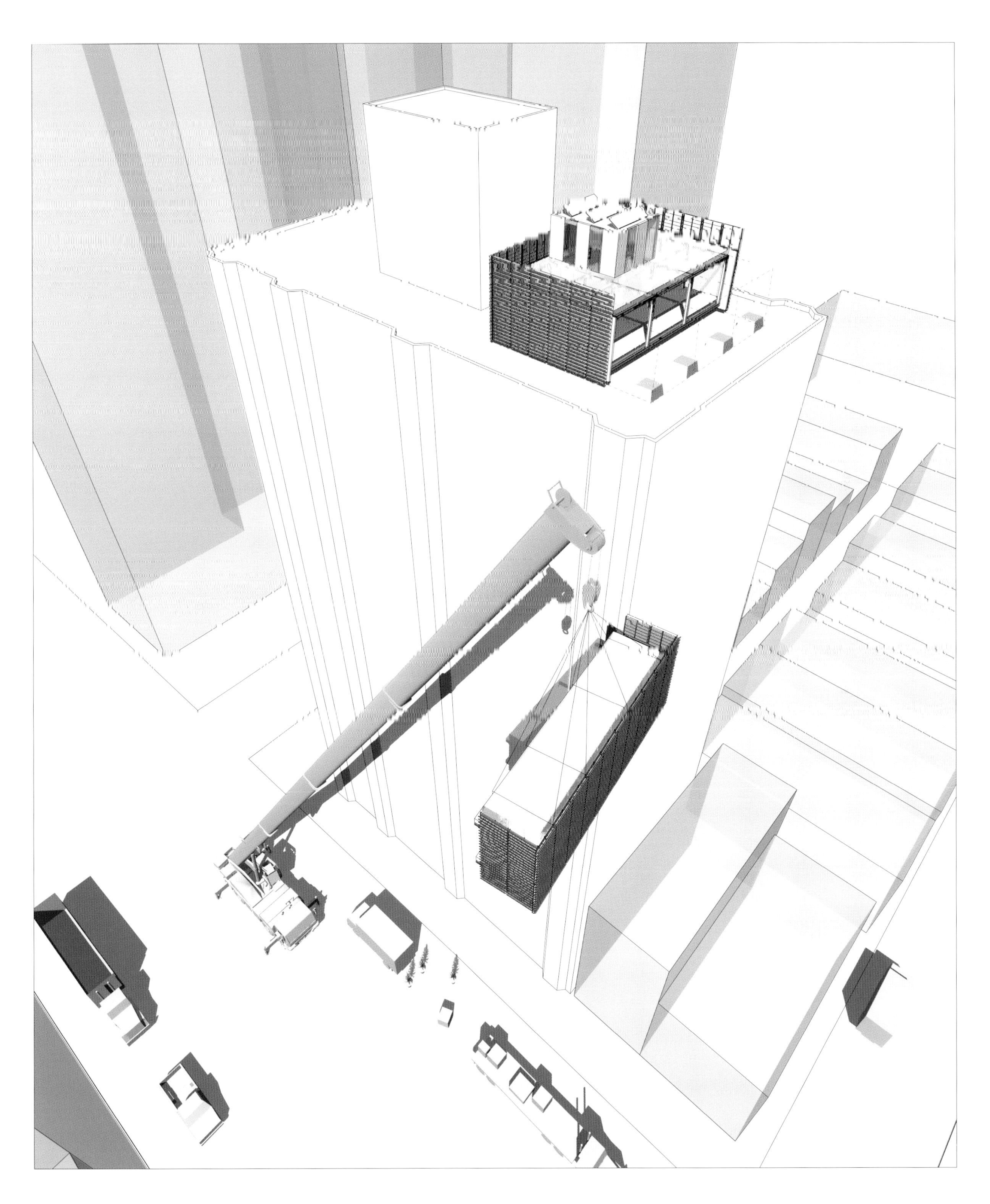

Organic Urban Living Armature (OULA)

San Francisco, California / 2005

This high-density mixed-use building complex is designed to be fabricated almost entirely off site using a proprietary system of steel and concrete composite panels to be manufactured in Canada and assembled at a staging site near Sacramento into nearly complete, truckable building units. Ground-level retail/restaurant, mechanical/utility, and parking areas will be swung into place as prefabricated hybrid panels similar to tilt-up construction. The upper-level residential spaces will arrive as preassembled and prefinished living units. The entire manufacturing and prefabrication process can be completed off site within a five-month, just-in-time delivery framework, overlapping with a total three-month onsite construction period. The rationalized, componentized manufacturing, delivery, and erection process provides tremendous cost savings and reductions in urban disruption and site pollution. The building materials are high-recycled-content concrete and steel, inert, healthy, and free of off-gassing chemical products. The surfaces are hard and robust concrete and plaster—there are no vinyl or drywall products. Parking is limited, while car-share, bicycle parking, and sheltered street frontage on light rail tracks all encourage positive urban life. At the same time, the project reconnects through-streets and provides typical urban street parking to enhance existing traffic logic and maximize retail use at the street level.

Using the vehicle of a highly economical, radically rationalized construction process accompanied by sustainable, green-technology materials and systems, primary design emphasis is placed on high-quality urban street life. The integration of organic indoor-outdoor living within a high-density residential community offers a variety of unit types to accommodate diverse families. The two city blocks are organized to optimize day-lighting, ventilation, and outdoor access to all living units and to carve out a semi-protected, semi-secret, quiet-but-active restaurant/retail alley stepping back to lower residential units and quiet rooftop gardens above. Most building faces are composed of generous balconies or sunrooms intended to enliven all street and community garden facades with active, populated, and densely planted outdoor living areas. The configuration of the blocks steps down and adjusts to the neighboring historic buildings and steps back at street level to encourage outdoor cafes and to gently draw pedestrian traffic from the street into the active restaurant alley. The building is conceived as a dense urban retail block, porous to light and air at the residential levels and carved out at its alley core to provide an urban surprise of quiet outdoor restaurants and family gardens in the air. The building itself is detailed as a simple, rational frame armature bringing the life of shops, apartments, and hanging gardens into the forefront as a primary image of the site.

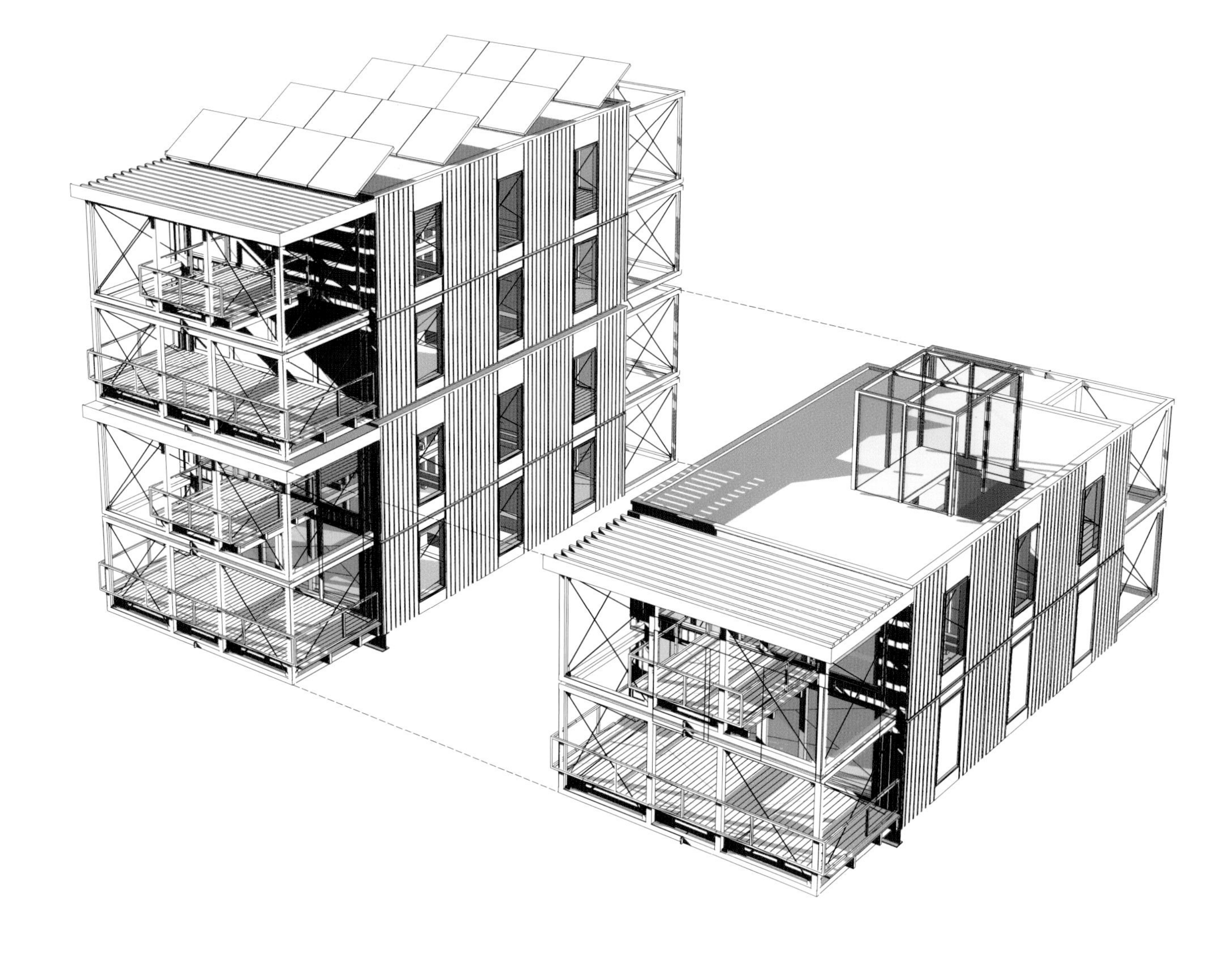

OPPOSITE
Detail of modules as configured for terrace rooftops and solar panel rooftops

Below: Rendered aerial view showing density of dwelling units, open spaces, and relationship with surrounding neighborhood density

Organic Urban Living Armature (OULA)

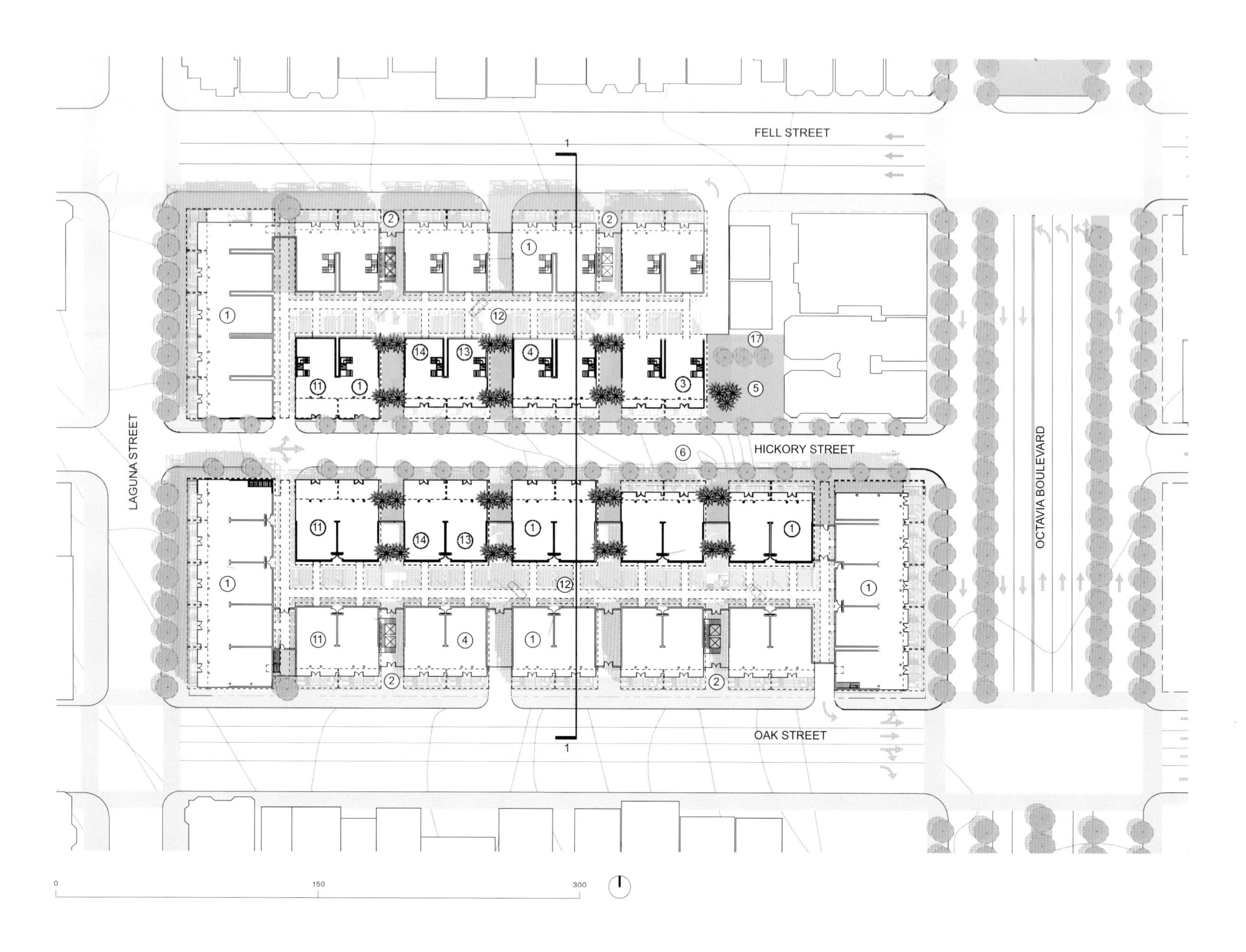

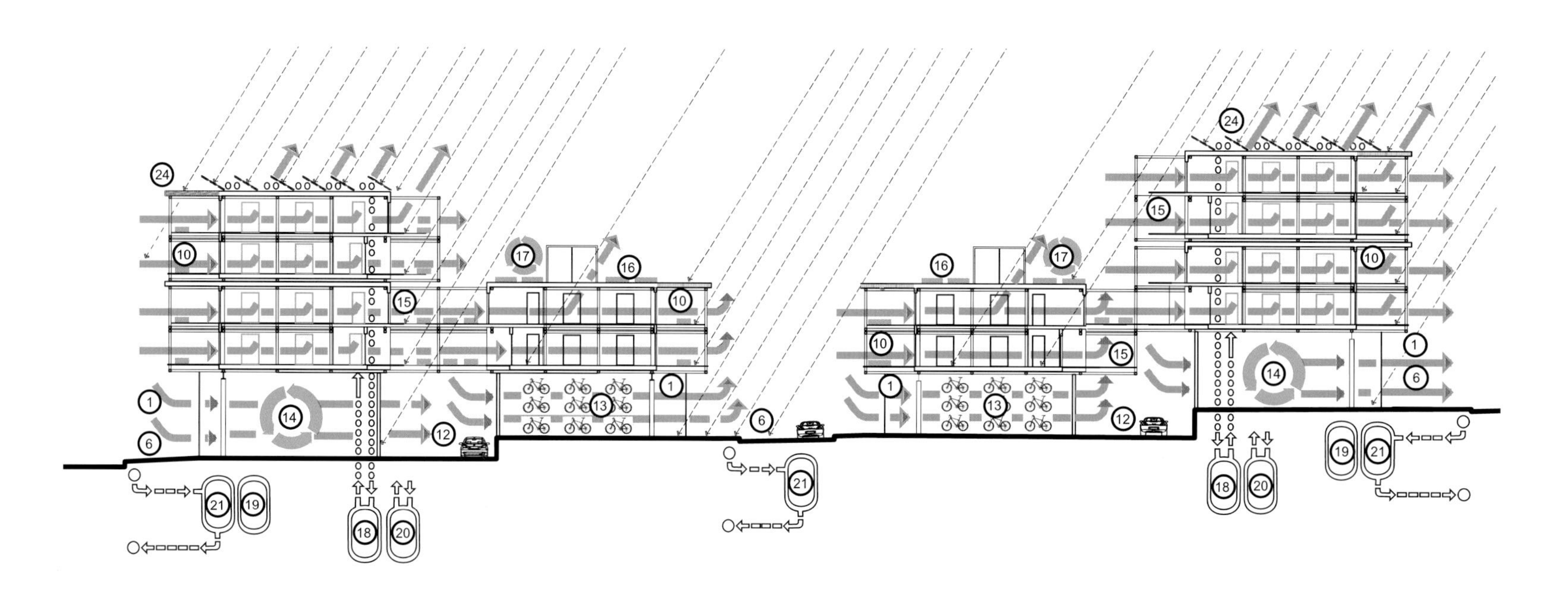

OPPOSITE
Top: Site plan
Bottom: Section
Key:
1. Flexible street-front retail/restaurant space
2. Residential lobby
3. Community center and day care
4. Mechanical spaces and utilities
5. Community playground
6. Outdoor gathering and outdoor restaurant space
7. One-bedroom residence
8. Two bedroom residence
9. Three bedroom, two-story townhouse residence
10. Residential balconies
11. Commercial office
12. Parking; electric car station; city car-share station
13. Bicycle parking garage
14. Recycling and trash center
15. Exterior residential hanging sidewalks
16. Roof gardens
17. Residential community compost center
18. Roof runoff catchment and filtration/recycling tanks
19. Black water sewer pretreatment tanks
20. Household gray water filtration and irrigation recycling tanks
21. Street runoff storm sewer pretreatment tanks
22. Community compost center
23. Residential community outdoor gathering space
24. Rooftop photovoltaic solar collection panels

Right:
2nd- and 4th-floor plans

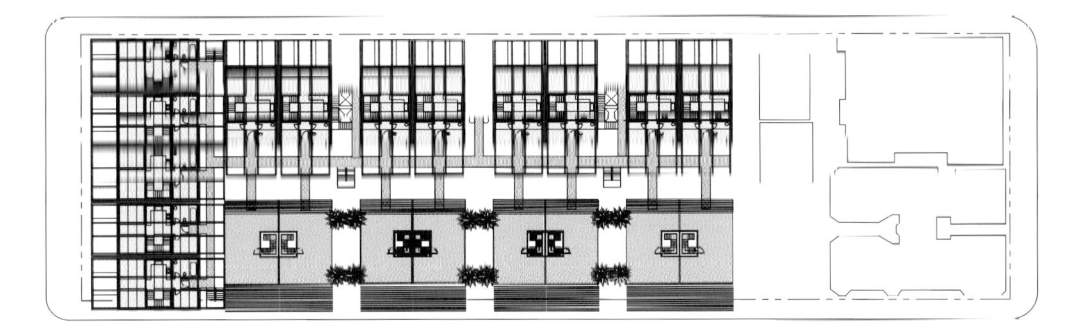

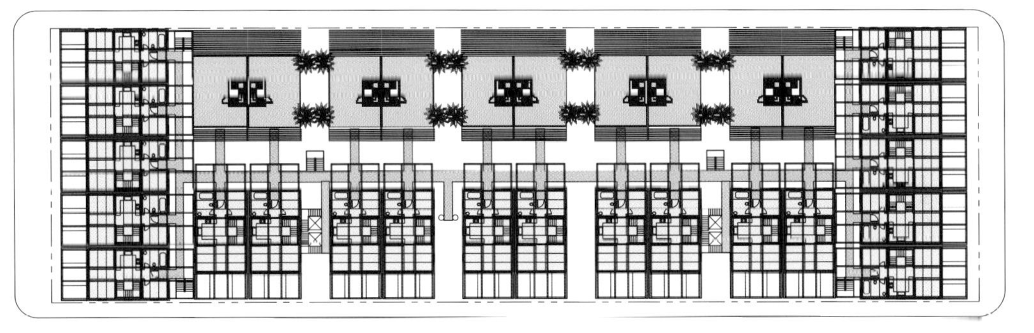

Fourth floor

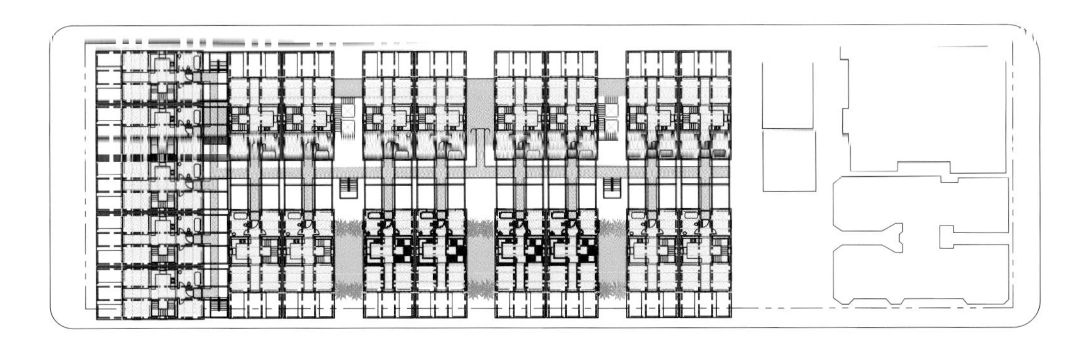

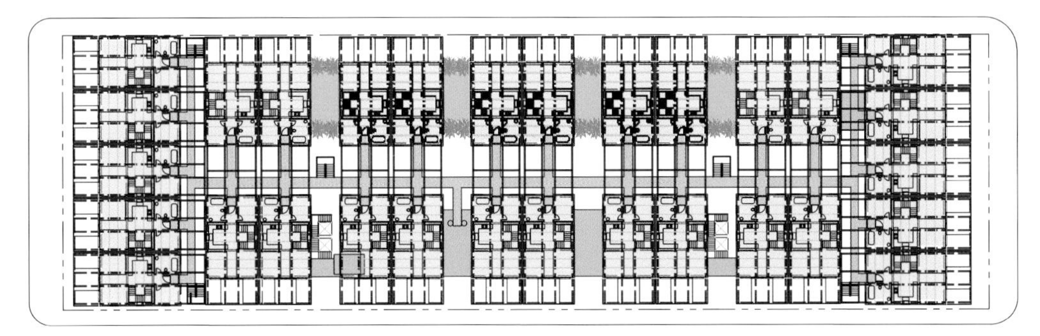

Second floor

Organic Urban Living Field (OULF)

Charlottesville, Virginia / 2005

This urban housing landscape will be fabricated almost entirely off site using a hybrid steel-frame/SIPs system. The individual building units will be efficiently manufactured off site in two road-legal halves per typical three-bedroom flat, and then stacked by crane as complete three-story, three-family units on top of semi-buried, prefabricated, composite concrete basement vaults. Earth excavated for building foundations is redistributed as rolling landscape berms, creating a unified outdoor common space flowing around the individual house blocks. The amount of earth cut and fill is balanced in order to minimize cost, energy expenditure, and existing community disruption, while simultaneously enhancing the rich symbolism of a community rooted in the local Jeffersonian earth. Dwelling units share a common geometric order defined by a superimposed agrarian orchard grid planted with fruit-bearing shade trees. Within the regular orchard grid, the slightly sliding house positions create a readably syncopated rhythm, allowing the common open space to shrink and swell across the rolling berms, creating variously sized outdoor gardening, picnic, and play areas. Market-rate dwelling units will be fully pre-assembled with finished interiors, while self-build units will incorporate homeowner and volunteer labor at both the factory and onsite construction stages. Self-build and volunteer labor construction process variations will accommodate differential cost structure, rather than overt distinctions in unit size, placement, or quality. Within a highly democratic common building language, a wide range of residential, retail, community gathering, and child-care spaces are included in the site planning and distribution of system modules, resulting in architectural, economic, and social diversity intertwining across the site. Community vegetable gardens, picnic, and play areas weave as continuously linked earth berms winding among the buildings and gently rolling down the cross-slope site, both defining internal community areas and flowing outward to the street edge as a welcoming integration with the larger neighborhood.

Primary design emphasis is placed on high-quality urban community life, applying a highly economical, energy-efficient, fair-wage manufacturing and construction process accompanied by sustainable land use patterns, optimized siting for solar and natural wind flow ventilation access and control, healthy, green-technology materials, and low energy-consumption mechanical and filtration systems. The cross-slope, hillside home siting is organized to optimize day lighting, ventilation, and outdoor access to all living units, and to carve semi-protected outdoor living and play spaces. Primary building faces are composed of generous balconies or sunrooms intended to enliven all street and community garden facades with active, populated, and densely planted outdoor living areas. The configuration of the housing blocks steps down and adjusts to the neighboring buildings and step back at street level to activate street frontage with outdoor cafes, retail, bus stops, and pedestrian traffic.

Underground environmental systems will be placed as entirely prefabricated utility vaults with primary plumbing and mechanical systems already integrated at the factory. The rationalized, componentized manufacturing, delivery, and erection process provides tremendous cost savings and reductions in urban disruption and site pollution. The building materials are high-recycled content concrete, steel, and recycled wood. The buildings are organized and detailed to provide maximum daylight and airflow to each unit, and all primary community spaces, stairways, and balconies are open air. All rooftops are designed for maximum photovoltaic energy production, and all roofs collect and filter rainwater for use as non-potable household water. Household gray-water will be filtered and recycled as garden irrigation. Black water and grade-level storm water will both be pre-filtered and partially treated prior to release into the respective city systems, in order to minimize the impact of increased density on existing city services. Primary street frontage allows for urban parallel parking and storefront commercial space to accommodate existing community traffic and maximize friendly retail, residential, and community center use at the street level.

Rendered exterior view showing retail street-front slat siding systems and operable window shutter elements

OPPOSITE
Rendered views showing indoor/outdoor living possibilities within high-density low-rise development on former trailer park site

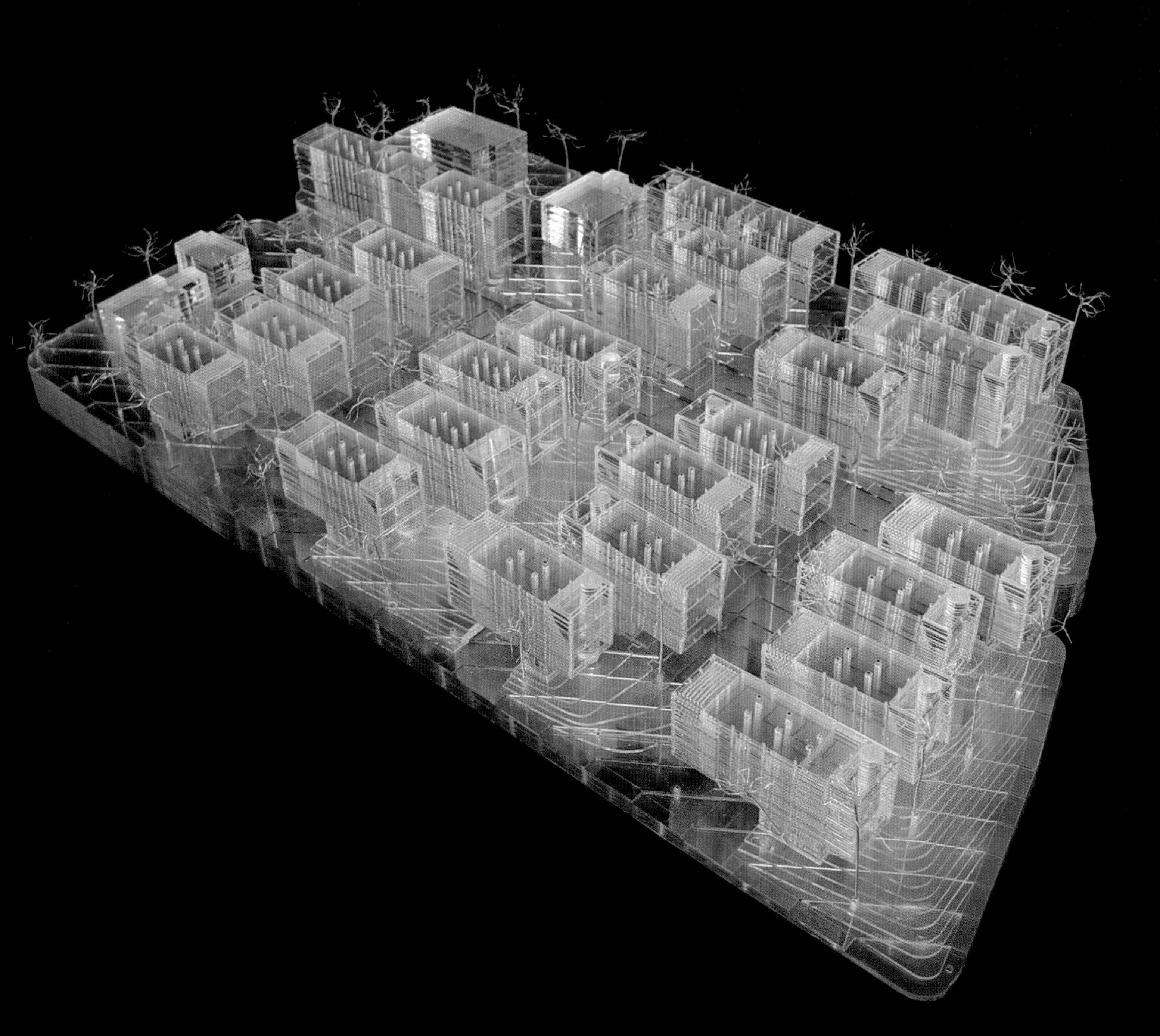

Site model showing distributed field of three- to six-unit buildings, staggered on site and following natural topography to provide greatest possible access to sunshine, breezes for natural ventilation, and private and community outdoor living spaces

OPPOSITE
Large scale model showing three-story modular steel frame with slatted exterior cladding, operable as shade and storm shutters at window locations

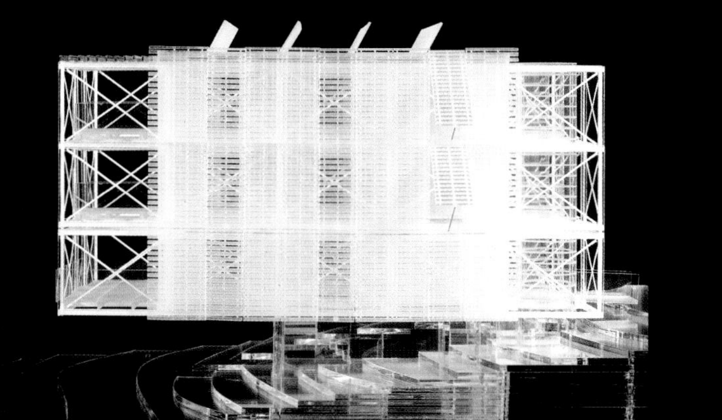

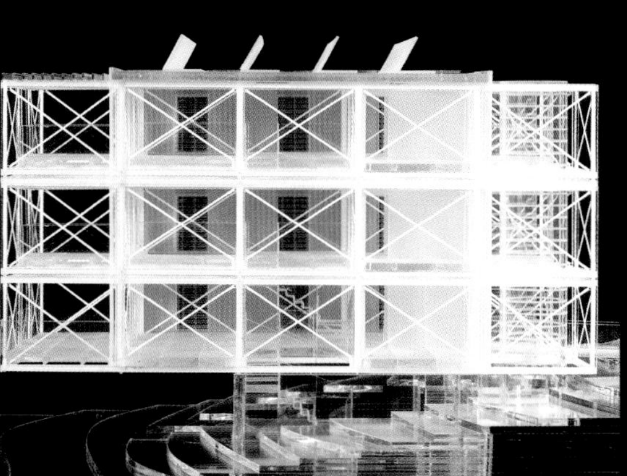

Organic Urban Living Field

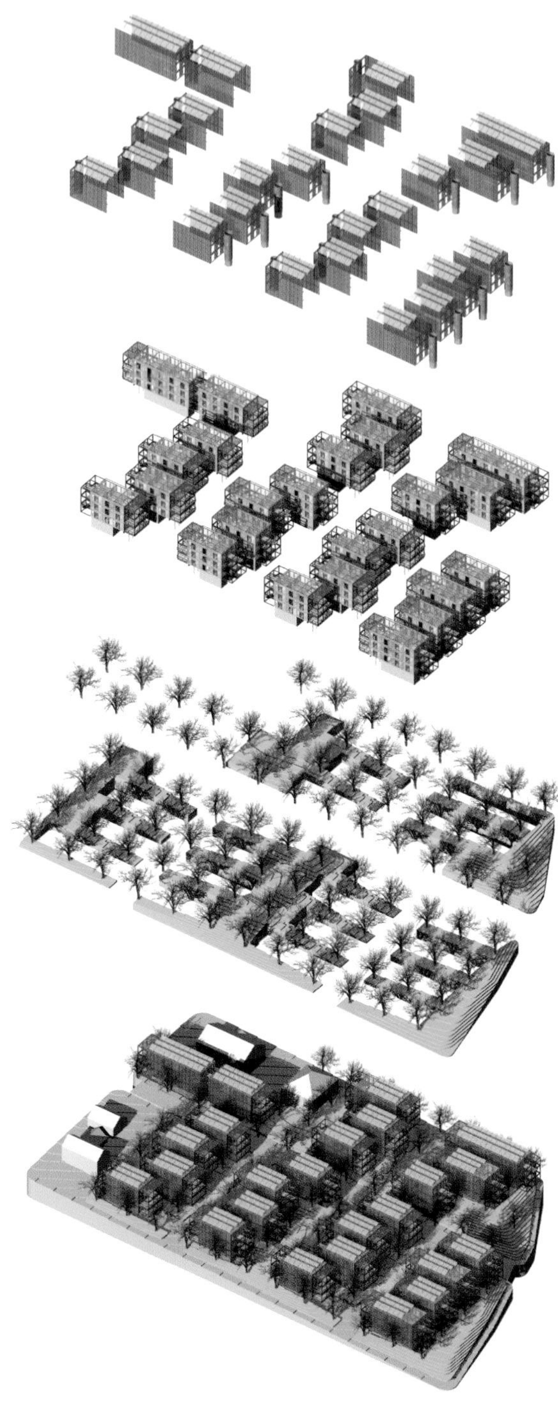

Left: Typical floor plans for different unit types, showing wide variety of configurations within standardized modular grid **Above:** Exploded view of primary site design elements, showing (from bottom to top) completed site; landscape berms and trees; structural frame modules; and exterior skin, solar panels, and other bolt-on elements.

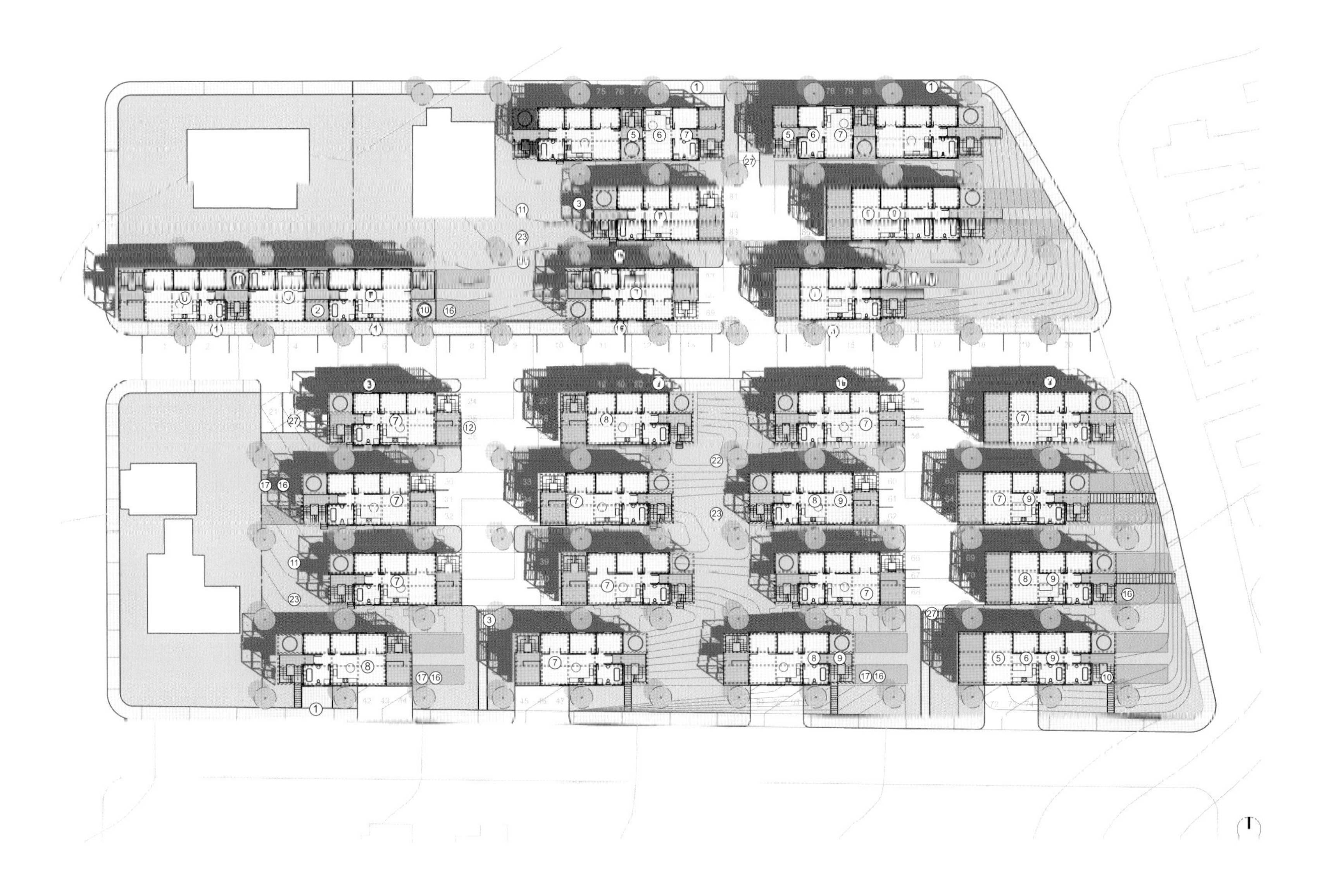

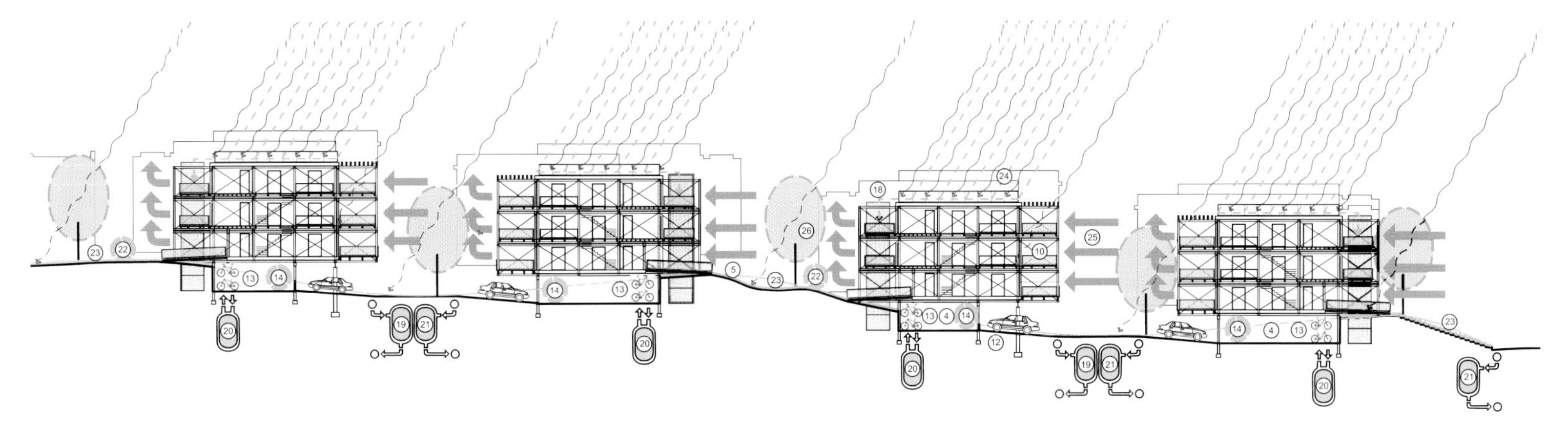

Key:

1. Flexible streetfront retail/restaurant space

2. Common residential stairs and balconies

3. Street-level accessible residences

4. Mechanical space and utilities

5. One-bedroom or studio residence

6. Two-bedroom, one-bath residence

7. Three-bedroom, one-bath residence

8. Three-bedroom, two-bath residence

9. Three-plus–bedroom, two-bath, two-story townhouse residence

10. Residential balconies

11. Community playground

12. Parking; electric car station; city car-share station

13. Bicycle parking

14. Recycling and trash center

15. Community center and day care

16. Community vegetable gardens

17. Community compost center

18. Rain barrel water catchment and filtration tanks

19. Black water sewer pretreatment tanks

20. Household gray water filtration and irrigation recycling tanks

21. Street runoff storm sewer pretreatment tanks

22. Balanced cut-and-fill rolling landscape berms

23. Community outdoor gathering space

24. Rooftop photovoltaic solar collection panels

25. Natural ventilation corridors

26. Fruit-bearing shade tree orchard

27. Community basketball hoops/hard surface play area

Organic Urban Living Field

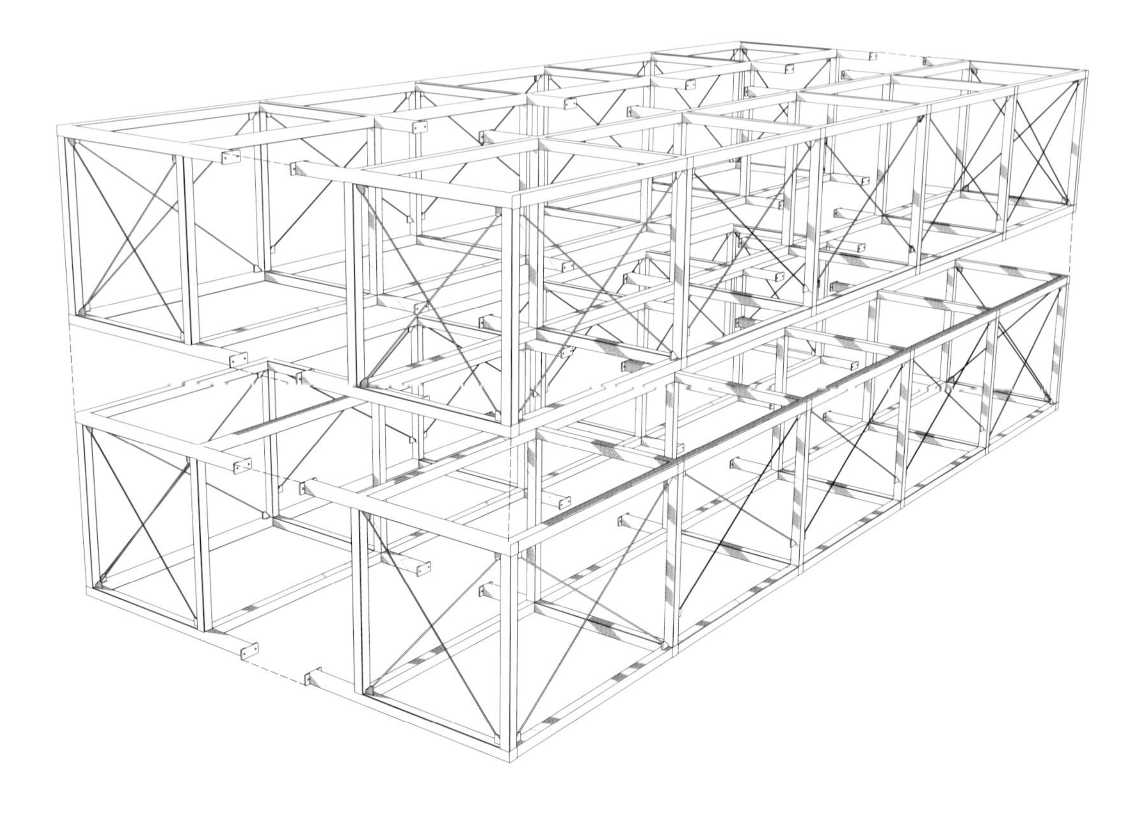

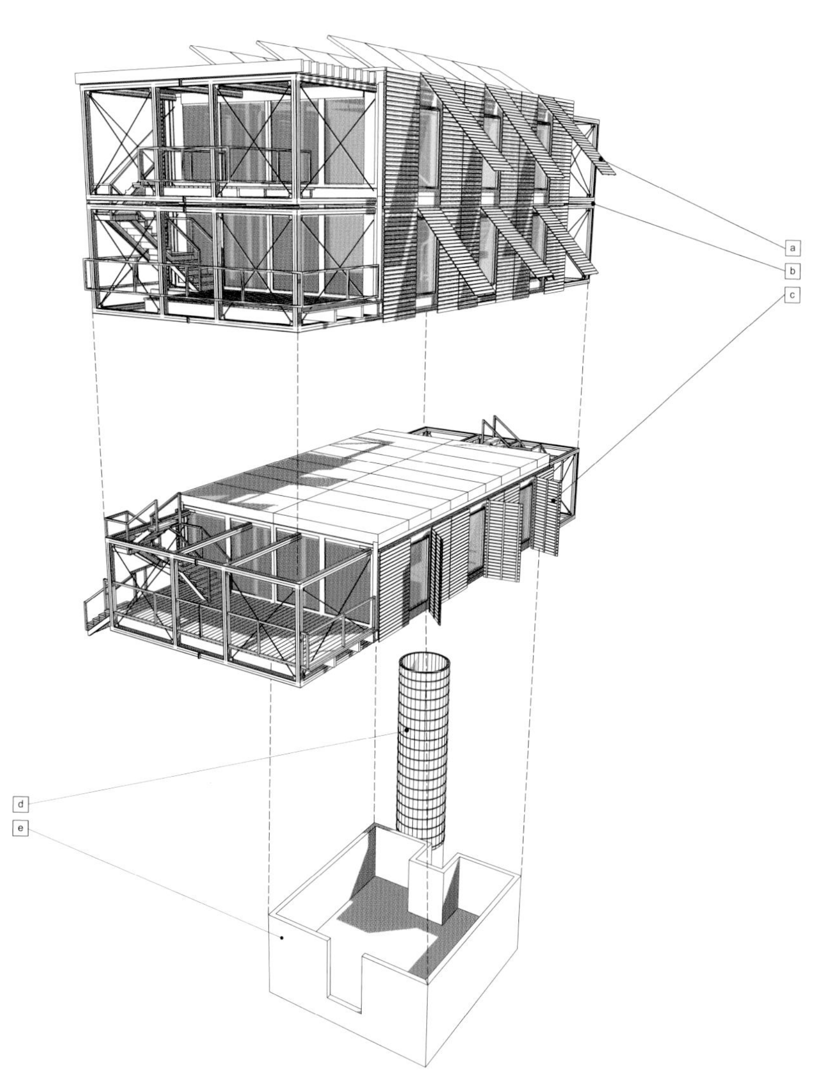

Top: Exploded assembly view of structural frame components prior to attachment of other elements
Bottom: Exploded view of primary building modules
Parts List:
a. Operable shade shutters
b. Two-story, four-bedroom, two-bath dwelling unit
c. One-story, three-bedroom, two-bath dwelling unit
d. Rain barrel roof water catchment and recycling tank
e. Prefabricated steel and concrete composite panelized basement unit with shared mechanical, storage, and recycling facilities

OPPOSITE
Top: Rendered interior view showing two-story space and operable window shutter elements
Bottom: Typical outdoor living space on upper-floor unit

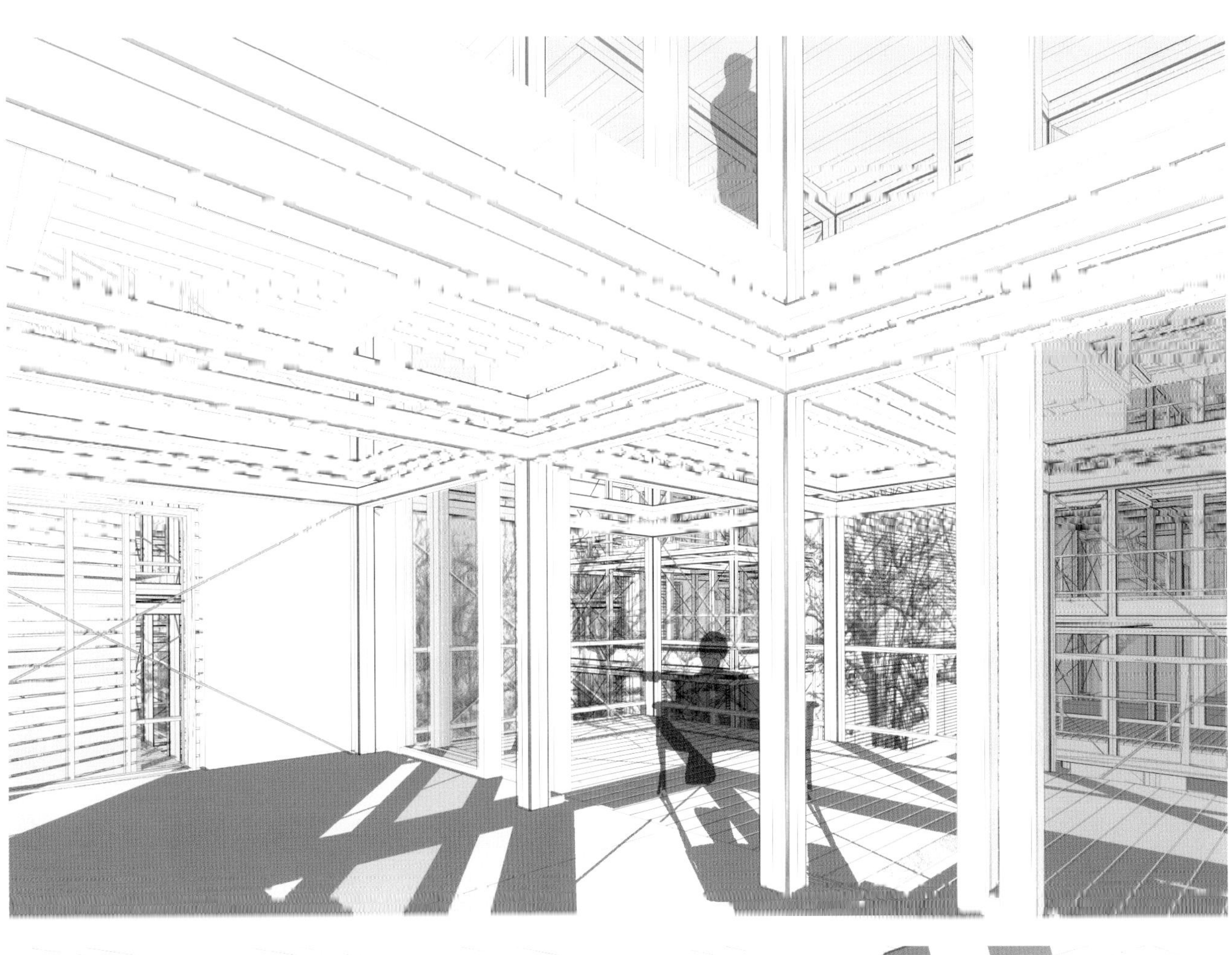

Organic Urban Living Field

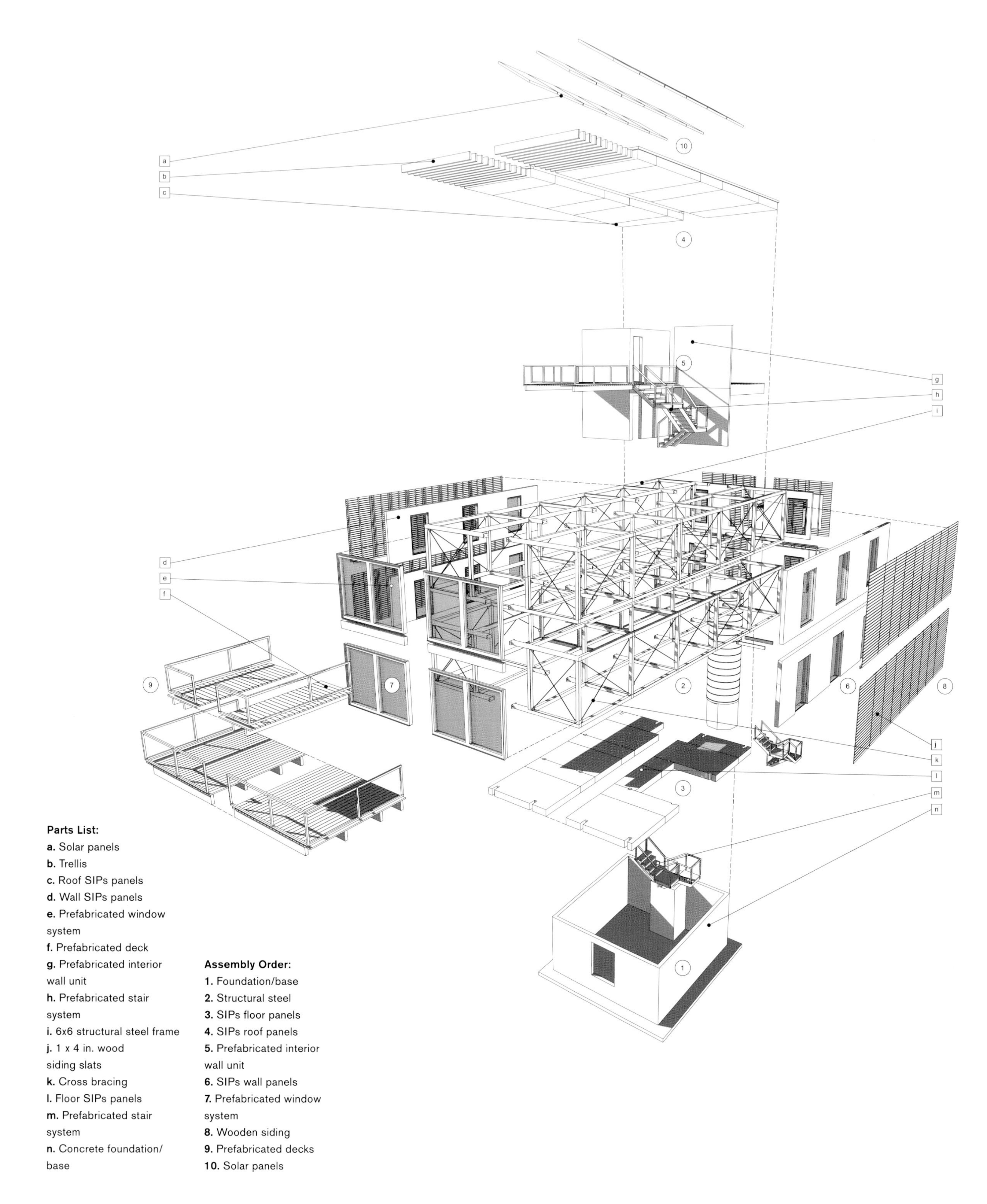

Parts List:

a. Solar panels
b. Trellis
c. Roof SIPs panels
d. Wall SIPs panels
e. Prefabricated window system
f. Prefabricated deck
g. Prefabricated interior wall unit
h. Prefabricated stair system
i. 6x6 structural steel frame
j. 1 x 4 in. wood siding slats
k. Cross bracing
l. Floor SIPs panels
m. Prefabricated stair system
n. Concrete foundation/ base

Assembly Order:

1. Foundation/base
2. Structural steel
3. SIPs floor panels
4. SIPs roof panels
5. Prefabricated interior wall unit
6. SIPs wall panels
7. Prefabricated window system
8. Wooden siding
9. Prefabricated decks
10. Solar panels

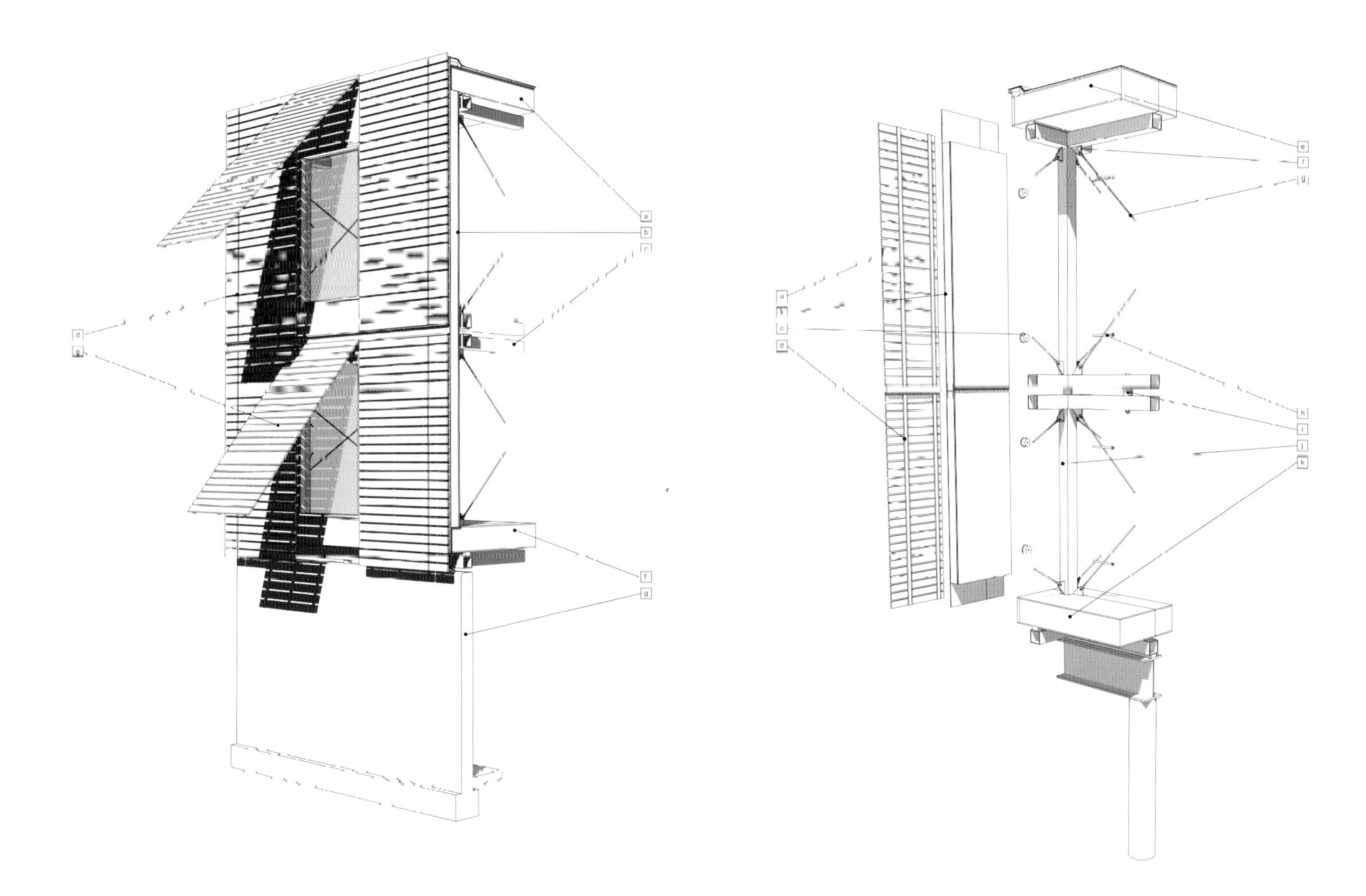

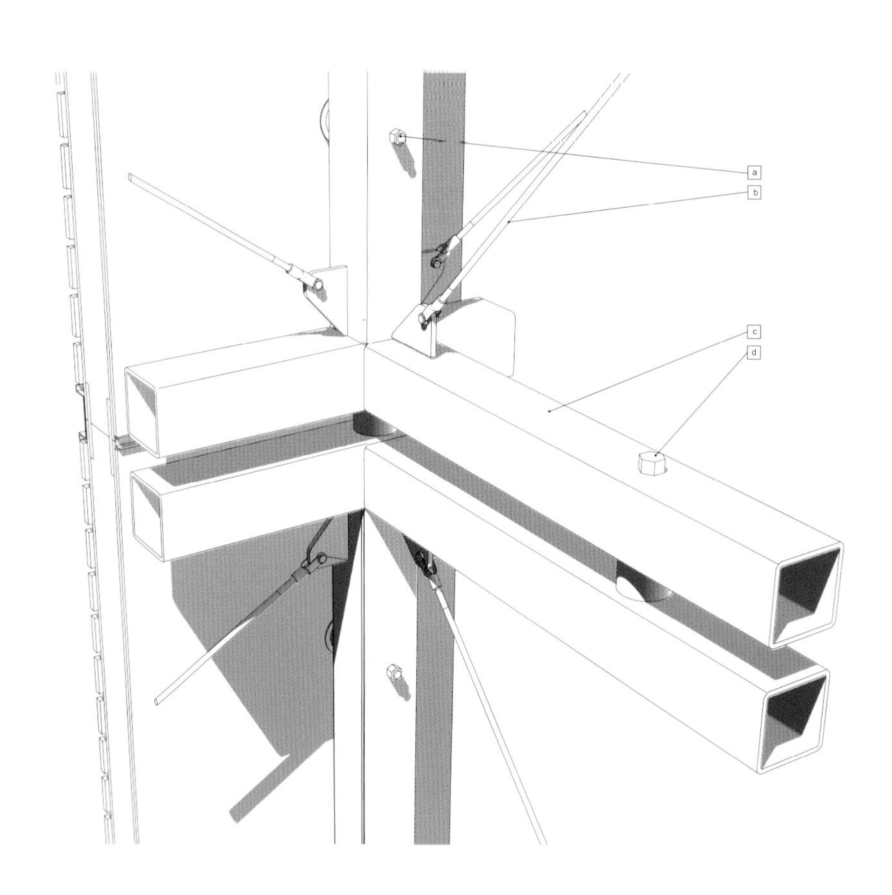

Top left:
Wall section detail perspective
Parts List:
a. R-50 structural insulated panel system
b. R-25 structural insulated panel wall enclosure system
c. Modular cross-braced hollow structural steel cantilever frame system
d. 1x4 wood slat siding
e. Operable horizontal or vertical swinging louvered window shutter system
f. R-50 12 in. structural insulated panel system
g. Prefabricated steel and concrete composite panelized foundation system

Top right:
Exploded corner detail perspective
Parts List:
a. 1x4 wood slat siding
b. SIPs wall panel
c. Steel wall spacer
d. 1 in. hollow steel tube support for skin
e. SIPs roof panels
f. Metal bracket for cross bracing
g. .5 in. diameter rod cross bracing
h. .5 in. bolt for wall spacer
i. Steel frame spacer
j. 6 in. hollow structural steel tube frame system
k. SIPs floor panel

Right:
Module frame connection detail
Parts List:
a. .5 in. diameter bolts connecting structural insulated panels to steel frame
b. .5 in. diameter steel rod tension cross-bracing
c. 6x6 hollow structural steel tube
d. 1 in. diameter bolts through 3 in. diameter steel tube spacers

Amber Block

Tulsa, Oklahoma / 2005

The City of Tulsa initiated major public projects and grant programs in 2005 intended to revitalize its downtown, opening new potential for an economically viable urban center focused on creating livable neighborhoods with attractive housing, workplaces, shopping, nightlife, and cultural activities at the heart of the city. At the east end of downtown on Third Street is the Amber Block, providing a mix of condominium residences and office and retail space and occupying the prominent corner where the north-south grid of east Tulsa makes a sharp bend into the diagonal grid of downtown, acting as a gateway from the outlying city into the density of the urban core.

Our design process involved extensive research and mapping studies analyzing historical and current social, economic, cultural, and urban architectural factors particular to Tulsa, viewed in context with successful national and international design approaches applied to other new urban housing neighborhoods. The site design and building massing are quite dense relative to the immediate community, intended to encourage adjacent development by providing concentrated open space, sidewalks, and a street-corner urban plaza that is enlivened by substantial retail frontage. Primary features of the architectural strategy are developed as modular systems that can accommodate a variety of options and alterations in project size, design, financing, and marketing approaches that are highly flexible as a prototype project for similar sites in Tulsa as well as in other American downtowns. The prefabricated design and construction system is a continued development of our cross-braced steel-frame modules, accompanied by modular development and financing strategies intended to allow great flexibility and site-specificity for various conditions while staying within a well-studied and well-structured overall framework.

The redevelopment of Third Street, with its numerous civic institutions and new convention center, is a major element in the city's plans for the revitalization of downtown. It is not until one arrives at the sharp bend of the Amber Block at Third and Kenosha that the skyline landscape of downtown unfolds, providing the perfect spatial opportunity to have this corner broaden into a small square. The plaza is not too large, as would be more appropriate in the center of downtown, but acts as a significant neighborhood center with its own identity and street life, with trees, a small fountain, urban-scale sculpture, overhanging residential gardens, and continuous street-level retail space on all street faces of the building. The extended street design incorporates significant existing buildings along Third Street also owned by the developer and provides simple continuities in street elements that will in the future complete a special outdoor public walking space contiguous with surrounding neighborhoods.

The opposite end of the Amber Block on Second Street, while less of a pedestrian corner than Third and Kenosha, is equally important to the image of the city. Second Street is a major one-way conduit from downtown, leading departing visitors to the airport and commuters to the freeways and to outlying neighborhoods to the east and south. This corner also offers streetfront sidewalk presence in the form of two-story live/work units, as well as high gardens and balconies overhead. Cars entering the city on First Street and leaving on Second will all view this building as a gateway corner for downtown Tulsa, and with the adaptive reuse of the existing cellular phone tower on the plaza at Third Street—as a neon-spiraled swizzle-stick skewering two gigantic, perforated-steel martini olives wedged against the building—the block will become a significant, memorable, and attractive point in the city, adding both substantial new housing development as well as a pocket of sidewalk shopping that contributes to the immediate neighborhood and to a symbiotic spirit of life and new development throughout the whole of downtown.

The project developers, Micha Alexander and his father, Mike Alexander, both own steel fabrication businesses and have substantial experience in systematic manufacturing of specialty steel components. This building is in many ways a celebration of steel fabrication and efficient production processes that harness the particular interests and expertise of the developers as well as the architects. An important component of the building design is its detailing for efficient prefabrication of many steel elements. A secondary yet extremely important urban and financial aspect of this project is the intention of the developers to set up a prefabrication process near the site, which will first employ local workers in this project's construction and will subsequently establish an ongoing building components fabrication facility with the potential to be a significant employer and revenue producer in the downtown community.

The building follows many of the same construction and planning strategies developed in our previous and related multi-family, mixed-use urban projects. The Amber Block is organized and detailed to provide maximum daylight and airflow throughout the dwelling units and community spaces, and the rooftops are designed to capture photovoltaic energy and rain water while offering a variety of shaded and sunlit rooftop garden areas serving multiple private, community, retail, and public functions. Trees and plantings are situated throughout the building structure to give added life and scale to the urban street walls and to shade the roof gardens, public sidewalks, and street-corner market square. The building skin is a perforated and slotted sheet steel siding designed to provide a unique richness of color and shadows, enlivening the street and adding to the texture of the building landscape at this edge of the city. The folded stainless steel skin panels are painted an intense amber color on their interior surfaces, creating a subdued coloration when viewed face-on from a distance, through the perforations and field of wheat-stalk-thin vertical slots, but increasing in color intensity as one approaches the building, due to the oblique upward view directly onto the painted surface itself. This shimmering variation in color and layered shadows, dependent on angle of approach and speed of passage, will create an endlessly shifting appearance from the surrounding streets and freeway under various light conditions.

Schematic perspective views of Amber Block public space. Urban plaza looking south toward skyline (top); plaza and street views (below)

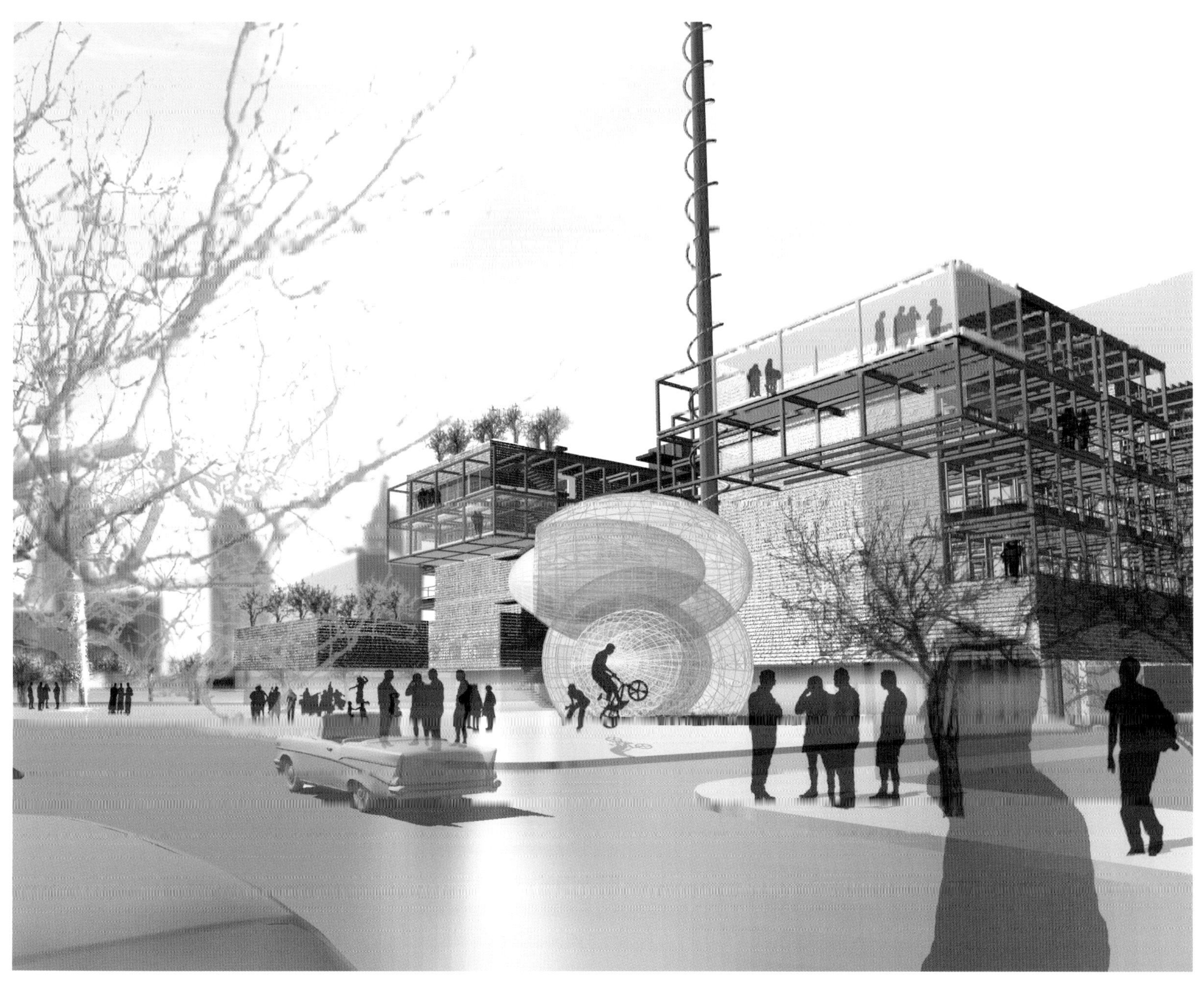

Amber Block

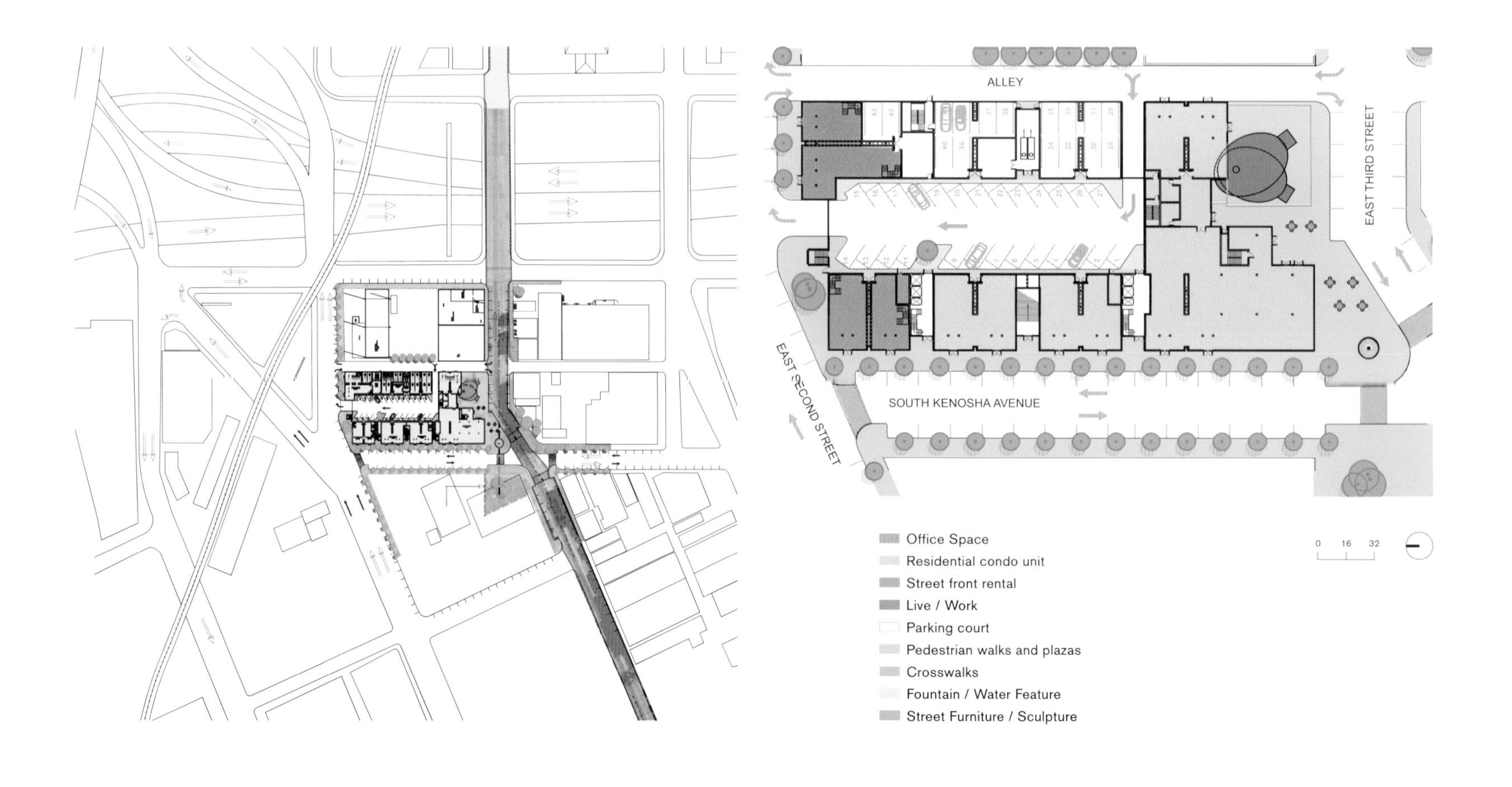

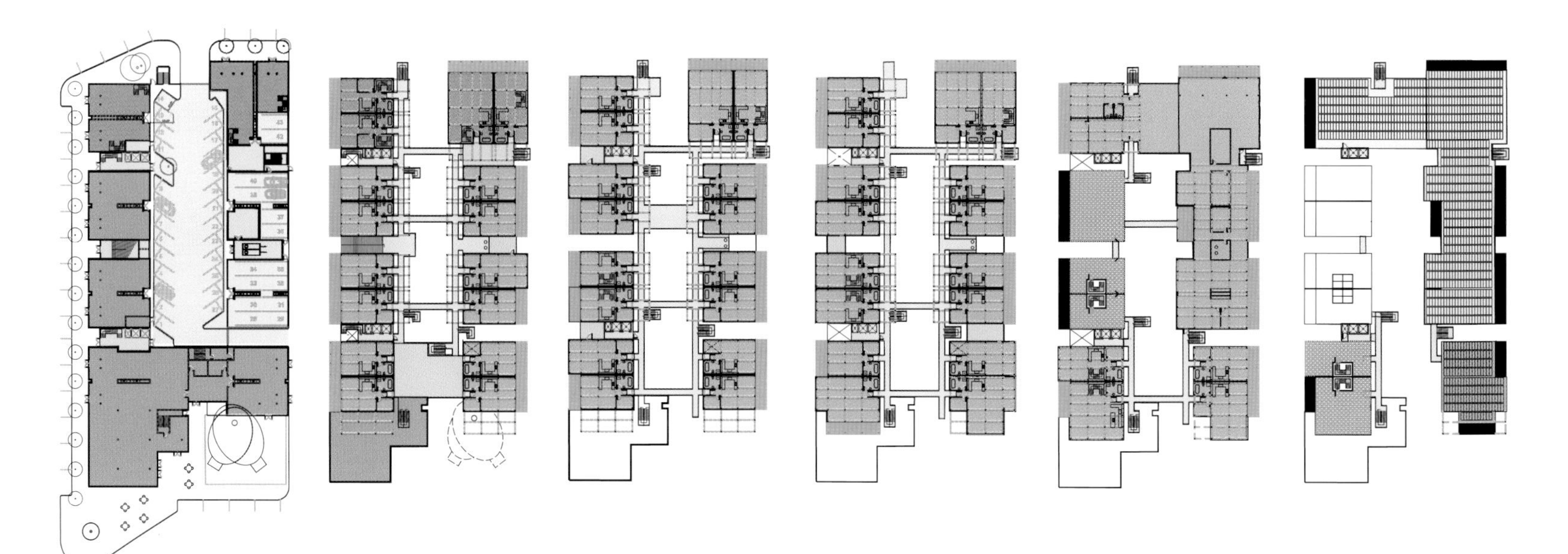

PROGRAM DISTRIBUTION

Retail / Office	32,399 sf
Residential	63,101 sf
Live / Work	(included in residential)
Exterior Terraces	2,679 sf
Service / Circulation	12,882 sf

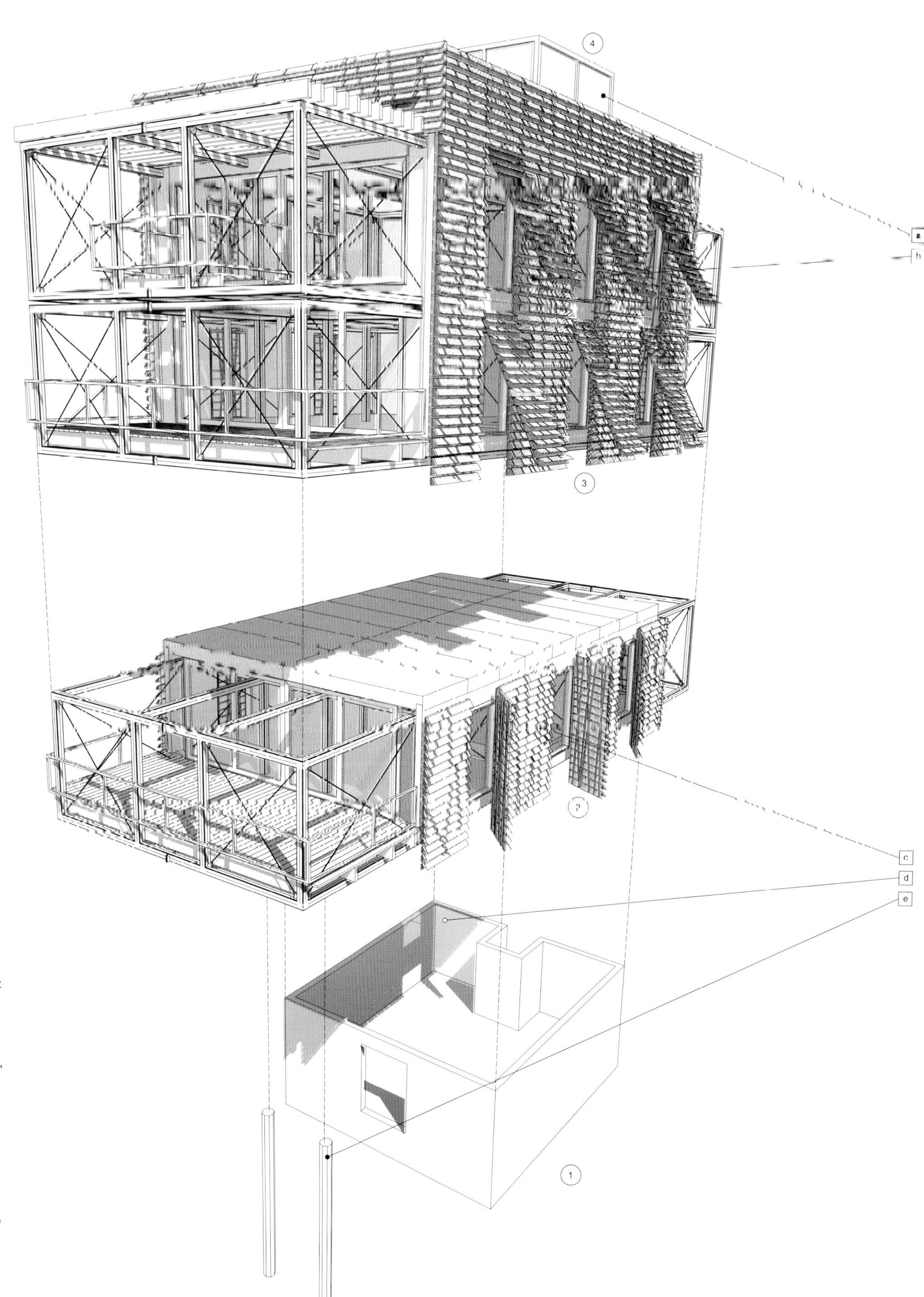

Parts List:

a. Roof terrace access unit

b. Two-story, three-bedroom, two-bath dwelling unit

c. One-story, two-bedroom, two-bath dwelling unit

d. Prefabricated steel and concrete composite panelized foundation unit with shared mechanical, storage, and recycling facilities

e. 12 in. diameter concrete support columns

Assembly Order:

1. Foundation/base

2. One-story unit

3. Two-story unit

4. Roof-access unit

Amber Block

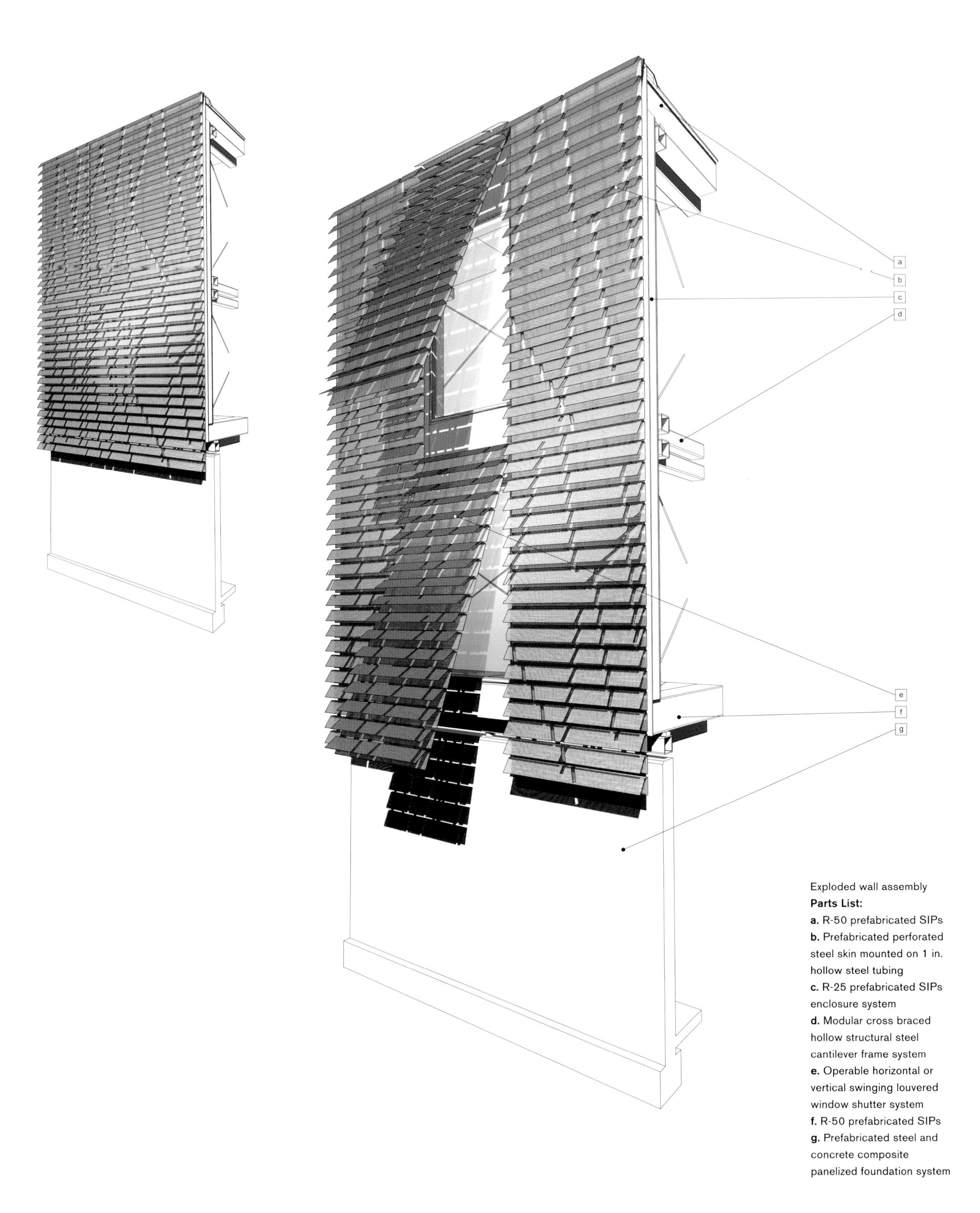

Exploded wall assembly
Parts List:
a. R-50 prefabricated SIPs
b. Prefabricated perforated steel skin mounted on 1 in. hollow steel tubing
c. R-25 prefabricated SIPs enclosure system
d. Modular cross braced hollow structural steel cantilever frame system
e. Operable horizontal or vertical swinging louvered window shutter system
f. R-50 prefabricated SIPs
g. Prefabricated steel and concrete composite panelized foundation system

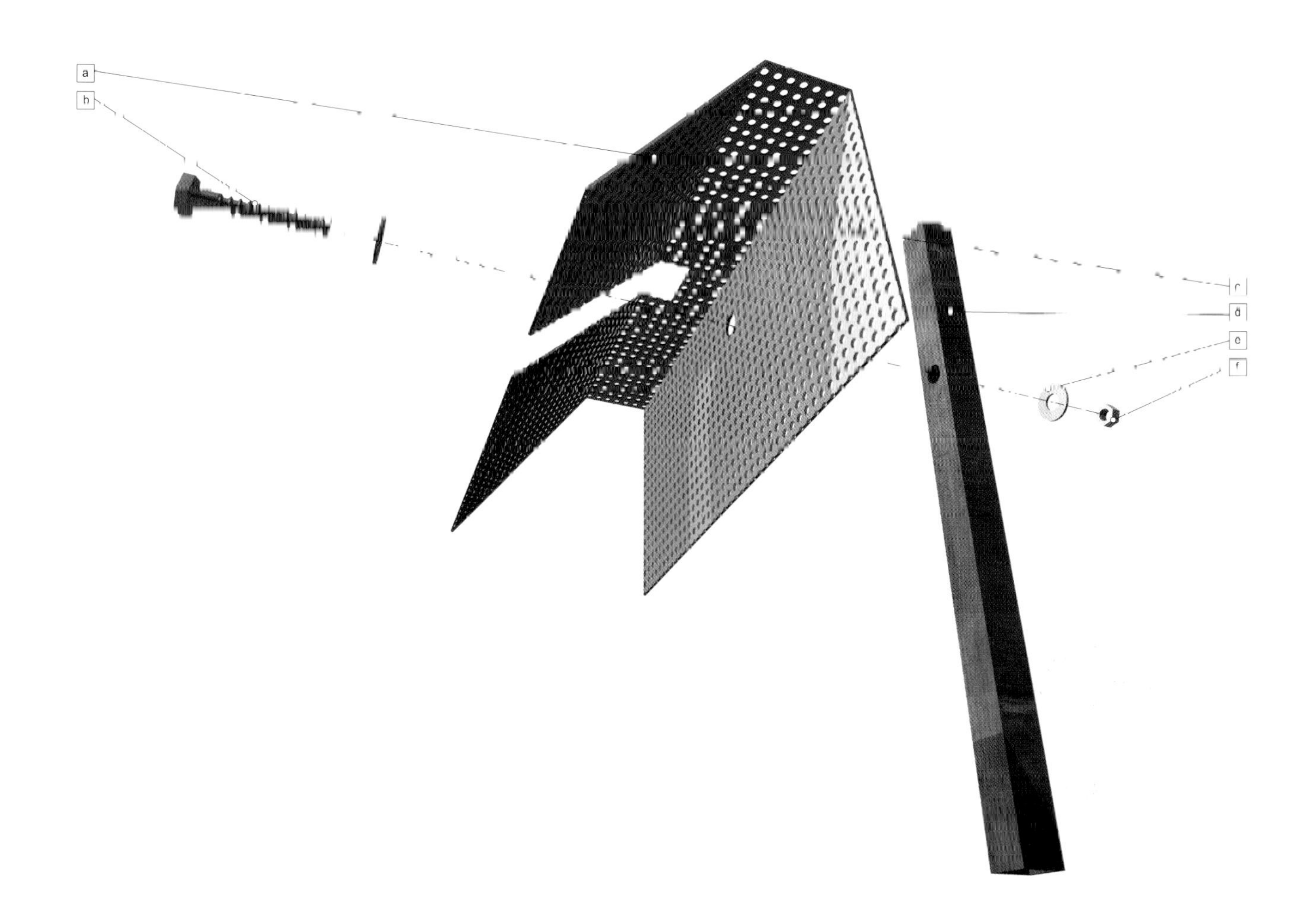

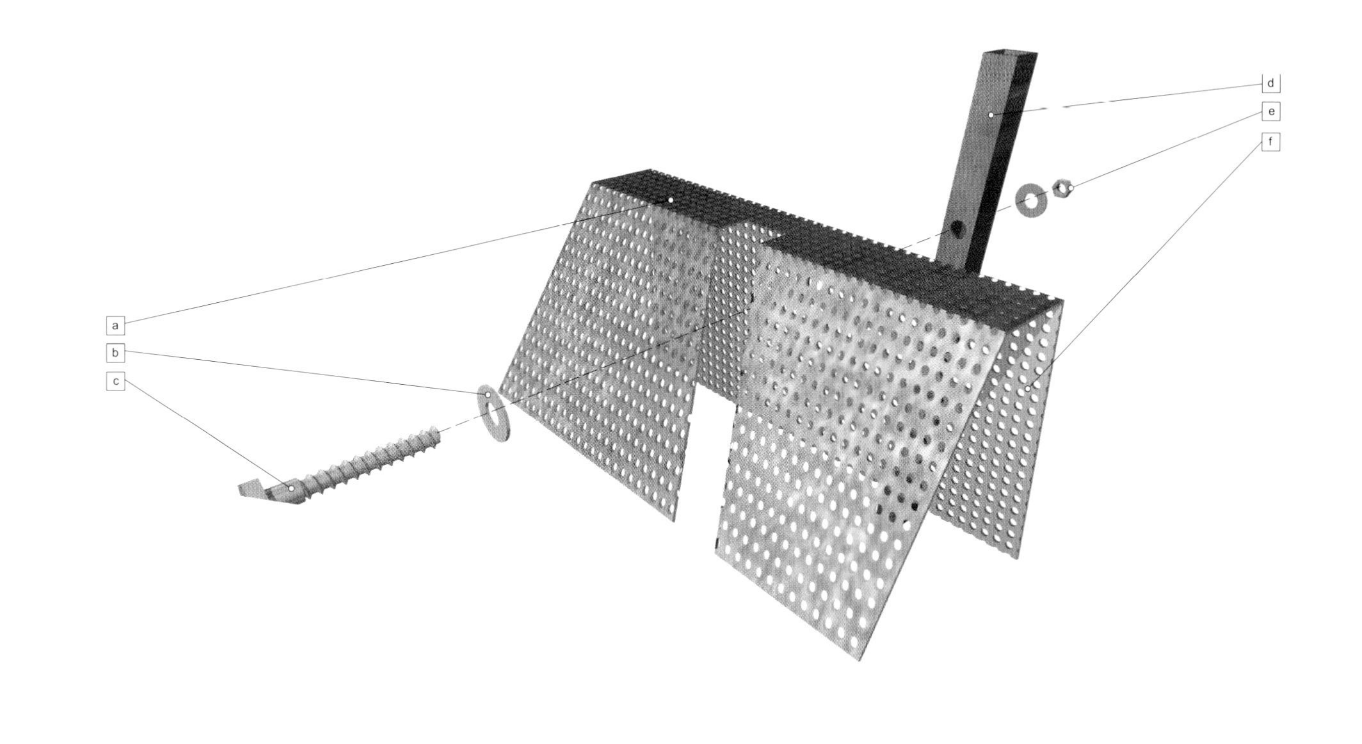

Stainless steel siding system details

Top:

Parts List:

a. Prefabricated perforated steel skin with natural silver surface

b. .5 in. bolt

c. Prefabricated perforated steel skin with interior painted surface

d. 1 in. square hollow steel tube spacer frame

e. .5 in. washer

f. .5 in. nut

Bottom:

Parts List:

a. Prefabricated perforated steel skin with natural silver surface

b. .5 in. washer

c. .5 in. bolt

d. 1 in. square hollow steel tube spacer frame

e. .5 in. nut

f. Prefabricated perforated steel skin with interior painted surface

Amber Block

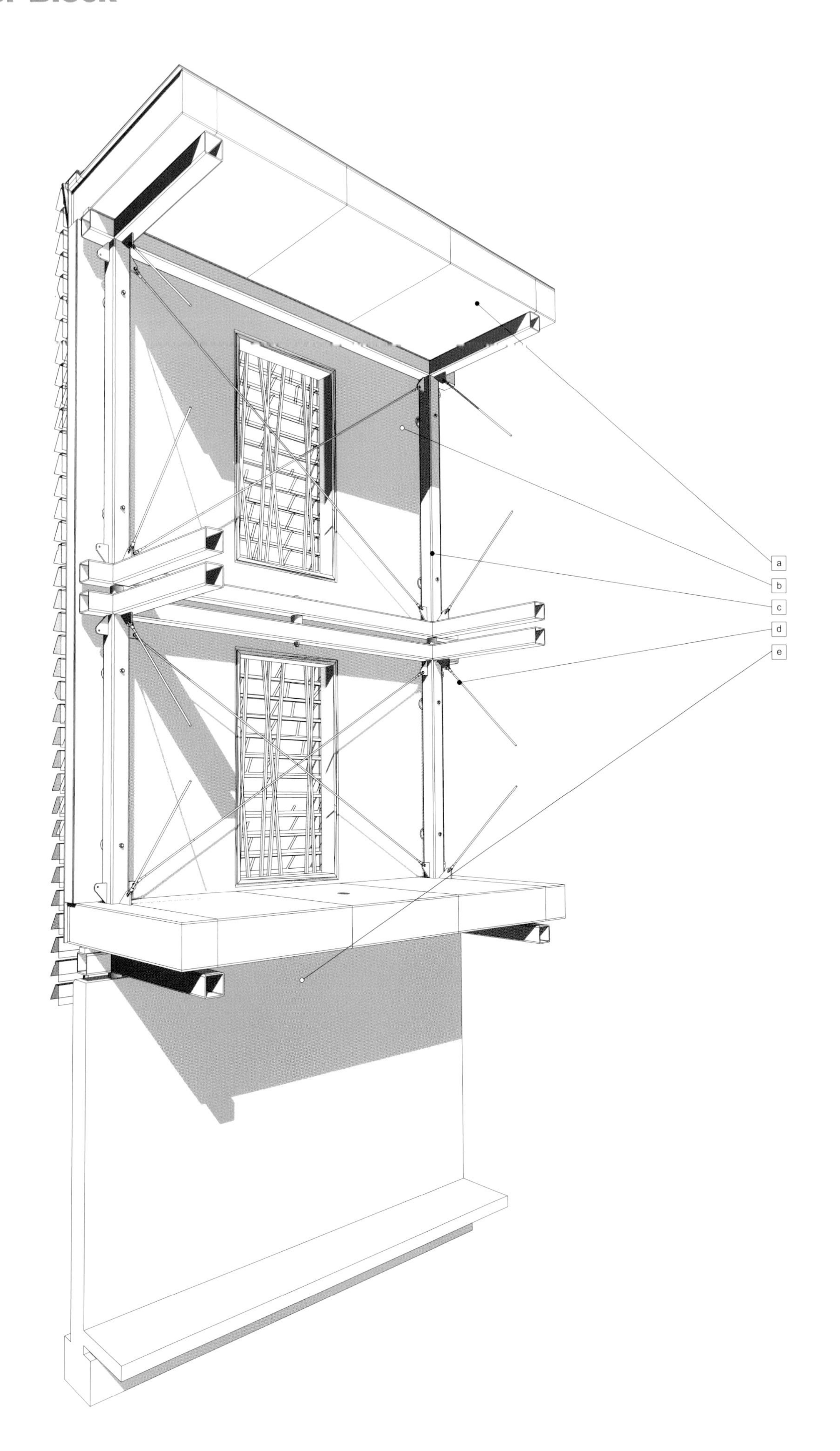

Typical foundation-to-roof wall section detail at double-height space

Parts List:

a. R-50 12 in. SIPs roof enclosure system

b. R-25 6 in. SIPs wall enclosure system

c. Hollow structural steel tube frame

d. .5 in. diameter diagonal tension bracing

e. Prefabricated concrete and steel composite panelized foundation system

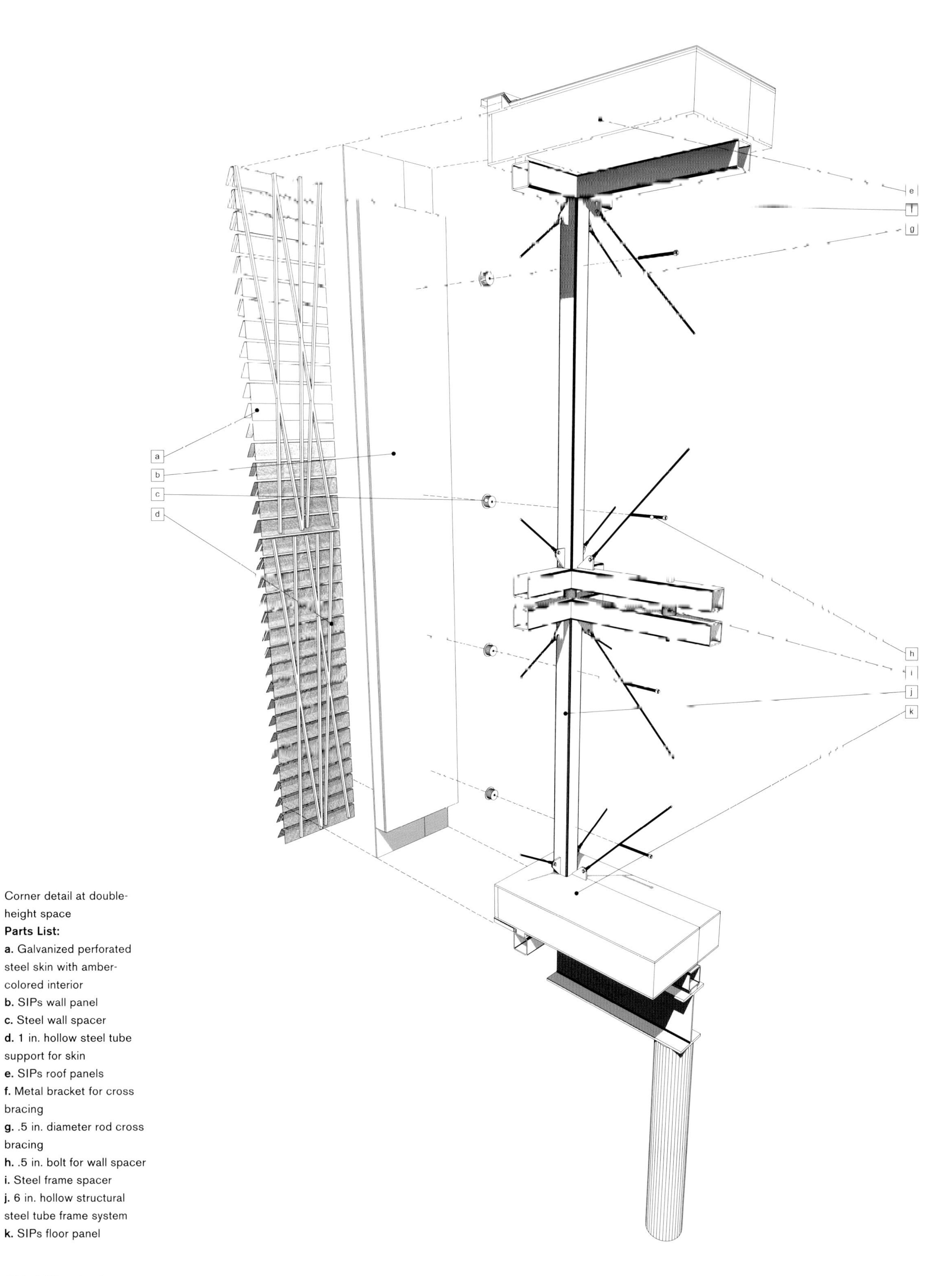

Corner detail at double-height space

Parts List:

a. Galvanized perforated steel skin with amber-colored interior
b. SIPs wall panel
c. Steel wall spacer
d. 1 in. hollow steel tube support for skin
e. SIPs roof panels
f. Metal bracket for cross bracing
g. .5 in. diameter rod cross bracing
h. .5 in. bolt for wall spacer
i. Steel frame spacer
j. 6 in. hollow structural steel tube frame system
k. SIPs floor panel

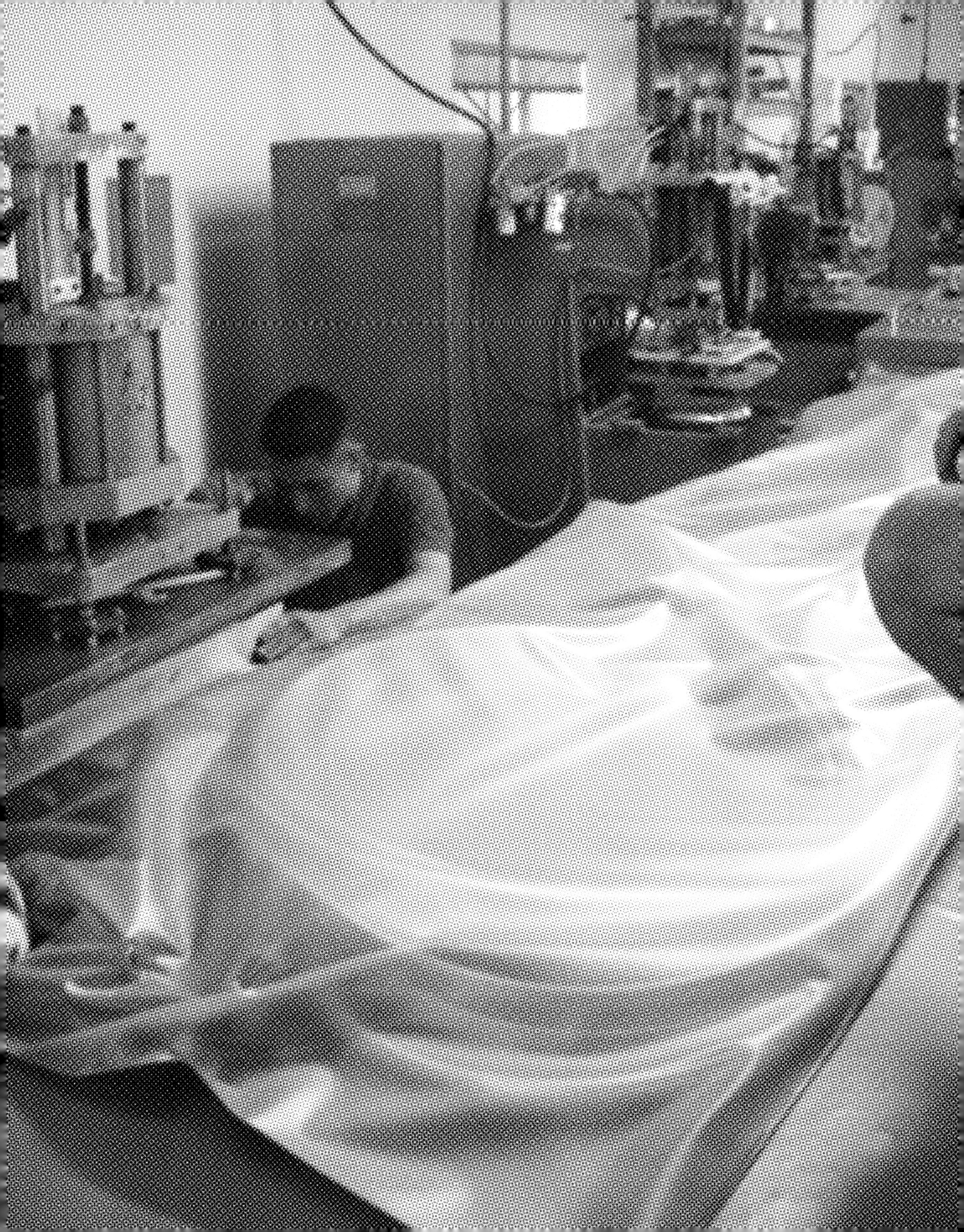

Chapter 7

Further Experiments

Prefabricated structures can be as solidly fixed and heavy-duty as any site-built structure. Most mainstream manufacturers take particular pains to demonstrate that prefabricated buildings may be as ordinary and everyday as any other type of building. In many minds, there has been the conventional understanding that prefabrication is necessarily cheap and repetitive, and thus the knee-jerk inclination for many manufacturers is to design products based on the objective of overcoming any sense of factory production, modularity, or portability, rather than celebrating these qualities. While many may instinctively dismiss factory production as cheap and plastic, there is still an aspect of prefabrication that also connotes and makes technically feasible far more imaginative, flexible, mobile, and otherwise experimental, celebratory, and exotic buildings.

This division of views regarding prefabricated buildings follows a salient historical division in culture generally: between those who welcome change, innovation, and progress, and those who fear it or at best find it vaguely unsettling. The dominant cultural voice in this division occasionally shifts from one side to the other, as histories of politics and culture clearly demonstrate. Although active in the debate and generally more progressive than society as a whole, architecture as a discipline is typically more conservative than the other arts, at least in recent centuries, and typically less inquisitive and experimental than the sciences, at least in organized ways. The construction industry, to the extent that it is most closely aligned with business rather than engineering and technical practice, is far more conservative still, primarily basing its decisions and planning on appeal to the largest number of customers and desiring to offend nobody. In a for-profit construction economy, in which land and buildings represent a primary collateral asset for the economy as a whole, the entire structure of real estate, banking, government regulation, manufacturing, construction, and individual business and homeowner finances conspires to preclude any particularly radical or experimental deviation from design and construction norms. The outlook and immediate cultural atmosphere of society may rise and fall with respect to the celebration of change and innovation, but as the largest and arguably most rooted and cumbersome pillar of the world economy, construction responds very slowly, so that at moments of seemingly great opportunity, as in California during the 1950s and '60s with respect to prefabrication, for example, the mainstream of construction cannot broadly react until long past the peak of any particular moment of opportunity and enthusiasm. We have already

Prairie Ladder: Weather Station. Perforated aluminum skin being lifted onto welded steel frame. Prairie Ladder is a series of landscape interpretive structures built for the Connemara Conservancy near Dallas, Texas, in collaboration with sculptor Cameron Schoepp.

discussed in broader terms the importance of digging into the construction industry at its complicated, tangled roots in order to meaningfully engage with any actual opportunity for structural progress. Although within this industry context the prototype building projects that we have discussed in this book are creative in ambition and substantively experimental, comprehensive design projects, realistically expecting approval from clients, financiers, and regulators can be frustratingly conservative affairs relative to more narrowly focused experiments, proposals, and studies that escape the bounds of satisfying every demand of permanent construction. Occasional breakthrough projects occur in architecture when talented and intensely focused architects create or happily encounter extraordinary opportunities with adequate financing and equally ambitious clients, communities, and project colleagues. Such breakthrough events do not emerge from a creative vacuum however, and in order for great events in architecture and construction to occasionally occur, there must be a strong and pervasive subculture of architectural ambition, inquiry, and investigation always pushing along in large and small ways, as opportunity and energetic engagement will allow, within schools, within firms, and in independent practices. Just as laboratories of pure and applied science and engineering are essential to the progress of knowledge and technological innovation in many fields, the independent creative inquiry of architects within experimental drawing, writing, and physical production drives the engine of progress in architecture. With notable exceptions in Japan and France in particular, and in some British and American research universities, there is relatively little basic construction research on the scale pursued in most other fields. Unlike other major industries, there is less consolidation and fewer immense companies with resources to devote to technical research. Unlike electronics or pharmaceutical industries, for example, construction competition is based largely on cost advantage rather than on having any perceivable technological edge. This may slowly change, particularly as issues of resource efficiency and sustainability increasingly require more sophisticated building solutions, but, for the most part, innovation in architecture and construction relies on the passionate, personal commitment of small groups of creative architects or small-scale building-craft or technology enthusiasts—in passive solar construction, or rammed-earth building, or prefabricated housing, for instance—to push forward ideas and opportunities in the built environment. The subculture of architecture inquiry pursued through design competitions, gallery installations, personal creative work, university design studios, and dedicated, critical experimentation within independent practice are the essential research well from which breakthrough buildings and ideas occasionally flow. To maintain this essential culture, a large group of architects must do more than practice in commissioned work or teach students in existing structures and technologies. The culture of inquiry and creative research must be continually engaged and supported. Most often, creative opportunities for experimentation and learning are not offered or readily available, and must be invented and painstakingly built with colleagues, collaborators from other fields, and with the next generation of architects in school. Although it would be nice to think that major corporations, governments, large architecture firms, and university research labs are all hard at work inventing new opportunities and ideas for the construction industry, the real work most often happens in smaller, individual ways, and these small projects support the occasional big event.

Further Experiments

Prairie Ladder: Earth Plane Sky Barge. Barge element is bolted onto rotating bearing plate at top of ladder.

This chapter draws together a number of such small, creative investigations pursued independently of commissioned work or within commissioned work freed of more typical project requirements, allowing deeper experimentation within a limited range of focus. Many of these projects are undertaken with students, others in collaboration with other artists or with creative and ambitious businesspeople. They are not the comprehensive building proposals of the previous chapters but offer potential insights into many issues of prefabrication and construction research that may subtly extend the boundaries or lay the groundwork for broader application.

With respect to industrial production and prefabrication, many architects during the 1960s in particular experimented with mobile structures, inflatables, plastics, and industrial processes conceived for mass production. Among the most intriguing were the fantastic works of the British pop architects working under the name Archigram, who designed balloon-borne cities and all manner of robotic, inflatable, and pop-industrial structures. During the past thirty years architect and industrial designer Gaetano Pesce has produced a remarkable body of work including buildings, furniture, and production objects experimenting with the capacity for serial uniqueness and random chance within tightly controlled mass-production processes. Much of his work has been in exciting new materials of the time, largely resins, plastics, and foams. His experimental work has been instrumental in engendering new creativity within furniture and industrial design, and his long-term approach to material investigations has now become a central theme in the vanguard of creative architecture. We first had the opportunity to work with Pesce as graduate students in the 1980s when we built one of his projects for an exhibition at Harvard, and we have been much influenced by his approach to investigating common production technologies and reworking them slightly to suit new creative purposes, while still producing functional, mass-produceable, and sometimes marketable works.

While it can be so difficult to effect quick change in many areas of building construction, there are always opportunities to experiment in smaller projects on the margins. In addition to our larger building design projects, we do a great deal of collaborative work with other artists on performance projects, installations, and public artworks. While prefabrication is not a driving conceptual issue in any of these projects, it is invariably essential as a pragmatic tool, and we often learn a good deal about flexible, modular construction in these smaller projects that can be applied to building construction, as much as our construction background contributes conceptually and practically to these art projects and experimental structures.

One of the opportunities we now have for experimental construction work is in directing academic research and design studio projects with students and other faculty members. These projects attempt a quadruple intention: to explore a spatial concept; to investigate a particular set of technical processes; to offer some pragmatic social contribution; and to offer students the opportunity to engage physically with exactly this sort of multivalent intellectual endeavor that should be at the heart of every architectural project.

We have built numerous large and small experimental projects with collaborating artists and with art and architecture students. Some are complete buildings, some temporary urban art events or gallery installations, and some have been primarily site-based projects as heavy and rooted as rammed earth. We have experimented with a range of inflatable, airborne, waterborne, suspended, and temporarily mounted mobile structures; temporary public art installations and event structures; stage sets and performance collaborations with musicians, choreographers, and film-makers; prototype mobile dwelling units and emergency shelters; highly practical mobile construction offices, construction shelters, and fencing; suspended forest canopy research structures; and even semi-permanent urban plaza walls. Most of these projects have been highly prefabricated experiments with new materials, new processes, and new program intentions—for instance, sucking up carbon dioxide from building exhausts and returning the carbon to the earth via living algae walls before it can escape into the atmosphere and deplete the ozone, as in our Thick-SkinGasHuffPhytoCurtain, a suspended steel cable, plastic bubble, and biomass structure built with students at the Tulane University School of Architecture in New Orleans, Louisiana; or offering an inflatable outdoor, solar-heated amphitheater, as in the Hot White Orange, built with students at the University of California, Berkeley. While each of the projects on the following pages represents specific highlighted areas of experimentation rather than comprehensive building proposals, each represents elements of programmatic or technical concerns that may continue to energize and expand the range and richness of future environments, harnessing prefabrication to achieve far greater and sometimes even irrational richness of experience, afforded through the economies and research logic of rationalized production.

S.W.E.L.L. Interior view from ramp entry. As a visitor ascends the ramp, the visitor's weight causes the ramp to descend, drawing the suspension cables together and swelling the interior volume. The more people who enter, the greater the weight on the ramp, and the larger the volume becomes.

S.W.E.L.L.

Berkeley, California / 2003

S.W.E.L.L. was designed and constructed as an intensive thirty-day collaboration with fifteen graduate architecture students at the University of California, Berkeley. The final stage of a two-project elective studio course, the construction of this dynamically articulated folding tube of inhabitable fabric, steel, and sticks followed a more conventional studio project in which each student worked independently on the design and construction detailing of a six-person, semi-mobile mountain shelter situated on a high Sierra glacier. Following some mountain travel and design experiments undertaken in small groups, the previous individual projects had yielded a range of innovative skin and structural system ideas. It was intended that this preliminary work should offer raw material and shared understandings of each student's design approach as the group entered into this final project as a large collaborative team.

The educational intention was two-fold: to provide a structured opportunity for hands-on construction of a creative work of the students' own design and to provide an intensive experience in the complex dynamics of collaborative design and construction administration. Most often in design studios, the primary focus is upon the development of a personally directed and deeply thoughtful design process directed toward the origination of well-structured ideas. Frequently this studio emphasis does not extend to the ambition of developing a design to the point of fully understanding and describing its constructability. This is perfectly fine in school, as no project should be built without first developing a deeply thoughtful approach to the underlying structure of ideas essential to a good work of architecture. It is very difficult to develop and describe a strong design concept as a well-integrated and deeply considered body of ideas. Teaching the conceptual foundations for strong architectural thought is the most difficult and fundamental obligation of a school of architecture, more so than completing comprehensive, technically detailed building designs.

At the same time, many architects have a fundamental creative desire to directly engage in making things. Many students share this interest and are increasingly enthusiastic to join studios that are bent on turning out a constructed product. This is exciting. One realistic approach to this ambition is to diminish the depth of the design investigation in order to have sufficient time and focus to spend on the later stages of detailing and construction process, essentially inverting the typical studio emphasis. To a certain extent this narrowing or switching of focus is inevitable in a one semester studio intending to build something. And yet this is exactly the difficult balance of time and focus faced by all architects engaged in practice. It is acceptable in school to minimize the emphasis on later stages of the design process leading up to and through construction in order to strongly develop fundamental spatial and conceptual skills. Once something is to be built, particularly in the practice of adding long-term elements of construction into the world, it is no longer acceptable to truncate any portion of the essential processes of design that may be expected to lead to a good work of architecture. A strong body of ideas must be conceived and developed, and this foundation needs to be carried with great care into and throughout the process of construction.

A construction project in school should emphasize the same high standards of design investigation that should be expected in practice. The S.W.E.L.L. students were not asked to proceed into construction with an abbreviated or prepackaged-by-the-instructor conceptual foundation, program, or design process. Instead, the instructor and students began together at the beginning, with only the clarity of intention that within the next thirty days we would conceive of something of value, develop the idea into constructability, figure out how to pay for it, obtain permission for the construction to have a site, and then build it; and finally, we would do all of this in cumbersome coequal collaboration as a large team of fifteen designers.

At first glance, S.W.E.L.L. resembles an undulating wooden basket, stretched with fabric, carrying long shadows and points of light. Stepping onto the slatted wooden bridge, a visitor traveling upward into the constricted tube will increasingly weigh down the suspended structure, gently lowering the bridge to the ground and causing the tube itself to elongate and swell into a voluminous, occupiable space. The counter-weighted bridge hangs from a larger "scissor" structure that folds downward under the application of the occupant's weight, simultaneously dragging the finer-grained scissor structure of the wooden-ribbed, fabric-skinned enclosure tube into an expansive arc. This action expands the interior space from a relatively minimal aperture into a translucent, lozenge-shaped room more than sixteen feet in diameter. When a visitor continues through the space, descends the bridge on the opposite end, and leaves the structure empty, S.W.E.L.L. pulls itself back to its resting position, raising the bridge and shrinking back to its most compact size. Elastic nylon fabric panels adjust to the changing shape, gathering and projecting light, shadow, and the silhouettes of bodies multiplied and floating through space.

Prefabricated architecture has frequently drawn inspiration from the ingenious traditions of tent-making. While this project is not specifically developed as a dwelling, it falls within the tradition of mobile, flexible architecture that may be compactly stowed, transported, and then expanded outward to inhabitable scale. The folding, scissorlike structural cage of S.W.E.L.L. folds down upon itself without any unbolting to become a long, thin, flexible tube only about sixteen inches (40.6 cm) in diameter, or just over 8% of the fully expanded maximum diameter. Fabricated by the students in their eighth-floor architecture studio, this remarkably light and mobile structure was squished down to its minimum diameter and maximum length, of about eighty feet (24.38 m), and then carried as a thirty-legged dragon under the arms of fifteen students as they snaked and spiraled down the twisting fire stair to the ground. In discussions with developers who are interested in affordable rooftop penthouse additions on tall buildings in New York beyond the reasonable reach of mobile cranes, this experience has greatly influenced our thinking about the possibility for lightweight, highly compressible structural and enclosure systems that might pack into a standard passenger elevator and then expand into a habitable dwelling upon arrival at the roof.

Further Experiments

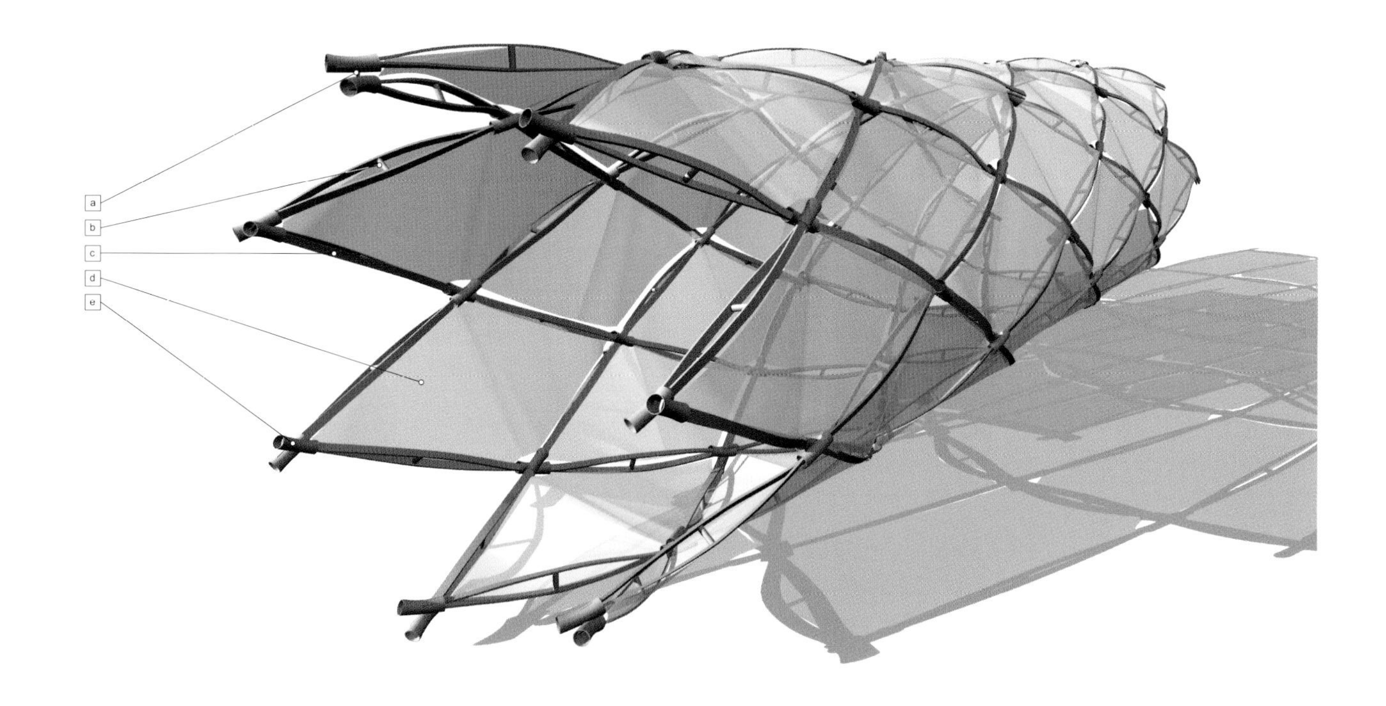

Top: Aerial view of S.W.E.L.L.
Bottom: Perspective drawing of S.W.E.L.L. showing elements of construction
Parts List:
a. Rotating pin connection
b. .5 in. conduit spacer
c. .5 x 2 x 48 in. ash structural skin frame
d. Nylon/spandex fabric skin
e. High strength rubber tubing connectors

S.W.E.L.L. models, testing mock-ups, and full-scale details under construction. Note compact form of structure when fully collapsed for transport.

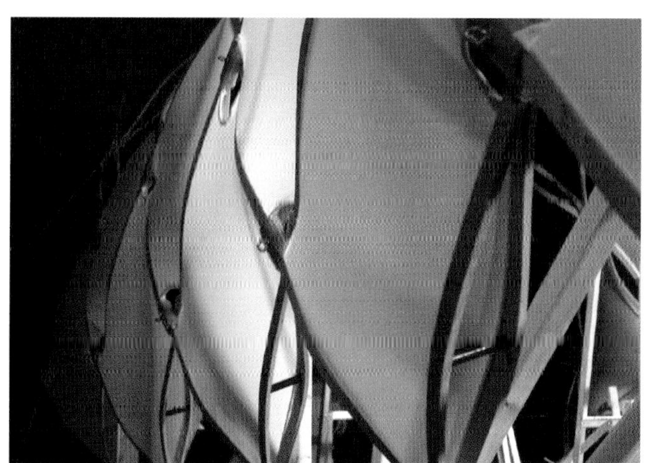

S.W.E.L.L. models, testing mock-ups, and full-scale details under construction. Note compact form of structure when fully collapsed for transport.

Clackety-Yak

Honolulu, Hawaii / 2001

This project is a large-scale steel cable and bamboo construction project designed and built with students of the School of Architecture at the University of Hawaii, Manoa. The structure was erected in the school's central courtyard and dealt with weather-related experience and social interaction affected by ambient sound, light, and shadow, as well as with essential issues of prefabrication, detailing, and construction process. This project developed in part from our particular interest in the highly standardized and rationally prefabricated systems available for cable suspension and wire rigging, a common structural system in many of our experimental projects. The cable suspension systems developed in the Clackety-Yak project have since been applied in fast-erection, prefabricated projects such as our Fire Island Pines Town Center design.

Wire rigging in the construction industry is primarily used for temporary rigging for crane lifting of slings of material or prefabricated panels. This is a well-standardized system of parts and cables that can be assembled quickly by a specifically skilled and common labor force using pre-engineered, highly tested, and well-documented components available as standard materials in large and small cities around the world. All of this, in addition to its light weight and compact storage and shipping parameters, makes wire rigging an especially cost-effective and fruitful area of consideration for affordable, prefabricated construction.

Clackety-Yak is an approximately 100 x 150 foot (30.48 x 45.72 m) grid of wire cables suspended at a height of about thirty-five feet above the architecture school courtyard, the singularly bleak and unshaded heart of the school, linking all of the classrooms and offices but offering no place of respite from the searing daytime heat and glare. This oddly blank outdoor stucco box separated the school from the otherwise lush tropical world surrounding it, and one could never scurry head-down across this blazing beige blankness without wondering what was happening in the real world of color and breezes outside. It was a constant point of discussion around the school, considering what was to be done about this space, and the students began to design a prosthetic means of registering the rich phenomena of the outside world back into their daily lives. Intriguingly, at approximately ten-foot (3.05 m) intervals around the courtyard rim were steel bolt-eyes embedded in the concrete. Clearly there had been some intention that something should happen up there, and the possibility of a steel cable structure was the logical solution, though there was no recorded documentation of what was originally intended, or what structural capacity these bolt-eyes were designed for. The students contacted the building's architect, who referred them on to the original engineer, who reluctantly agreed to meet with the class and appraise their proposal in light of his knowledge of the alluring bolt-eyes.

The students documented their proposal with elaborate and comprehensive drawings, models, and film detailing their intended construction system and their interest in registering the sun and wind phenomena of gently waving shadows down from the sky and into their courtyard. The proposal was simple and elegant, with a field of uprooted, upside-down bamboo trees suspended from the cable grid with one third of their length jutting into the sun and wind above the courtyard, and two thirds of the trunk and bushy leaf hanging down just above head level, waving in gentle circles generated by the wind blowing through the upper stalks, and throwing dancing shadows onto the suddenly occupiable courtyard amid the murmuring brush sounds of the bamboo and the distant clackety-clack of the occasionally colliding roots up in the air. What a great presentation, we thought, but the engineer turned out to be a grump, and proceeded to explain to the students how silly and ridiculous and unlikely their project was, and why the school would never allow them to do it anyway. Aside from the crime of schooling children in unimaginative and uninspiring buildings and landscapes, there is little that can generate anger like seeing the imagination of students stepped upon, and there are more than a few projects we have pushed forward from exactly this impetus. With the gauntlet thrown down, all that was left was to figure out how to pay for it, where to find a hillside of bamboo trees volunteered for harvest, and how to get twelve hardworking students safely up into the air.

Clackety-Yak under construction. Primary cable suspension structure with suspended upside-down bamboo trees draws the wind motion from above down into enclosed courtyard

Further Experiments

Fire Island Pines Town Center

Fire Island Pines, New York / 2004

The concept of prefabrication is most often associated with the construction of projects that will remain permanent once fixed in place, but it is equally appropriate for ephemeral constructions that require similar capabilities. In a recent design proposal for the rehabilitation of an existing town center on Fire Island in New York, we have selected a combination of fixed and changeable prefabricated components that will allow the renovation of the facilities to be incrementally phased while they are in constant use.

An open web of .25 inch (6.35 mm) diameter stainless steel aircraft cable will be stretched taut between existing buildings and across the entire site atop a new grid of thin, twelve-foot-high (3.66 m) ironwood columns planted strategically according to the existing and proposed logic of the site. An inexpensive, quick, easily predictable, and largely offsite-fabricated construction system, this first stage of cable construction will serve as an armature for the immediate transformation of the site within a single season. With this web armature and column field in place, translucent white nylon curtains will be hung vertically from each line of the overhead cable grid, dividing the site into an interwoven lattice palace of mysterious passages, peek-a-boo private chambers, great outdoor halls, and swimming pavilions.

In the imagination of visitors sailing in toward the island, the curtained complex will first appear on the horizon as a great billowing windjammer sailing forth from the harbor. Passing into the complex full of flickering shadows, flashes of color, and distant snatches of laughter and murmuring conversation, the visitor will next feel that he or she has entered an Arabian tent city, a gossamer Casbah, thick flight of butterflies, or perhaps the memory of a childhood theater created from nothing but white sheets strung from a backyard clothesline.

As a construction process, the existing buildings will remain at first much the same, but over time, from season to season, the buildings and outdoor restaurant decks will be remodeled with offsite, prefabricated modules that may be quickly installed on site during the off-season without disrupting the affected businesses. The initial clothing of the site in cable and curtains will already provide an entirely new and highly memorable image and experience, while the incremental addition of prefabricated additions and building repalcements will continue to add new surprises and new comforts for returning visitors to discover within this dramatic cloak of spatial continuity.

While gathering the existing hodge-podge complex of the town center into an imageable continuity, at the same time this gauzy skin offers a continually fresh and constantly shifting quality of light, color, and spatial experience, varying with the weather and with daily and seasonal cycles of human mood and density. The town of Fire Island Pines has a primary constituency of weekend homeowners, their guests, and occasional returning visitors. To engage this group, the new design offers ever-changing experiences and events for the regulars, including the continual attraction of interesting casual visitors drawn here by the frisson of something new, delightful, and curious in its never quite tangible mutability. To build such a place at reasonable cost requires an approach of creative theater, harnessing not only the ephemeral imagery of the theater but also the highly developed practicality, scheduling efficiency, and cost-effectiveness of the stage. With some experience as stage carpenters and as designers of objects and spaces for theater and exhibition, as well as a good deal of experience in the scheduling of building construction and prefabrication process, we are very interested in the technical mechanics and practicalities of sequenced projects such as this, and also in the scripting of seasonal architectural and social theater in unique places and communities. The potential for architects to more broadly apply prefabrication and modularity not only to the conception of the final object, but more powerfully still to the design of sequenced events and the social theater of construction and change over time, offers the most exciting avenues for the harnessing of existing industrial systems to enliven urban architecture and bring the popular design concept of prefabrication into a broader understanding of the modern role of urban architecture in relation to popular, staged media events and the mass theater of the city itself.

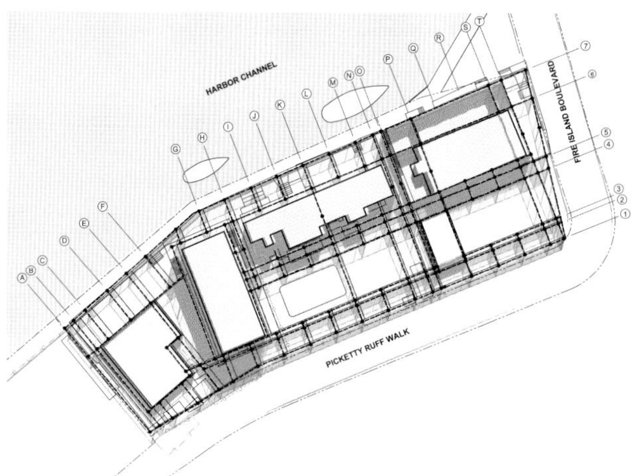

Aerial view and site plan

OPPOSITE
Model views of curtain system integrating the existing hodge-podge of buildings

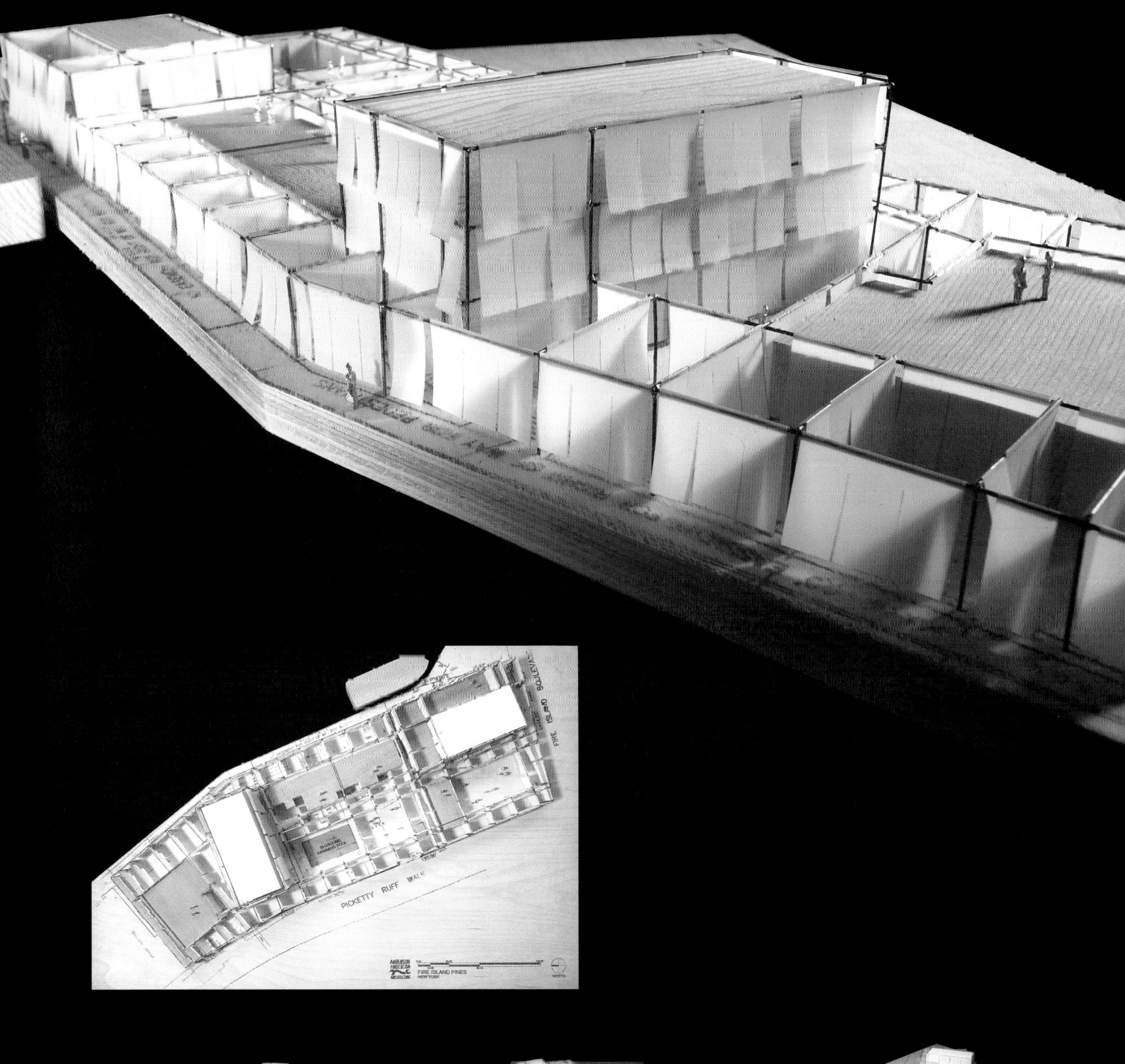

FIRE ISLAND BOULEVARD
IN GROUND SWIMMING POOL
PICKETTY RUFF WALK
ANDERSON ANDERSON ARCHITECTURE
FIRE ISLAND PINES
NEW YORK
NORTH

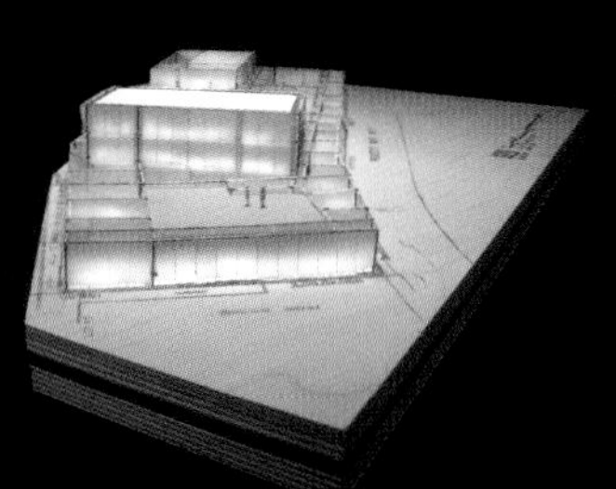

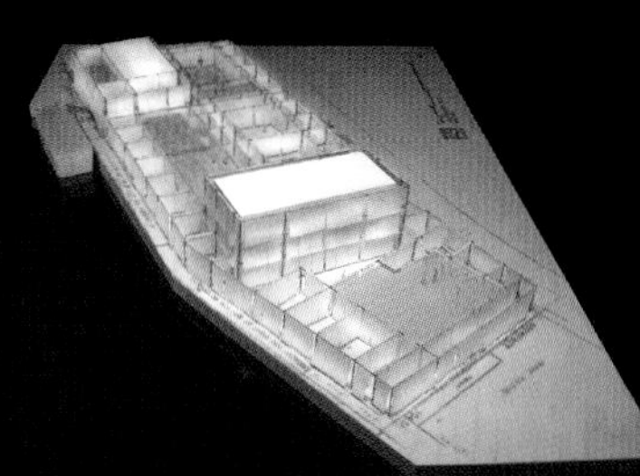

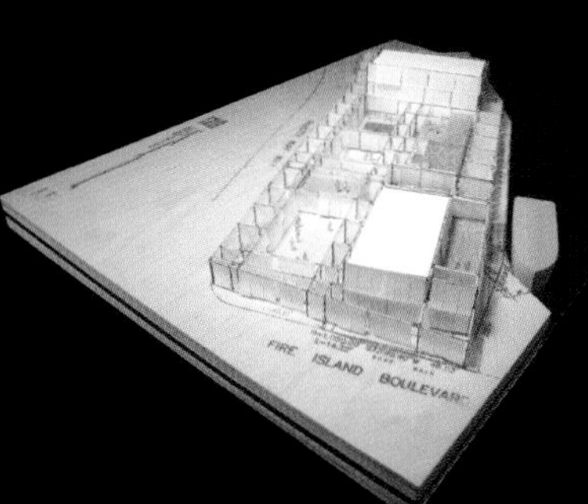

FIRE ISLAND BOULEVARD

Fire Island Pines Town Center

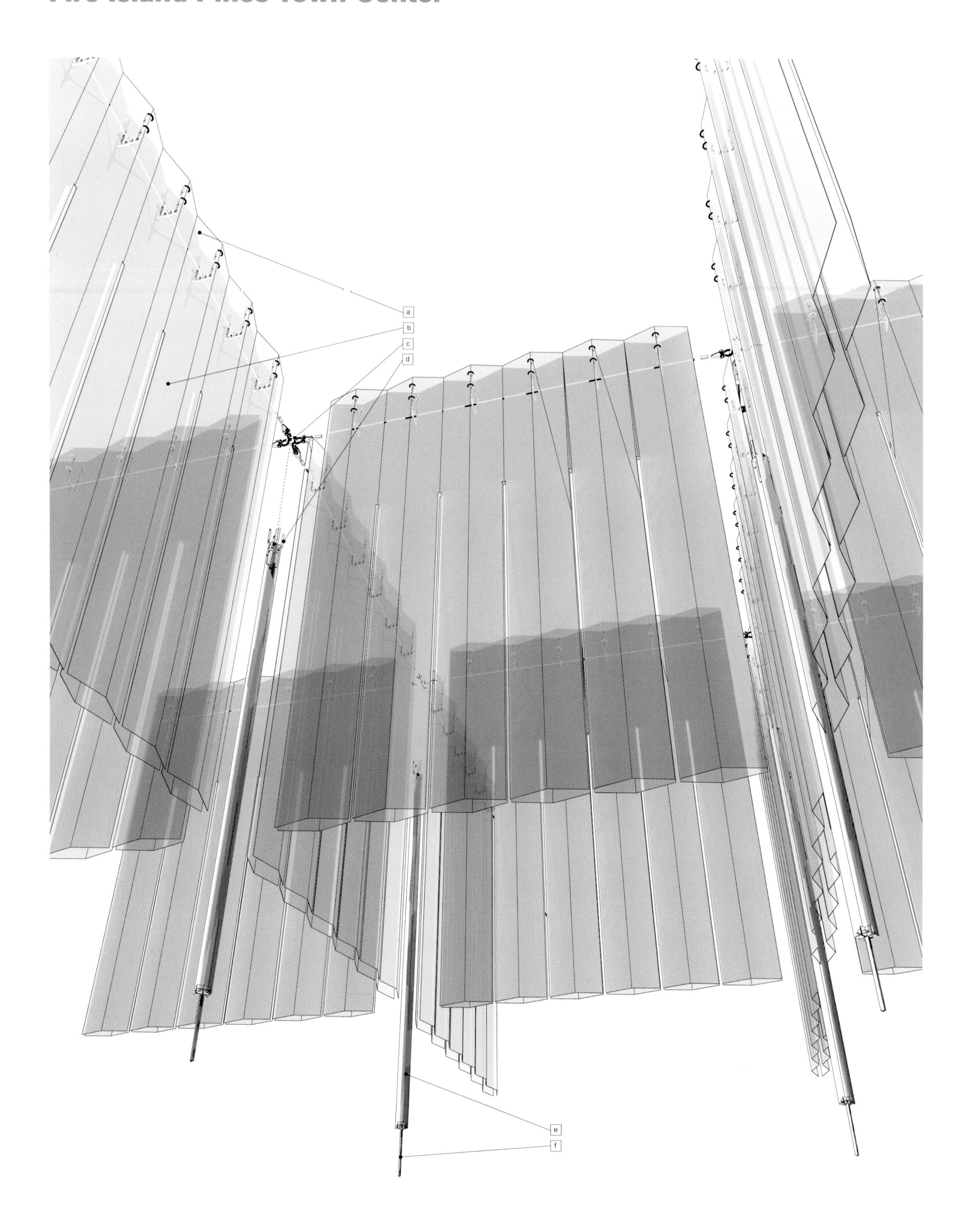

Further Experiments

OPPOSITE
Perspective view
from below
Parts List:
a. Metal cable
b. Translucent fabric curtain
c. Rigging assembly
d. Rigging support assembly
e. Glue laminated post
f. [illegible] pipe

Below:
Perspective view from above
Parts List:
a. .5 in. diameter stainless cable and fittings
b. .63 in. steel pipe saddle
c. 1 in. steel pipe
d. .25 in. steel plate
e. 3.5 x 3.5 in. glue-laminated timber column
f. Translucent fabric curtain

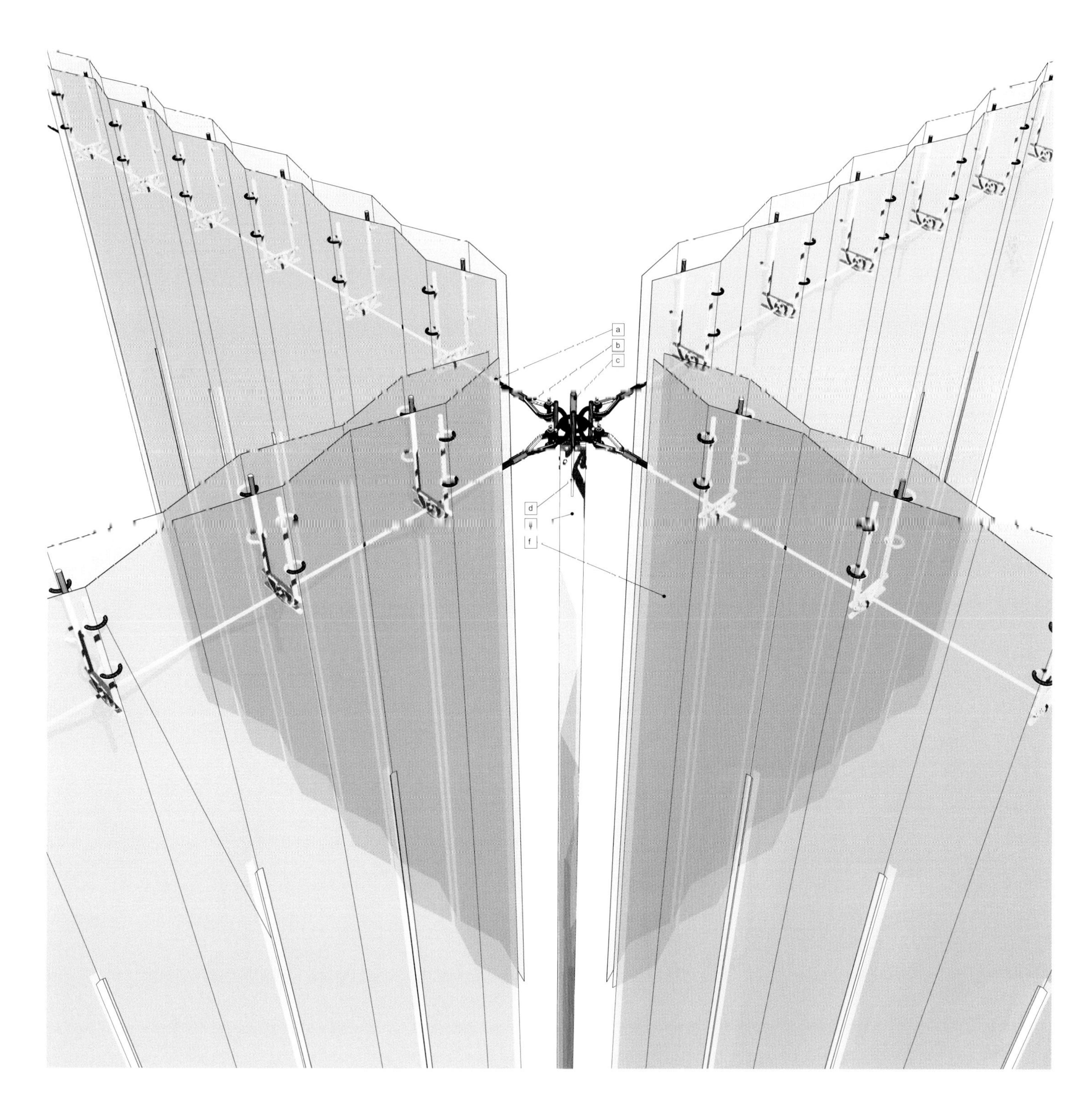

Hot White Orange

Berkeley, California / 2005

Hot White Orange is a portable, solar hot water–heated outdoor amphitheater designed and built with architecture students at the University of California, Berkeley. The intention was to explore design and production communication and fabrication processes within a theoretical investigation of sensual space-making and constructed experiential phenomena. The group included twenty-two students divided into a number of separate teams responsible for design and production of specific components of the project.

To maintain integration and coordination, the form and general approach was established in one group meeting at the beginning of the project. To minimize gross-scale design negotiation, the formal structure was established as precisely that of an orange. With this parameter, a simple three-dimensional digital model was created as the common base geometry for the project, within which all further design and fabrication issues would be negotiated. The size of the object was established as a ten foot (3.05 m) diameter sphere, the functional purpose of which would be the amphitheater seating of at least thirty people for viewing outdoor films. This program purpose generated all other functional criteria for weatherability and outdoor seating comfort.

As in many projects involving prefabrication, a great majority of the time was spent in designing, sourcing, and commissioning the fabrication of the many elements that would ultimately make up the completed work, and a dramatically short time in the actual assembly process. This stands in contrast to other design-build and hands-on projects that we and others have done together with students, where the concept of "making" concentrates on the physical part of the process. This also stands in contrast to standard onsite construction processes, and closely models the future of designing for and within manufacturing, with rapid installation on site.

As useful as it is to learn to work with tools and the job-site realities of the construction process, we focused here on the process of prototyping components through drawings, models, and mockups—using to the greatest extent possible the same CAD/CAM technologies that would be used for the full-scale fabrication—then documenting and communicating the information learned to the professional fabrication shops that would collaborate on design refinement and make the parts. The assembly process, once the many parts were delivered by or picked up from the different suppliers, was more about evaluating the proof of concept than it was about learning to put things together.

The project has a steel frame exoskeleton, air-filled bladder internal structure, hydronic heating coils supplied by pump from a satellite solar heating bladder and wrapped around water-filled thermal ballast blankets beneath the external vinyl skin. The orange fruit serves as both formal geometry paradigm as well as inspiration for the complex interior structure and mechanical systems of a living, pulsing, vascular bladder architecture. The project generated tremendous new insights into rich alternative worlds of fabrication technology involved in complex structural and material processes outside of, and yet readily accessible to, the more typical world of building construction.

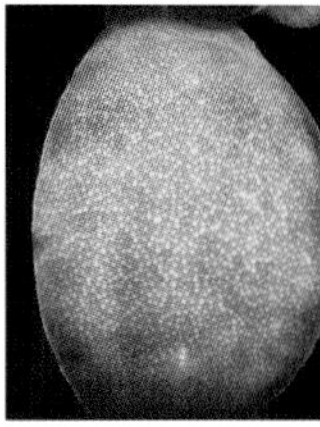

Early lighting study models with orange fruit and skin

OPPOSITE
Nighttime installation view of completed project

Hot White Orange

Top row: Steel frame drawings for CAD/CAM bending (left) and welding process (right)
Middle row: Mechanical plumbing systems (left) and installation of hydronic heating hose on fabric skin panels (right)

Bottom row: Elevation and plan views of steel frame in both open and closed positions; and Interior lighting system

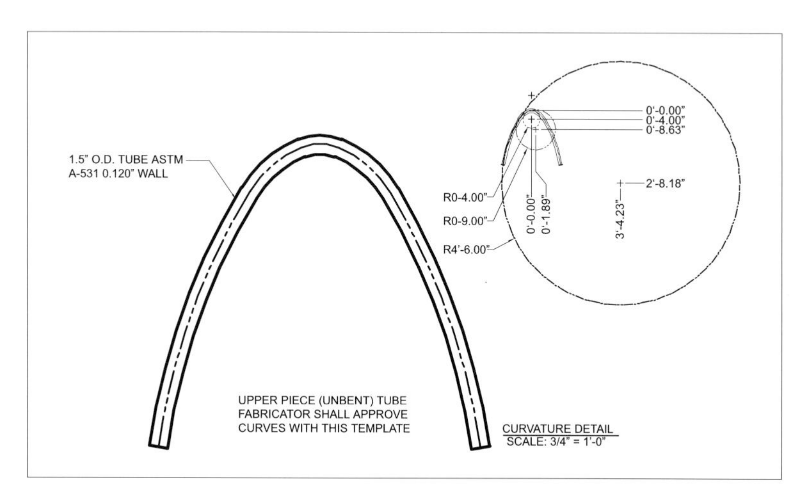

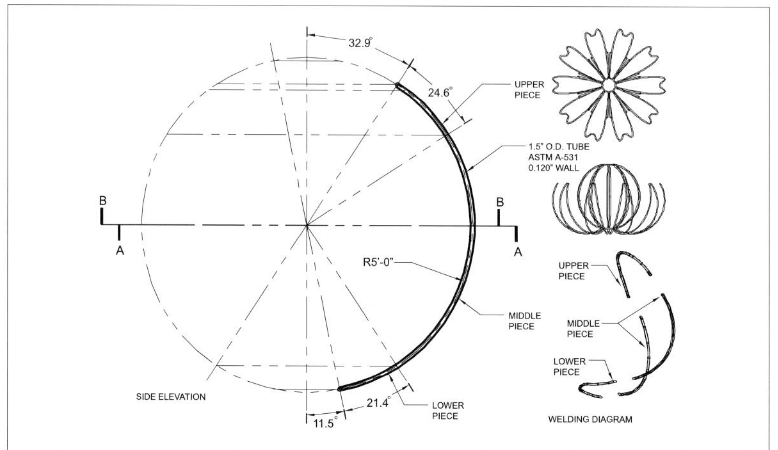

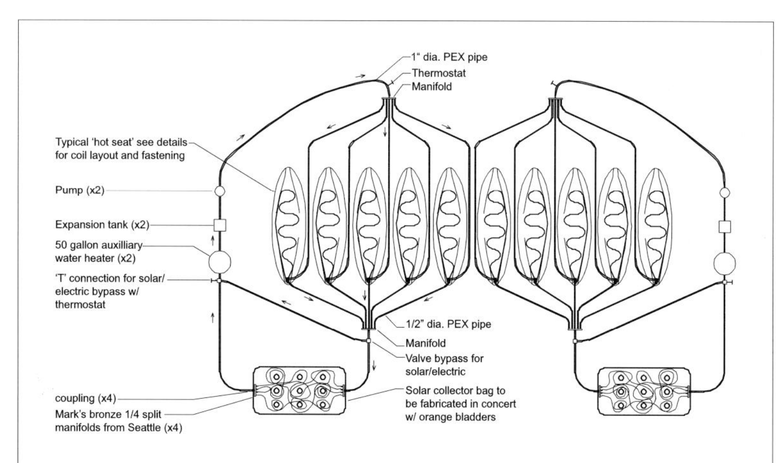

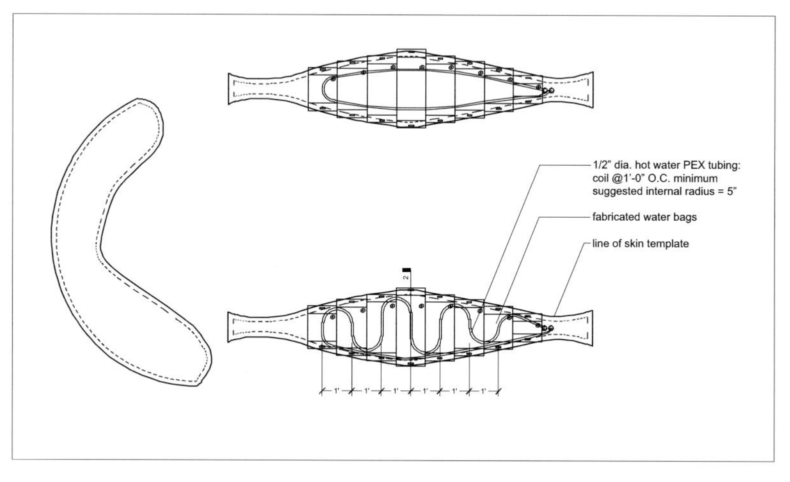

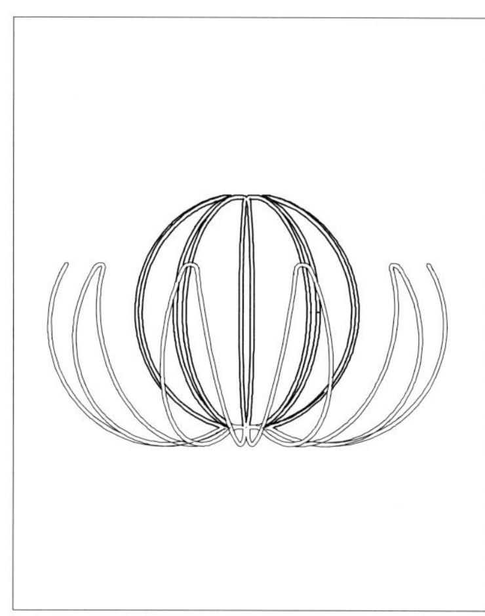

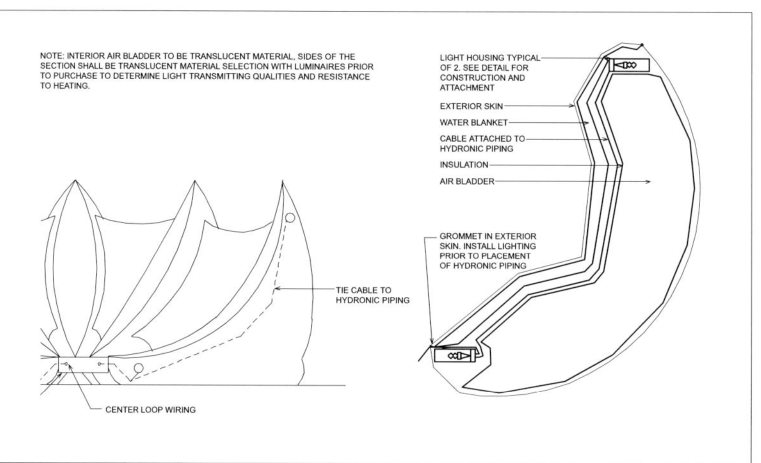

Further Experiments

Top: Construction document drawings showing relationship of bladder system to fabricated steel frame
Bottom: Construction documents detailing fabricator addresses

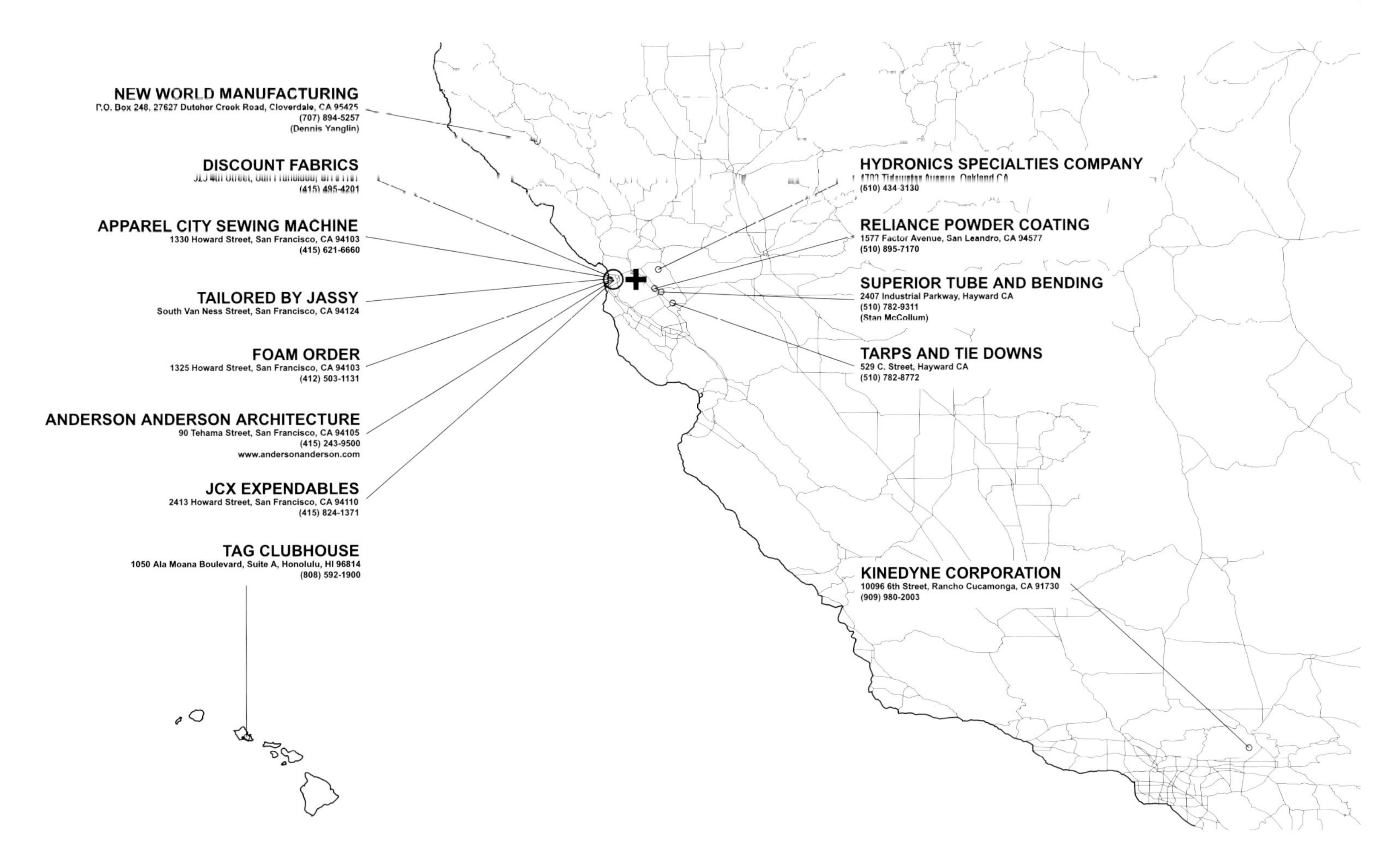

Hot White Orange

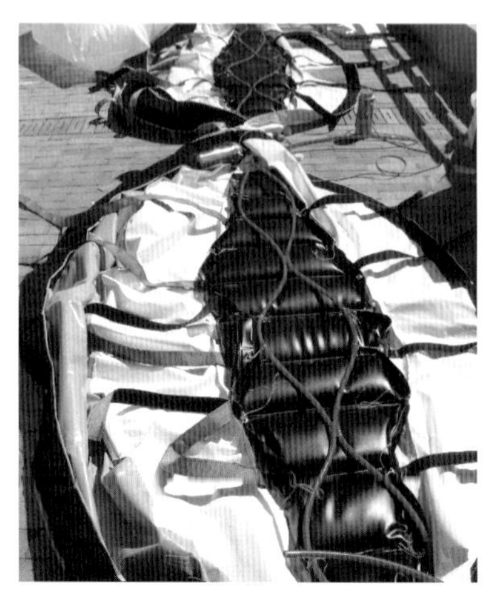

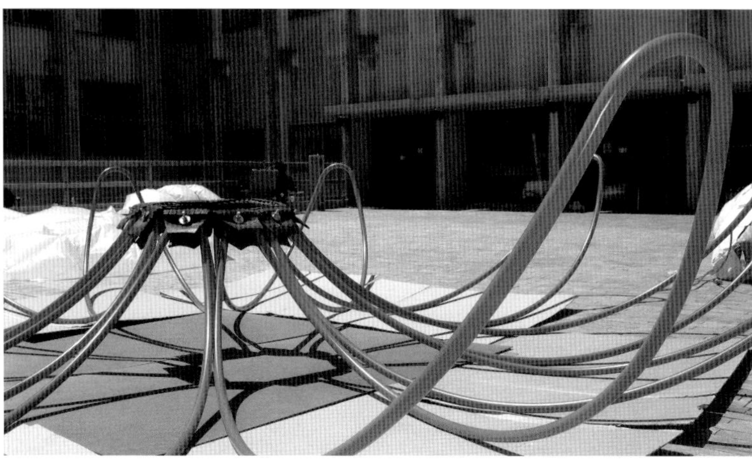

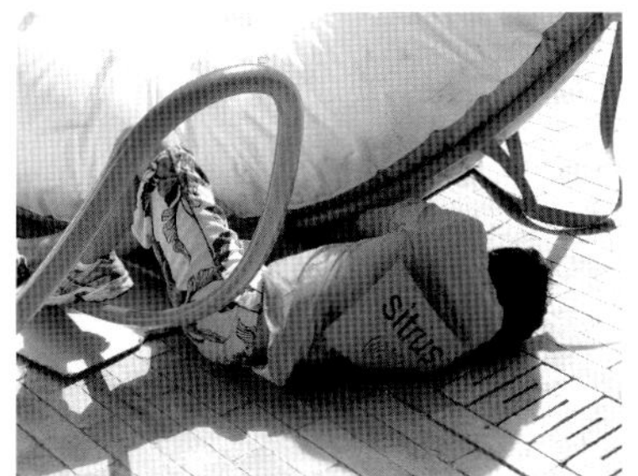

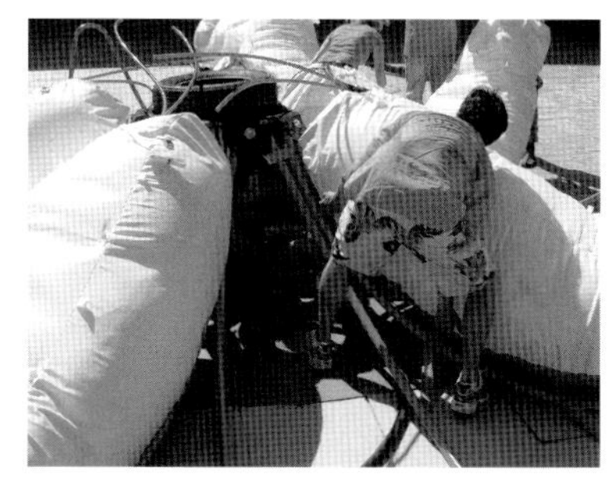

Hot White Orange under construction. Numerous layers of internal bladders, skins, straps, piping, and wiring are enclosed within the outer fabric and exposed steel frame.

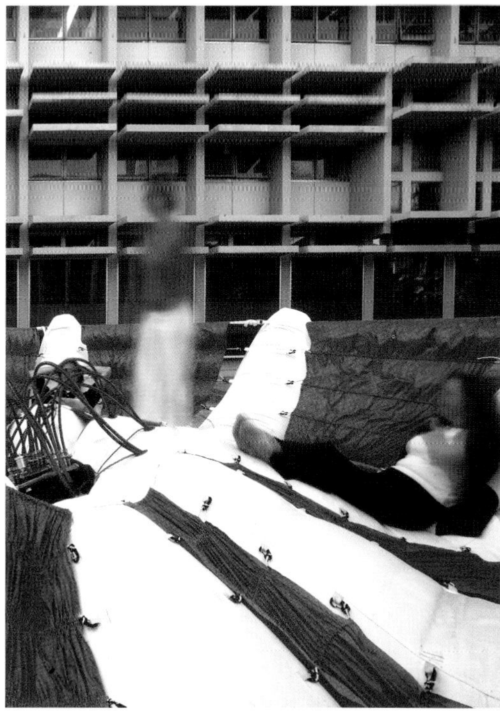

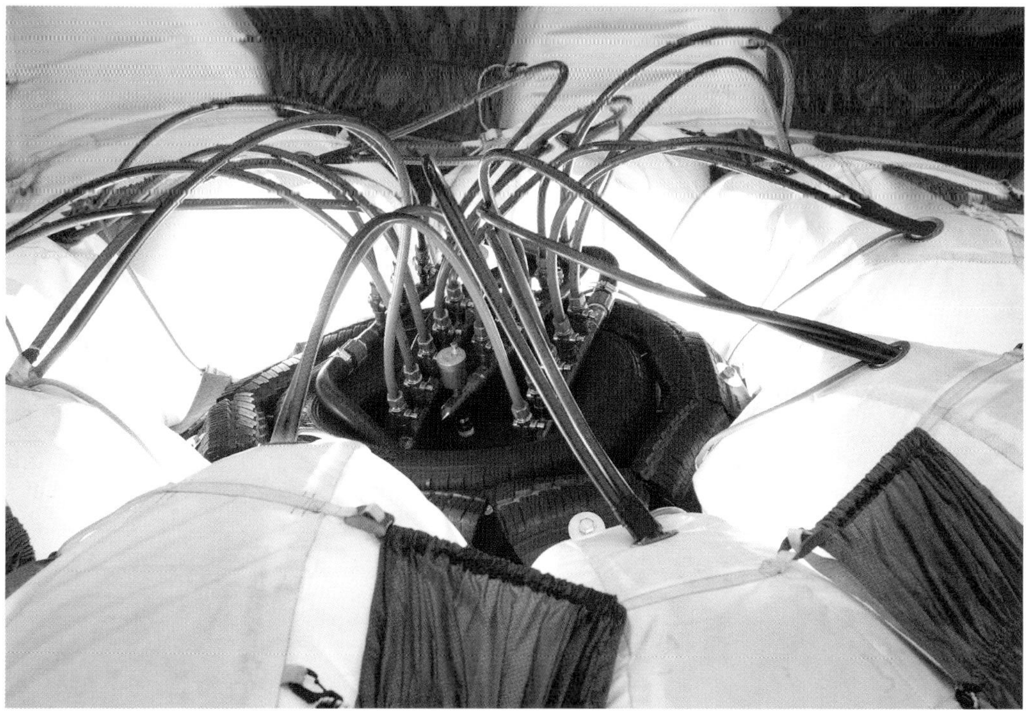

Completed Hot White Orange. Note central plumbing manifold that distributes solar-heated hot water into the interior heating bladders of each orange segment

Hot White Orange

Further Experiments

Nighttime view of solar-heated amphitheater in use

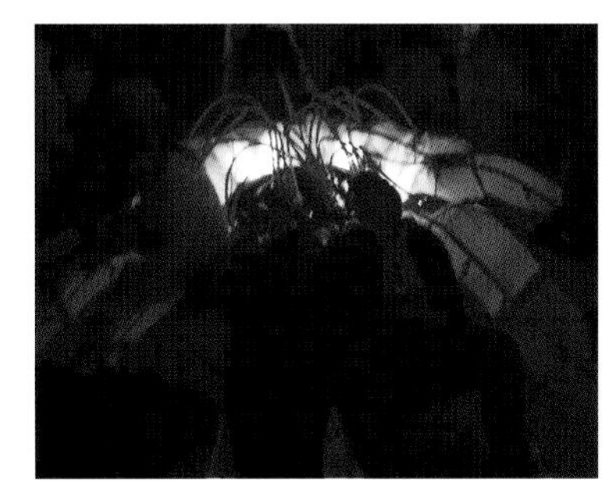

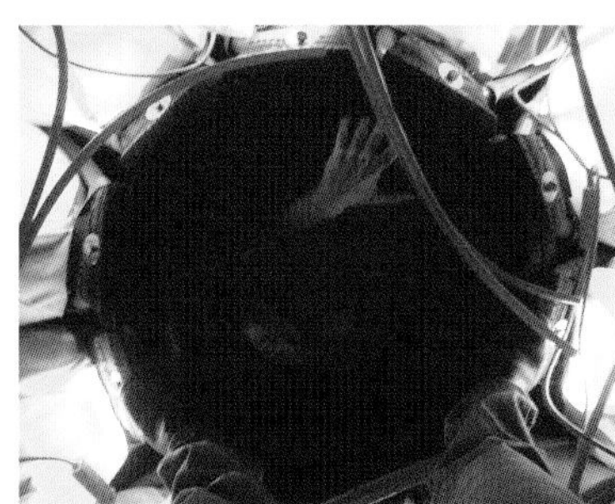

Repurposed Plastic Building Modules

Berkeley, California / 2003

As one in a series of experimental construction courses at the University of California, Berkeley, we researched the potential for modular construction assemblies to be developed from the encapsulation and reuse of otherwise nonrecyclable plastic products. As is frequently encountered as a considerable nuisance when attempting to recycle plastic containers, plastics are produced as a number of incompatible materials that cannot be intermixed in reprocessing, and many plastics cannot be recycled at all, at least not economically. For many people concerned with the environment and healthy living, plastics in general are the root of all evil. Frequently derived from nonrenewable petroleum resources and commonly associated with toxic processes and effects, there are many reasons to exercise caution and concerned restraint in designing for work in plastics. Still, plastics offer irreplaceable advantages in many construction situations, and on top of everything else, their potential for luminous translucence, vibrant color, and sensual tactility make them hard to forswear. We love plastic. Perhaps then as an act of atonement, we worked with our students to design new modular building products that would encapsulate nonrecyclable plastics into practical components, giving these seductively evil materials something useful to do, keeping them off the streets and out of the oceans and landfills.

Many intriguing proposals developed for products ranging from modular sound walls on freeways to interior partition walls, furniture, and glazing systems. Two products particularly relevant to building prefabrication are shown in these pages in more detail. The team of Victor Chu and Manuel Cordero-Alvarado experimented with polyethylene and polycarbonate juice bottles partially sliced open and folded into interlocking cellular structures that could be indefinitely linked three-dimensionally into strong wall elements with substantial insulative potential. Composed of variously colored, clear, opaque, and translucent bottles randomly acquired from campus dumpsters, these prototype walls developed incredible richness of color and luminosity. Polyethylene is frequently problematic as a construction material because it resists almost any method of glue fastening, limiting it to joining by strictly mechanical means such as bolts or screws. Further complicating the assembly process was the intermixing of bottles that were formally similar yet produced with very different chemical compositions—an even greater complication for recycling, and an important reason for finding an effective reuse strategy that did not require extensive sorting or identification of exact material composition. Although the color, clarity, and chemical composition of the bottles was quite varied, the general form and size (based on volume of the packaged product) and the bottle caps were relatively standardized. After experimenting with a variety of fasteners, it soon became clear that the best way to bolt the unfolded bottles together was through the use of their own threaded screw tops. We have experimented with using bottle tops as bolt fasteners in the past, and this system completed the bottle wall assemblies as a very strong, simple to produce system that made complete use of all the bottle qualities and parts and left nothing to be thrown away.

Approaching the project from a different direction, focusing on a tool process first and searching for the appropriate materials second, Mathew Stromberg developed a beautiful, high-strength modular partition wall and insulation/glazing system based on a composite sandwich vacuum-forming method, using three common, difficult to recycle materials: clear polystyrene sheets, closed cell polyisocyanurate insulation foam, and that bright orange polyethylene mesh typically used as temporary construction fencing and increasingly found half-buried on every empty lot and highway shoulder. A set of dies was fabricated to give structural shape and interlocking snap connections on the panel edges; a layer of pink insulation foam was sandwiched on both sides by a layer of the orange mesh serving as tensile structure, followed by sheets of polystyrene serving as finish surfaces and structural skin. Using a vacuum forming machine to heat up and then suck this sandwich of plastics against the dies, an extremely rigid, unbreakable, and relatively well-insulating modular structural panel was created. Translucent and brilliantly colored, the panels can be easily snap-connected together to create large, flat, or gently curving wall surfaces. Using some of the most mundane and environmentally problematic materials, an inexpensive product was developed that simultaneously encapsulated typical landfill mass into a productive new use. It is exactly this sort of creative invention and problem solving—allied with rich spatial, conceptual, and formal imagination—that may facilitate a comprehensive agenda for architecture and industry, leading to an overall ecology of invention and production that seeks to integrate and resolve the full complexity of a rationally structured, experientially rich world of humane construction.

Detail views of recycled plastic building modules
Top three rows: Completed and construction views of interlocking wall system by Mathew Stromberg
Bottom two rows: Views of plastic bottle structural panel by Victor Chu and Manuel Cordero-Alvarado

Further Experiments

Afterword

Reuse, Recycle, Prefabricate

The use of recycled materials in construction offers the potential to lower raw material processing and transportation costs with subsequent reductions in building costs, while at the same time contributing to a healthier environment and greater economy at many levels of production, transport, consumption, and disposal. The recycling of construction waste at manufacturing plants and at building job sites contributes to this return of raw materials into the remanufacturing cycle and, just as importantly, contributes to worksite efficiency by rationalizing procedures and reducing costs for waste stockpiling, transport, and dis-posal—some of the most vexing strategic problems in production staging and work flow, especially for tight urban work sites and efficiently sized factories. One of the strongest arguments for the maximization of offsite fabrication is the vastly larger efficiencies that may be brought to stockpiling, recycling, energy use, and waste reduction in specialized and continually reoptimizing factory procedures, offering symbiotic and therefore powerful incentives in the form of both economic and environmental savings.

As recycling efficiency has gained increasing attention in material production and waste management for the construction industry, increased attention is also directed toward the downstream recyclability of obsolete buildings themselves and to the potential for the wholesale reuse of construction materials and assemblies with minimal reprocessing. This objective arises from the recognition that recycling itself consumes a great deal of energy and generates substantial waste product, and therefore the greater efficiency is in optimized reuse of already manufactured products more than in secondary reprocessing.

This sounds very good in theory, but as builders with a packrat bent, we know that the real-world potential for reuse of parts offers many limitations. From time to time we have purchased whole warehouse loads of misfit windows and doors and other high-value building products that manufacturers have wished to close out of their inventories, and when scrounging through Boeing Surplus or scrap steel yards for specifically necessary parts, we always run across large lots of immediately unnecessary material that is just too special to pass up. To accomplish this high-minded acquistion for intended reuse, we have had to rent large trucks and take our construction crews off more productive projects in order to pick up, transport, restock, and catalog vast arrays of not-quite-matching yet beautifully crafted stuff. The intention has always been to purchase high-quality products at a nickel on the dollar—products that we could never afford to use in our projects if ordered new—and then to design buildings around these wonderful parts. Again, the theory is wonderful, but the burdens of stockpiling and managing inventories quickly accumulate until the actual costs begin to outweigh the advantages over ordering new: there are the truck and labor costs, the headaches of drawing up and maintaining detailed inventories sufficiently complete to allow design work to proceed smoothly without needing to climb around in the warehouse, moving loads of material to check measurements, and then turning around in this three-dimensional warehouse puzzle to gain access to the next part to measure. We have built several small buildings and a house for our parents this way, and we have occasionally managed to provide a good deal of donated material to various worthy causes such as Habitat for Humanity and other community and student building projects, but for the most part, for all of the magnificent construction components and tools that we have scrounged and saved over the years, we have been able to give away or make use of relatively little. Even when clients or community groups are scraping by on a limited budget and would love to get something for free, the percentage of parts that will actually work for a particular site and project are always frustratingly few, and it is almost always necessary to obtain most parts and materials and many tools newly manufactured.

There are many hard-earned lessons related to issues of prefabrication that can be learned from such simple experiences in purchasing, shipping, warehousing, and building. Since beginning to work on highly structured construction research projects in Japan and then teaching at UC Berkeley and becoming involved with faculty, researchers, and students from engineering and construction management fields, we have learned that the school-of-hard-knocks lessons we encountered in the everyday business of construction has, at an entirely different scale, engendered large fields of academic, government, and industry research inquiry. Supply chain studies of manufacturing, construction, and business processes bring sophisticated insights into the flow of ideas, design, financing, permitting, material sourcing, purchasing, warehousing, fabrication, marketing, sales, shipping, onsite erection, project completion, billing, and aftermarket servicing that lie at the heart of the success or failure of any construction project and particularly demonstrate the need for increased system rationality represented generally by concepts surrounding the idea of prefabrication.

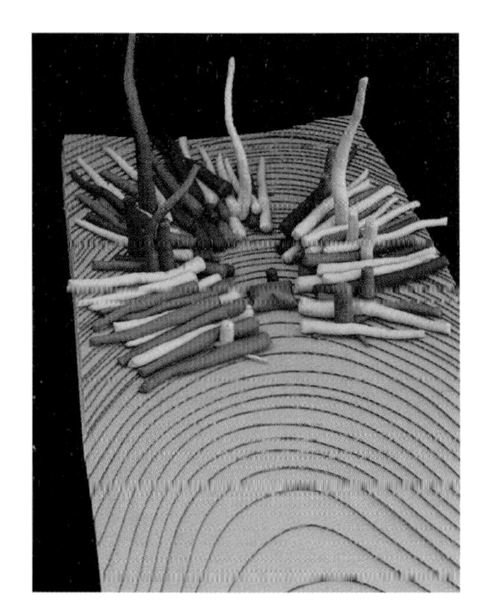

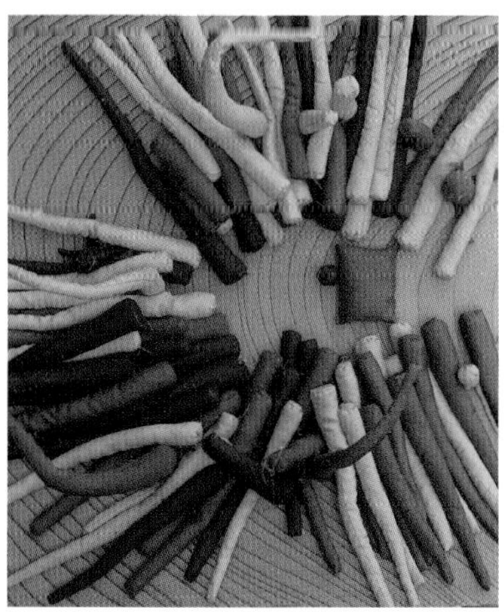

Study model of Moss Pit Amphitheater, Arboretum of the Cascades. The system is constructed of discarded tires currently littering the site, cabled together into long, log-shaped strands, covered in growing medium–filled plastic fabric mesh skins, plumbed with solar-heated warm water and nutrient-laden drip irrigation, and quickly grown over with soft moss and lichens to create a soft, warm theater in the forest.

Although designing for the reuse of products is in many respects more complex than issues of recycling, there are many reasons why this frustrating goal deserves attention. There are substantial environmental, economic, and even creative sociocultural benefits that may be obtained from imaginative repurposing, reapplication, and reuse of previously designed and/or previously used products. To make this sort of reapplication reasonable requires substantial integration of production concepts that consider serialization, repetition, modularity, standardization, and broadly shared scientific inquiry into the properties and behavior of materials. All of this relates directly to prefabrication and other fundamental issues of standardization in materials, processes, and concepts, which incidentally links ideas of prefabrication directly to some of the most prominent conceptual threads in post-enlightenment intellectual and economic history.

Bibliography

Abel, Alan O., et al. *The Industrialized House: A Study of Its Development and Prospects.* Cambridge, Mass.: Harvard University Graduate School of Business Administration, 1952.

Adamson, P., and M. Arbunich. *Eichler: Modernism Rebuilds the American Dream.* Salt Lake City: Gibbs Smith, 2000.

"Aesthetics and Technology of Preassembly." *Progressive Architecture* vol. 76 (1995).

Albert Farwell Bemis Foundation and B. Kelly. *The Prefabrication of Houses.* Cambridge, Mass., and New York: The Technology Press of the Massachusetts Institute of Technology and Wiley New York, 1951.

Anderson, M. and P. *Anderson Anderson: Architecture and Construction.* New York: Princeton Architectural Press, 2001.

———. "Playing in Traffic." *Offramp* 7, "Detours and Dialogues" (May 2000): 110–28.

———"Trains in the City." *Design Matters*, vol. 5, no. 1 (Feb. 1997): 4–11.

Anderson, P. "The Future of Wood Construction in China." *U.S. China Building Materials* (Mar. 2003): 3.

Anderson, S. *Peter Behrens and a New Architecture for the Twentieth Century.* Cambridge, Mass.: MIT Press, 2000.

———. *Eladio Dieste: Innovation in Structural Art.* New York: Princeton Architectural Press, 2003.

Arieff, A. "Dwell Home Invitational." *Dwell* vol. 3, no. 7 (July/Aug. 2003): 3.

Arieff, A., and B. Burkhart. *Prefab.* Salt Lake City: Gibbs Smith, 2002.

Baker, E. *Great Inventions, Good Intentions: An Illustrated History of Design Patents: 1930–1945.* San Francisco: Chronicle Books, 1990.

Baldwin, D. "Prefab Designs: Homes for the Target Generation." *New York Times*, Sept. 26, 2002.

Barnett, R. S. "Better Building Blocks." *Building Design & Construction* vol. 40, no. 2 (Feb. 1999): 82 (1).

Bauer, C. *Modern Housing.* Boston: Houghton Mifflin, 1934.

Bemis, A. F. *The Evolving House, vol. 3: Rational Design.* Cambridge, Mass.: Technology Press, Massachusetts Institute of Technology, 1936.

Bemis, A. F., and John E. Burchard. *The Evolving House.* Cambridge, Mass.: Technology Press, Massachusetts Institute of Technology, 1933.

Bender, R. *Selected Technological Aspects of the American Building Industry: The Industrialization of Building.* Prepared for the National Commission on Urban Problems. New York: R. Bender, 1968.

Benson, A. C. "Early Example of Prefabrication." *Royal Institute of British Architects Journal* (June 1963): 251.

Benton, T. *Form and Function: A Source Book for the History of Architecture and Design 1890–1939.* London: Crosby Lockwood Staples, 1975.

Bernhardt, A. D. *Building Tomorrow: The Mobile Manufactured Housing Industry.* Cambridge, Mass.: MIT Press, 1980.

Bogardus, J., D. D. Badger, et al. *The Origins of Cast Iron Architecture in America.* New York: Da Capo, 1970.

Boone, A. R. "Look, No Nails: House Hooks Together." *Popular Science* (Nov. 1950): 138–39.

Brown, P. L. "Modern Houses, Futurist Ideals." *New York Times*, Nov. 16, 2000.

Bruce, A., and H. Sandbank. *A History of Prefabrication.* New York: John B. Pierce Foundation, 1943.

Building Block Modules, Inc. Introduces a New Concept in Low Cost/Quality Housing. Oakland: Building Block Modules, Inc., 1968.

"Building for Defense." *Architectural Forum* 75 (Aug. 1941): 107.

Burchard, J. "Research Findings of Bemis Industries, Inc." *Architectural Record* vol. 75, no. 1 (Jan. 1934): 3–8.

Carreiro, J., New York State College of Home Economics, Dept. of Housing and Design, et al. *The New Building Block: A Report on the Factory-Produced Dwelling Module.* Research Report no. 8, Center for Housing and Environmental Studies. Ithaca, N.Y.: Cornell University, 1968.

Cherner, N. *Fabricating Houses from Component Parts.* New York: Reinhold, 1957.

Coupland, K. "Dream Homes on Demand: A New Brood of Architects Rely on Prefabrication, Green Materials and Techniques, and Computer Aided Design to Imagine the House of the Future." *Acura Style* (Winter 2004).

Cramer, N. "Winning Projects from American Wood Council Awards." *Architecture Magazine* vol. 85, no. 4 (Apr. 1996).

Creighton, T. H., and Princeton University. *Building for Modern Man: A Symposium.* Princeton, N.J.: Princeton University Press, 1949.

Darlington, I. "Thompson Fecit." *Architectural Review* 124 (Sept. 1958): 187–88.

Davidson, J. "Prefab-ulous: A Cadre of High-Style Architects Brings Designerly Ambitions to the Mid-priced, Kit-built Home." *Newsday*, Nov. 24, 2003.

Delman, A. S. "Modular Buildings: A Perfect Fit for Education." *American School & University* vol. 64, no. 4 (Dec. 1991): 54 (3).

"Department Store Subdivision Contains Eight Preassembled Houses." *Architectural Forum* 84 (Feb. 1946): 7–8.

Diamant, R. M. E. *Industrialized Building 2.* London: Iliffe Books, 1965.

Bibliography

Dietz, A. G. H., and L. S. Cutler, eds. *Industrialized Building Systems for Housing*. Cambridge, Mass.: MIT Press, 1971.

Eichler, N. *The Merchant Builders*. Cambridge, Mass.: MIT Press, 1982.

"The Emphasis is on "Pre" in Fabrication of Buildings." *Modern Hospital* vol. 66 (Apr. 1946): 46–49.

European Productivity Agency. *Prefabricated Building: A Survey of Some European Systems; Project no. 226*. Paris: Organisation for European Economic Cooperation (OEEC), 1958.

"Expansible Prefab House for Post-War. Walter Gropius, Architect." *Architectural Record* 96 (Dec. 1944): 69.

Fetters, T. T. *The Lustron Home: The History of a Postwar Prefabricated Housing Experiment*. Jefferson, N.C.: McFarland, 2002.

Final Report on the State of the Art of Prefabrication in the Construction Industry to the Building and Construction Trades Department, AFL-CIO. Columbus: Battelle Memorial Institute, 1968.

Fisher, H. T. "Prefabrication: What Does It Mean to the Architect?" *AIA Journal* 10 (Nov. 1948): 219–27.

Fitch, J. M. *Walter Gropius*. New York: Braziller, 1960.

"Flexibility Through Standardization/Ezra D. Ehrenkrantz, John D. Kay." *Progressive Architecture* vol. 38 (July 1957): 105–15.

Franciscono, M. *Walter Gropius and the Creation of the Bauhaus in Weimar*. Urbana, Ill.: University of Chicago Press, 1971.

"French System of Prefabrication." *Architects' Journal* 103 (27 June 1946): 487–91.

"Fresh Approach to Housing: Steel Prefabrication, Flexible Size, and Igloo-Like Design: Martin Wagner's House." *Architectural Forum* 74 (Feb. 1941): 90–97.

Fry, E. M. "Walter Gropius." *Architectural Review* vol. 117, no. 619 (Mar. 1955): 155–57.

"General Panel Corp." *Techniques et architecture* 5 (May–June 1945): 134–35.

"Germany Now has Copper Houses Made in Sections Easily Set Up." *Copper and Brass Bulletin* (1931): 2–3.

Giedion, S. *Walter Gropius: Work and Teamwork*. London: Architectural Press, 1954.

Gordon, A. "Putting the Fab in Prefab." *New York Times*, Sept. 26, 2002.

Gould, J. *Modern Houses in Britain, 1919–1939*. London: Society of Architectural Historians, 1977.

Gropius, W. "The Formal and Technical Problems of Modern Architecture and Planning." *Journal of the Royal Institute of British Architects* 19 (May 1934): 679–94.

———. *Scope of Total Architecture*. New York: Harper, 1955.

———. *The New Architecture and the Bauhaus*. London: Faber and Faber, 1956.

———. "Gropius at Twenty-Six." *Architectural Review* 130 (July 1961): 49–51.

———. "True Architectural Goals Yet to be Realized." *Architectural Record* 129 (June 1961): 147–52.

———. "Tradition and Continuity in Architecture." *Architectural Record* 135 (May 1964): 131–36.

———. "Tradition and Continuity in Architecture." *Architectural Record* 136 (June 1964): 133–40.

———. "Tradition and Continuity in Architecture." *Architectural Record* 136 (July 1964): 151–56.

Gropius, W., and I. F. Gropius. *Apollo in the Democracy: The Cultural Obligation of the Architect*. New York: McGraw-Hill, 1968.

Gropius, W., and L. Moholy-Nagy. *Rebuilding our Communities*. Chicago: P. Theobald, 1945.

"The Gropius Exhibition." *The Architect and Building News* (1934): 181.

Hardless, T., ed. *Europrefab Systems Handbook: Housing*. London: Interbuild Prefabrication Publications, 1969.

Harrison, D. D., J. M. Albery, et al. *A Survey of Prefabrication*. London: Ministry of Works Directorate of Post War Building, 1945.

Haskell, D. "Prefabrication: Assembly Lines Reach Out for Markets." *Architectural Record* 93 (June 1943): 62–69.

Herbers, J. *Pre-Fab Modern*. New York: Harper Collins, 2004.

Herbert, G. *The Synthetic Vision of Walter Gropius*. Johannesburg: Witwatersrand University Press, 1959.

———. *Pioneers of Prefabrication: The British Contribution in the Nineteenth Century*. Baltimore: Johns Hopkins University Press, 1978.

———. *The Dream of the Factory-Made House*. Cambridge, Mass.: MIT Press, 1984.

Herrey, H. "At Last We Have A Prefabrication System Which Enables Architects to Design Any Type of Building with 3-Dimensional Modules." *Pencil Points* 24 (Apr. 1943): 36–47.

Hodes, D. A., G. F. Jensen, et al. *The Case for Industrialized Housing—Reexamined*. n. p.: Housing Research Incorporated, 1973.

Holden, A. C. "Prefabrication and the Architect." *Architect and Engineer* vol. 149 (May 1942): 27–29, 32.

"House in 'Industry': A System for the Manufacture of Industrialized Building Elements by Konrad Wachsmann and Walter Gropius." *Arts and Architecture* 644 (Nov. 1947): 28ff.

"House of Many Parts. Building Reporter." *Architectural Forum* vol. 127, no. 4 (Nov. 1967): [78]–[81].

"Houses for Defense." *Architectural Forum* 75 (Nov. 1941): 321–26.

"Houses for Defense... Defense Houses at New Kensington, Pa. Gropius and Breuer, Associated Architects." *Architectural Forum* 75 (Oct. 1941): 218–20.

"Housing Revolution Unfolds." *Popular Mechanics* vol. 178, no. 1 (Jan. 2001): 20.

Howarth, D. *Tahiti: A Paradise Lost*. New York: Viking Press, 1984.

Hutchings, J. F. *Builder's Guide to Modular Construction*. New York: McGraw-Hill, 1996.

"The Industrialized House: General Panel Corporation." *Architectural Forum* 86 (Feb. 1947): 115–20.

Jeffries, N., and Keates, N. "Perfecting Prefabs." *Everett (WA) Herald*, June 13, 2004.

———. "Upwardly Mobile Homes: Can a Prefabricated House be 'Architecture'?" *Wall Street Journal*, Jan. 9, 2004.

Keith, N. S. *Politics and the Housing Crisis since 1930*. New York: Universe Books, 1973.

Kellogg, C. "Again, Architecture Discovers Prefab." *New York Times*, May 29, 2003.

Kolarevic, B. *Architecture in the Digital Age: Design and Manufacturing*. Oxford: Routledge, 2005.

Kronenburg, R. *Transportable Environments: Theory, Context, Design, and Technology: Papers from the International Conference on Portable Architecture, London, 1997*. London and New York: E & FN Spon, 1998.

———. *Portable Architecture*. Oxford and Burlington, Mass., Architectural Press, 2003.

Kronenburg, R., J. Lim, et al. *Transportable Environments 2*. London and New York: Spon Press, 2003.

Lafranchi, G. *Urbanomad*. Vienna and New York: Springer-Verlag, 2004.

Landecker, H. "Export Housing for Japan." *Architecture Magazine* 85 (Oct. 1996): 171–75.

Lane, B. M. *Architecture and Politics in Germany, 1918–1945*. Cambridge, Mass.: Harvard University Press, 1968.

Le Corbusier and F. Etchells. *Towards a New Architecture*. London: Architectural Press, 1946.

"A List of Prefabricators." *Architectural Record* 93 (June 1943): 75–79.

Llewellyn, F. T. "Steel in Residence Construction." *Architectural Record* 73 (June 1933): 438–42.

Low-cost Prefabricated Wooden Houses: A Manual for Developing Countries. Vienna: United Nations Industrial Development Organization, 1992.

Lynn, G. "Greg Lynn: Embryologic Houses." *Architectural Design* vol. 70, no. 3 (June 2000): [26]–35.

"Manifestos of the Modern House." *A+U* 3 (Mar. 2000, special issue): 56–61.

Manning, F. "Selling the Prefabricated House? Here are Some Unsolved Problems." *American Builder* 66 (Feb. 1944): 96, 98–100.

Marzolf, K. "Production Line Architecture: A Warning from Scandinavia." *American Institute of Architects* vol. 48, no. 4 (Oct. 1967): [62]–70.

Massachusetts Institute of Technology. Dept. of Architecture, Master's Class. *Housing Systems; Seven Studies of Factory-produced Steel and Concrete Modular Units*. Cambridge: Dept. of Architecture, Massachusetts Institute of Technology, 1970.

Maynard, N. "Some Assembly Required: Is Modular Construction an Answer to the Affordable Housing Problem?" *Residential Architect* (Nov.–Dec. 2002): 2.

McCoy, E. *Modern California Houses: Case Study Houses, 1945–1962*. New York: Reinhold, 1962.

———. *Case Study Houses, 1945–1962*. Los Angeles: Hennessey & Ingalls, 1977.

McKee, B. "The Impossible Dream?" *Metropolitan Home* (May/June 2002): 2.

McKennee, O. W., and the Staff of The Housing Institute, Inc. *Prefabs on Parade*. New York: Housing Institute, 1948.

McQuade, W. "An Assembly-Line Answer to the Housing Crisis." *Fortune* vol. 79, no. 5 (May 1, 1969): [98]–[103], 136–40.

"Metal Building Systems Dominate Low-Rise, Non-Residential Market." *ENR* vol. 241, no. 9 (Sept. 7, 1998): S14 (1).

"The Modern Houses of the Century of Progress Exposition." *Architectural Forum* 59 (July 1933): 51–62.

Moholy-Nagy, S. "The Diaspora." *Journal of the Society of Architectural Historians* vol. 24, no. 1 (Mar. 1965): 24–26.

Moltke, W. V. "Prefabricated Panels for Packaged Buildings." *Architectural Record* 93 (Apr. 1943): 50–53.

"The Month in Building... News." *Architectural Forum* 79 (Sept. 1943): 65–68.

Moore, R. "An Early System of Large-Panel Building." *RIBA Journal* vol. 76, no. 9 (Sept. 1969): 383–86.

Naylor, G. *The Arts and Crafts Movement: A Study of Its Sources, Ideals, and Influences on Design Theory*. Cambridge, Mass.: MIT Press, 1980.

Neuhart, M., and J. Neuhart. *Eames House*. Berlin: Ernst & Sohn, 1994.

"Newcomers: General Panel is First to Benefit from Everything Government Has to Give." *Architectural Forum* 86 (Jan. 1947): 15–16.

"The New Look in Builder Houses: A New Heating System and a New Idea in Prefabrication." *Architectural Forum* 91 (July 1949): 90–91.

Noakes, E. H. "Housing and Prefabrication." *Royal Architectural Institute of Canada* 23 (Sept. 1946): 207–11.

North, A. T. "Prefabricated Buildings Will Bring Lower Costs." *American Architect* vol. 141, no. 2607 (May 1932): 66–67, 90.

Bibliography

Park, J. H. "Anderson Anderson Architecture." *POAR* (Seoul) 3 (Apr. 1999): 58–67.

Pehnt, W. "Gropius the Romantic." *Art Bulletin* (Sept. 1971): 379–92.

Perkins, N. S., and Douglas Fir Plywood Association. *Plywood: Properties, Design and Construction*. Tacoma: Douglas Fir Plywood Association, 1962.

Peterson, C. E. "Early American Prefabrication." *Gazette des Beaux-Arts* 6th ser., no. 33 (1948): 37–46.

———. "Prefabs for the Prairies." *Journal of the Society of Architectural Historians* vol. 11, no. 1 (Mar. 1952): 28–30.

———. "Prefabs in the California Gold Rush, 1849." *Journal of the Society of Architectural Historians* vol. 24, no. 4 (Dec. 1965): 318–24.

Peterson, J. L. "History and Development of Precast Concrete in the United States." *Journal of the American Concrete Institute* vol. 25, no. 6 (Feb. 1954): 477–96.

Pople, N. *Experimental Houses*. London: Laurence King, 2000.

"Portable Steel and Copper House Developed in Germany." *Iron Age* (1931).

"Prefabricated Buildings." *California Architect and Building News* vol. 215 (Mar. 25, 1959): 373–98.

"Prefabricated Construction." *California Architect and Building News* vol. 222, no. 17 (Aug. 15, 1962): 226–44.

"Prefabricated Houses." *Architectural Forum* 78 (June 1943): 89–96.

"Prefabrication, for and against." *California Architect and Building News* vol. 168 (Dec. 19, 1941): [169].

"Prefabrication Gets Its Chance." *Architectural Forum* 76 (Feb. 1942): 81–88.

"Prefabrication: Modulok." *Architectural Forum* 79 (Sept. 1943): 65–68.

"Prefabrication Up-to-Date." *American Builder* 65 (Jan. 1943): 44–45, 77–79.

"Prefabs in the California Gold Rush, 1849." *Journal of the Society of Architectural Historians* vol. 5 (1945–46): v.: ill., maps.

Rand, A. "This Factory-Built House Came True." *Los Angeles Times*, Dec. 17, 1950.

"Recreational Center Proposed for Key West, Florida. Walter Gropius, Architect, Konrad Wachsmann, Associated." *Architectural Forum* 77 (Aug. 1942): 83–85.

"Review of Current Techniques." *Prefabricated Homes* (Nov.–Dec. 1946): 26–27.

Reidelbach, J. A. *Modular Housing—1971: Facts and Concepts*. Boston: Cahners Books, 1971.

Report of the Seminar on Prefabrication of Houses for Latin America. New York: United Nations, 1972.

Robertson, H. "The First Nazi Exhibition, Berlin." *Architect and Building News* (1934): 193–95.

Royal Institute of British Architects, Industrialized Building Study Teams. *The Industrialization of Building: An Appraisal of the Present Position and Future Trends*. London: Royal Institute of British Architects, 1965.

Russell, J. S. "Pre-engineered by Design." *Architectural Record* vol. 179, no. 10 (Oct. 1991): 128 (8).

Sanders, W. *Manufactured Housing: Regulation, Design Innovations, and Development Options*. Chicago: American Planning Association, 1998.

Sharp, D., and Architectural Association (Great Britain). *Sources of Modern Architecture: A Bibliography*. London: published for the Architectural Association by Lund Humphries, 1967.

Shiffman, J. "The Building Industry in Palestine." *Palestine and Middle East Economic Magazine* 7–8 (1933): 288.

"Shop Fabrication from Architects' Plans." *Architectural Record* 82 (Aug. 1931): 32–33.

"Site Prefabrication in Housebuilding." *Prefabrication* 9 (May–June 1948): 10–11.

Smith, E. A. T., E. McCoy, et al. *Blueprints for Modern Living: History and Legacy of the Case Study Houses*. Los Angeles and Cambridge: Museum of Contemporary Art and MIT Press, 1998.

Sears, Roebuck. *Sears, Roebuck Home Builder's Catalogue: The Complete Illustrated 1910 Edition*. New York: Dover Publications, 1990.

———. *Small Houses of the Twenties: The Sears, Roebuck 1926 House Catalogue*. An Unabridged Reprint. New York: Dover Publications, 1991.

Seehrich-Caldwell, A. *Starterhäuser*. Stuttgart, Germany: Karl Kramer Verlag, 1997.

"Steel Houses." *Architectural Forum* 58 (Apr. 1933): 327–31.

Stevenson, G. *Palaces for the People: Prefabs in Post-war Britain*. London: Batsford, 2003.

Stowell, K. K. "Housing the Other Half." *Architectural Forum* 56 (Mar. 1932): 217–20.

"Structural Systems and Designs." *Architectural Record* 80 (July 1936): 63–73.

Sullivan, B. J. *Industrialization in the Building Industry*. New York: Van Nostrand, 1980.

Sumichrast, M. *Housing and Money Markets, 1950–1966*. Washington, D.C.: National Association of Home Builders of the United States, 1966.

"Systems Building: What it Really Means." *Architectural Record* 145 (Jan. 1969): 147–54.

"Technical News and Research: New Housing Designs and Construction Systems." *Architectural Record* vol. 75, no. 1 (Jan. 1934): 12–34.

Tengbom, A. "The Growth of Prefabrication in Sweden." *Prefabricated Homes* 6 (May 1946): 17, 24.

Thomson, D. *Europe since Napoleon*. New York: Knopf, 1962.

"Timber Houses, Prefabricated Construction." *Builder* vol. 154 (Feb. 18, 1938): 360.

"Timber Prefabricated Houses." *Architects' Journal* vol. 98 (Dec. 31, 1944): 485–86.

Towne, C. A. "Housing and Prefabrication in Germany, Great Britain, and the U.S.A." *Progressive Architecture* 27 (May 1946): 89–93.

U.S. Department of Commerce, Iron and Steel Division. *Construction of Steel Houses Increasing in Western Germany*. Washington, D.C.: U.S. Government Printing Office, 1928.

U.S. Department of Housing and Urban Development. "Housing Systems Proposals for Operation Breakthrough." Washington, D.C.: U.S. Government Printing Office, 1971.

Vale, B. *Prefabs: A History of the UK Temporary Housing Programme*. London and New York: E & FN Spon, 1995.

"Variety of Houses from Identical Prefabricated Units of General Panel Corp.: Designed by Harvard Students." *Pencil Points* 24 (Dec. 1943): 76–77.

Wachsmann, K. *The Turning Point of Building: Structure and Design*. New York: Reinhold, 1961.

Ward, Jr., R. "Konrad Wachsmann: Toward the Industrialization of Building." *AIA Journal* (Mar. 1972).

Weidenhoft, R. V. *Workers' Housing in Berlin in the 1920s: A Contribution to the History of Modern Architecture*. New York: Columbia University, 1971.

White, R. B., and Building Research Station (Great Britain). *Prefabrication: A History of its Development in Great Britain*. London: H.M.S.O., 1965.

Whitely, P. "Off-the-Shelf Architecture." *Sunset* 195, no. 4 (Oct. 1995): 130.

Williams., A. "Absolutely Pre-fabulous: Prefabrication Seems to be Coming into its Own at Last." *Architects' Journal* vol. 215, no. 22 (June 6, 2002): 38–39.

Wilson, F. V. J. "New Approach to 'Packaged' House Recognizes Architects." *Architectural Record* 82 (Aug. 1931): 86–87.

———. *Tomorrow's Homes*. Trenton: Homasote Co., 1939.

Winchell, P. "The Dwelling of Tomorrow: An Economic Study in Residential Construction Showing Why a Growing Use of Iron and Steel is Inevitable." *Iron Age* 117 (1927): 613–15, 686–87, 766–68, 840–41, 930–31, 992–93.

Windsor, A. *Peter Behrens, Architect and Designer*. London: Architectural Press, 1981.

Wingler, H. M., and J. Stein. *The Bauhaus: Weimar, Dessau, Berlin, Chicago*. Cambridge, Mass.: MIT Press, 1969.

Wurster, C. B. "The Social Front of Modern Architecture in the 1930s." *Journal of the Society of Architectural Historians* vol. 24, no. 1 (1965): 48–52.

Zieger, M. "Prototype Prefabricated Design by Anderson Anderson." *Dwell* vol. 1, no. 4 (Apr. 2001): 10.

Project List

Anderson Anderson Architecture

Bay Pacific Construction

Projects dealing significantly with prefabrication are in bold.

2006

Nassau Street Habitat for Humanity Housing, Charlottesville, VA

New Orleans Modular Mixed-use Block, New Orleans, LA

2005

Amber Block, Tulsa, OK

Davis Residence, Los Angeles, CA

Tower House 2, Oakland, CA

Pope Allen Residence, Healdsburg, CA

Schach Residence, Pacifica, CA

Simpson Residence, Los Angeles, CA

Sunrise Trailer Court Redevelopment, Charlottesville, VA

2004

Alta Way House, Mill Valley, CA

Irish Beach Residence, Irish Beach, CA

Joshi-Owens House, San Carlos, CA

Marrone House Addition, Gig Harbor, CA

Private Airplane Hangar and Gallery, Durango, CO

San Bruno Mountain Houses, Brisbane, CA

Sherwin Residence, Arcadia, FL

Tang House Addition, Mountain View, CA

Tinker's Workshop, Berkeley, CA

Cheng Tower House, Oakland, CA

2003

Agnus Dei Lutheran Church, Gig Harbor, WA

Cal Sailing Clubhouse, Berkeley, CA

Experimental Theater/Artists' Retreat, Martinez, CA

Abiquiu House, Abiquiu, NM

Marrowstone Island House, Marrowstone Island, WA

Sierra Mountain House, Sierra National Forest, CA

Kelly-Yuoh Residence, Yonkers, NY

Orchard House, Sebastopol, CA

Maybelle Avenue Community, Oakland, CA

McQuistin-Smith Residence, Dash Point, WA

Mount Whitney Trailhead, Inyo National Forest, CA

Cantilever House, Arlington, WA

S.W.E.L.L., Berkeley, CA

Slice House, Pittsboro, NC

2002

Berkeley Office Interior, Berkeley, CA

Chameleon House, Leelanau County, MI

Cascades Arboretum Visitor Center, Preston, WA

Fonte Ferrata Chapel Interior, Castagneto Carducci, Italy

Malibu House, Malibu, CA

Marrone Residence, Mazama, WA

Nassiri-Merzenich Residence, Oakland, CA

Office Interior, Seattle, WA

Pigozzi-Casey Residence, Oakland, CA

Pillifant Multifamily Townhouses, Anchorage, AK

Scarbrough Residence, Gig Harbor, WA

Venice Lagoon Project/Barene Drawings, Venice, Italy

Wurster Workshop, UC Berkeley, CA

2001

Bloome Gallery, Seattle, WA

Clackety-Yak ScumBag Bamboo, Honolulu, HI

Cooper Residence, San Francisco, CA

Evans Multiplex, San Diego, CA

Hawaii State Capitol Lighting Renovation, Honolulu, HI

Judson College Library, Chicago, IL

Judson College Art and Architecture Building, Chicago, IL

Yanosan Headquarters Building, Bothell, WA

2000

East Side Community Pavilion, Detroit, MI

Hart Residence, Seattle, WA

Obata Residence, Kosai, Japan

ScumBagDirtClodGasHuffPhytoCurtain, New Orleans, LA

Lillegard Residence, Seattle, WA

1999

Iki Island Retreat Center, Iki Island, Japan

Nagao Residence, Kosai, Japan

Rian Residence, Honolulu, HI

Roe Ramsey Vacation Home, Hood Canal, WA

1998

Curry County Forest Canopy Park, Gold Beach, Oregon

Forest Canopy Research Center, Evergreen State College, WA

Obata Zebra Prototype, Washizu, Japan

Peninsula Lutheran Church Addition, Gig Harbor, WA

Stick Bladder, Seattle, WA

Tonn Residence, Tacoma, WA

1997

Chiba Mixed Use Prototype, Chiba Prefecture, Japan

Church's Shoe Store, Seattle, WA

Colbert Infrared Services Building, Seattle, WA

Cloud Room (Sitewerks Office), Seattle, WA

Enlow Residence, Cle Elum, WA

Obata Showroom Office, Washizu, Japan

Scofield Mixed Use Development, Gig Harbor, WA

Seattle Office Interior, Seattle, WA

Trains in the City, Seattle, WA

University of Hawaii Courtyard Competition, Honolulu, HI

World House 2020 Prototypes, Tokyo, Japan

1996

Construction Technology Seminars, Tokyo, Japan
Hamilton Library Remodel, Seattle, WA
Iwama Guest House, Mt. Fuji, Japan
Kansai-kan National Diet Library, Kyoto, Japan
Morimoto Residence, Setagaya, Tokyo, Japan
Nedderman Dock/Deck/Stair, Gig Harbor, WA
Obata Corp. Office and Residence, Shizuoka, Japan
University Apartments, Nagoya, Japan
Scofield Corp. Leavenworth Project, Leavenworth, WA
Sophia Elderly Housing Prototypes, Tokyo, Japan
Block House Addition, Fort Worth, TX
Sumitomo/Nichi-ha Test House, Bellevue, WA

1995

Amerikaya Garden Villa Houses, Tsuruga, Japan
Ceccanti Residence, Tahuya,WA
Chiyo New Town Affordable Housing, Fukuoka, Japan
Cowboy House Prototype 4, MT
Cowboy House Prototype 5, MT
Cowboy House Prototype 6, MT
Construction Technology Seminars, Kyushu, Japan
Ho Residence Remodel, Gig Harbor, WA
HotPlateColdPlateMudMapSnowBlind-BladderBladder, Anchorage, AK
Industrial Furniture, Seattle, WA
Import Housing Report for CTED, Kobe, Japan
Osaka Residence for I. K. Estem, Osaka, Japan
Ishida Ferrari Museum and Residence Odawara, Japan
Klein Residence, Tahuya, WA
Kobe Community Center, Kobe, Japan
Loeken Residence, Raft Island, WA
Markewitz Residence Remodel, Gig Harbor, WA
Obata Residence 1, Kosai, Shizuoka, Japan
Prado Museum Expansion, Madrid, Spain
Prastka Garage Addition, Harstine Island, WA
Prastka Residence, Harstine Island, WA
Rudin Residence, Port Angeles, WA
Richter/Wiener Ranch, Bend, OR
William & Zimmer Factory/Showroom, Lihue, HI

1994

Amerikaya Affordable Home Design I, Tsuruga, Japan
Big Ivy Affordable Home Prototypes, Sendai, Japan
Blodgett/Gullen Residence, Gig Harbor, WA
Carlberg Residence, Allyn, WA
Cowboy House System, MT
Cowboy House Prototype 1, MT
Cowboy House Prototype 2, MT
Cowboy House Prototype 3, MT
Ell Residence, Gig Harbor, WA
Ess Residence, Wollochet Bay, WA
Office Building Remodel, Gig Harbor, WA
Herron Apartment Project, Nagoya, Japan
Industrial Office Furniture, Seattle, WA
Lakebay Lumber Retail Facility, Vaughn, WA
Paly Residence, Gig Harbor, WA
Prairie Ladder: Exhibition, Phillips Gallery, Fort Worth, TX
Prairie Ladder: Lithograph Suite, Seattle, WA
Pioneer Square Loft Interior, Seattle, WA
Richards Residence, Gig Harbor, WA
Shinohara Residence, Tsuruga, Japan
Seattle Office Interior, Seattle, WA
Apartment Building Prototype, Tokyo, Japan

1993

Anderson Residence, Gig Harbor, WA
Bankson Residence, Wollochet Bay, WA
Finley Residence, Gig Harbor, WA
Kennedy Prototype Panelized House, Gig Harbor, WA
Marontate Residence, Gig Harbor, WA
Mulligan Residence, Harstine Island, WA
Affordable Housing, Sendai, Japan
Smart Technology House, Gig Harbor, WA

1992

Devita Residence, Gig Harbor, WA
Gig Harbor Office Remodel, Gig Harbor, WA
International Institute of Wood Construction, Tokyo, Japan
Apartment Building, Kanazawa, Japan
Exhibition Display, Kobe, Japan
Klein Residence Addition, Gig Harbor, WA
Prairie Ladder: Weather Station, Vaughn, WA
Shanghai Office Tower Development, Shanghai, China
Woodframe Laboratory, Tokyo, Japan

1991

Anderson Studio, Gig Harbor, WA
Bertelsen Residence Addition, Tacoma, WA
Burley Lagoon Properties, Burley, WA
Ceccanti Residence, Tahuya, WA
Clover Park Clinic, Tacoma, WA
Davie Art Studio, Vaughn, WA
Hammar Retail Interior, Seattle, WA
Klein Residence, Tahuya, WA
Mayer/Scheidt Residence, Olympia, WA
McCracken Residence, Spanaway Lake, WA
Mayer Residence Addition, Lakewood, WA
Courtyard House, Gig Harbor, WA
Canterwood Speculative House (2), Gig Harbor, WA
Steilacoom Inn Condominiums, Steilacoom, WA
Stuen Gallery, Pacific Lutheran University, Tacoma, WA
Sullivan Residence, Gig Harbor, WA

Project List

1990

Aiken Residence, Gig Harbor, WA
Camp Residence, Gig Harbor, WA
Chakerian Residence, Gig Harbor, WA
Johnson Vacation Home, Harstine Island, WA
Kraft Residence, Gig Harbor, WA
Meyers Residence, Port Orchard, WA
Ripperton Residence, Gig Harbor, WA
Schoepp Sculpture Studio, Fort Worth, TX

1989

Buder Residence, Gig Harbor, WA
Chiao Residence, Gig Harbor, WA
YSIT/Sam & Alice Dance Theater Set, University of Washington, Seattle, WA
Installation, Fairbanks, AK
Gig Harbor Office Renovation, Gig Harbor, WA
Prairie Ladder: Earth Plane Sky Barge, Connemara Conservancy, Dallas, TX
Thomson Residence, Gig Harbor, WA

1988

Anderson Dock and Greenhouse, Gig Harbor, WA
Harstine Sculpture Installation, Mason County, WA
Holden Residence, Harstine Island, WA
Marrone/Kresge Residence, Gig Harbor, WA
Shadowmaker Sculpture Installation, Carpenter Center for the Arts, Harvard University, Cambridge, MA
Torso Performance/Installation, Radcliffe College, Cambridge, MA
Wentzel Residence, Gig Harbor, WA
Fielding Addition, Wollochet Bay, WA

1987

Cowan Residence, Gig Harbor, WA
Lemke Residence, Wauna, WA
Lights Orot Sculpture Exhibition, New York, NY
Metzdorf Speculative House, Gig Harbor, WA

1986

Filmer Residence (Spec House), Gig Harbor, WA
Frangilo Residence (Spec House), Gig Harbor, WA
Fort Worth Sculpture Installation, Fort Worth, TX
Houston Sculpture Installation, Houston, TX
Speculative House, Gig Harbor, WA
Torrens Smash Colors House, Wauna, WA

1985

Architectural Exhibit, Harvard University, Cambridge, MA
Jerke Vacation Home, Harstine Island, WA
Marrone Residence Design, Gig Harbor, WA
Nelson Residence Addition, Gig Harbor, WA
Screaming Box Installation, Harvard University, Cambridge, MA
Wauna Speculative House, Wauna, WA

1984

Foster Residence Addition, Castagnetto Carducci, Italy
LeMarchand Addition, Vosves, France
McMenamin Residence Addition, Wauna, WA
North End Addition, Tacoma, WA
Private Residence Addition, Tacoma, WA
Rowlands Residence Addition, Tacoma, WA
Torrens Residence Addition, Wauna, WA
Torrens White Fish House, Wauna, WA
Williams Residence Addition, Tacoma, WA

1983

Anderson Cabin Workshop Addition, Harstine Island, WA
Cambridge Sculpture Installation, Harvard University, Cambridge, MA
Nordby Residence Addition, Tacoma, WA
Sculpture Installation, Tacoma, WA
Tonn Sunroom Addition, Tacoma, WA
Torrens Sculpture Studio, Wauna, WA

1982

Cox Residence Addition, Tacoma, WA
Eyler Residence Remodel, Gig Harbor, WA
MacBride Addition, Tacoma, WA
Rasmussen Residence Addition, Tacoma, WA
Tonn Residence Garage Addition, Tacoma, WA

1981

Anderson Residence Addition, Tacoma, WA
Commercial Fire Hydrant System, Gig Harbor, WA
Fonte Ferrata Farm Projects, Castagnetto Carducci, Italy

1980

Litzenberger Addition, Gig Harbor, WA

1979

MacLachlan Residence, Olympia, WA
Wishing Well Water Company, Gig Harbor, WA

1978

Litzenberger Vacation Home, Harstine Island, WA

1977

Anderson Vacation Home, Harstine Island, WA
MacLachlan Residence Addition, Lacey, WA
Reigstad Addition, Puyallup, WA

Design and Construction Credits

Anderson Anderson Architecture

Bay Pacific Construction

Design & Construction Team 1984–2005:
Mark Anderson
Peter Anderson
Tonita Abeyta
Phil Auge
Paul Baker
Scott Baker
Doug Barnes
Scott Bauerman
Hitasha Bhatia
Tom Benzenberg
Maria Bianchi-Lastra
Zachary Bischoff
Chris Bell
Hannah Brown
Jason Chai
Byron Chang
Carmello Echanis
Oliver Dering
Minako Domen
Lawton Eng
Mike Fish
Dean Flora
Kelly Forseth
Allen Frost
Rick Fuller
Teresa Funk
Rick Gagliano
Camila Garrido
Phillip Habell
Alexander Herter
Cathi Ho
Rich Holt
Kyle Hughes
David Jerke
Jeff Johnson
Matt Johnson
Jean Jonas
Jeff Jordan
Eric Jorgenson
Masatoshi Kasai
Kari Kimura
Suzanne Knapp
Aaron Korntreger
Dave Larsen
Youngchae Lee
Jarred Lowrey
Christopher Luthi
Francesco Maccarone
Martin MacDonald
Rex Manzano
Claudio Martonffy
Kaouru Matsumura
Kylie Moss
Erik Mott
Glenn Newton
Titian Niosi
Jeff Nye
Ken Olkonen
Dennis Oshiro
Todd Ottmar
Homero Nishiwaki
Olivier Pennetier
Casey Pritchett
Shaun Roth
Greg Rowe
Justin James Rumpeltes
Ed Sauerlinder
Phoebe Schenker
Cameron Schoepp
George Sharp
Margaret Sledge
David Sorey
Matt Stevens
Brent Sumida
Megumi Tamanaha
Mike Torgerson
Tom Trineer
Linda Uehara
Scott Wagner
Tim Walker
Todd Walker
Keenan Widrig
Carol Wilkinson
Joel Williams
Napier Wright
Bonnie Vesperman
Gary Yoshimura

Engineers
Jim Hands
Terry Nettles
Chris Peck
Roger Rached
René Abi-Rached
David Strandberg
Dave Wilson
Stan Wu

Contractors
Anderson Anderson/ Bay Pacific Construction
Comstock Construction Company
Daniel Martin Construction
Drew Allen General Contractor
Keever and Associates, Contractors

Acknowledgments

We wish to thank the many individuals and groups that have been involved in the production of this book and in the production of the research, design, and construction work presented within it. A list such as this is only the tip of the iceberg, however, when considering all the work by so many people that comes together to make a work of architecture, or a book. Although we cannot bring forward for special mention everyone who has contributed to these projects, there are many who deserve special acknowledgment.

First of all, we wish to thank the core group who has worked with us to produce this book. Within our office during this period, Brent Sumida, Dennis Oshiro, Aaron Brumo, Rita Sio, Ji Young Chung, Lawton Eng, Alan Owings, Hannah Brown, Chris Campbell, Chia Chieh Lee, Hitasha Bhatia, Carla Dominguez, and Camilla Garrido have among them worked on all aspects of drawing, designing, researching, building, archiving, sorting, critiquing, and image editing. Architect and photographer Anthony Vizzari has been instrumental in bringing some of our most important models and buildings to life in these pages. The librarians and staff at the University of California, Berkeley, College of Environmental Design Library, especially Elizabeth Byrne and Sue Koskinnen, have contributed greatly to our literature research. Above all we wish to thank our editor, Scott Tennent, who has been an outstanding and sympathetic colleague in the conception and writing of this book, as well as Paul Wagner, the book's designer, publisher Kevin Lippert, and the entire editorial, design, and production staff of Princeton Architectural Press.

Every building project requires the support of clients and collaborators who engage in the research, design, and construction process as a creative dialogue. In this role we have been fortunate to collaborate with outstanding individuals, businesses, and public institutions. We wish to especially mention the following client, research, and design collaborators:

Individuals Dan and Sue Brondyk, Tom and Rosalind Buffaloe, Lorn Dittfeld and Tara Schraga, Birgitte Ginge and Madeline Williamson, Sean Nassiri and Marghi Merzenich, Tuhin Roy, Phil and Carla Sherwin, Scott Stafne, Ben Kinmont and Naomi Hupert, Chris and John Jones, Mark and Sharon Bloome, Mike Obata, Bunnosuke and Junko Nagao, Kenji Shinohara, Cameron Schoepp and Terri Thornton, Jim and Ruth Mulligan, Melissa Kennedy, Itaru and Miho Ishida, Alka Joshi and Brad Owens, Leslie McQuistin and Sterling Smith, Heidi and Joe Davis, Leslie and James Simpson, David and Rossana Schach, Thomas Pope and Stewart Allen, Michael Marrone, Michael Scarbrough, Micha Alexander, Julie Molzahn, Jeffrey Bailey, Harrison Fraker, Mark Calhoon, Robert Yamazaki, Ivan Eastin, and Paul Boardman.

Additionally, the organizations and institutions Sumitomo Corporation; Nichi-Ha Corporation; Japan Ministry of Construction; City of Kitakyushu Departments of Housing, City Planning and Construction; United States Departments of Commerce and Agriculture; United States Forest Service; Amerikaya Construction Corporation; University of California, Berkeley; California College of the Arts; University of Hawaii at Manoa, School of Architecture and Construction Process Innovation Laboratory; International Laboratory for Architecture and Urban Design (ILAUD); Tulane University School of Architecture; University of Washington College of Forest Resources and Center for International Trade in Forest Products; Weyerhaeuser Corporation Research Division; Boeing Corporation; Microsoft Corporation; San Francisco Museum of Modern Art; Sophia Homes Corporation; Next Homes Corporation; Garden Pacific Homes, Japan; Obata Construction; Meiwa Construction; Jet Construction; *Dwell* Magazine; American Hardwoods Export Council; National Association of Homebuilders Research Council; Smart House Technology Consortium, American Plywood Association; Global Trek Corporation; Washington State Department of Trade and Community Development; US Bank; Bank of America; Washington Mutual Bank; Columbia Bank; Center on Contemporary Art, Seattle; Alaska Design Forum; Hyogo Prefectural Government; State of Hawaii Department of Accounting and General Services; State of Hawaii Foundation on Culture and the Arts; The Evergreen Partnership; Evergreen State College; Charlottesville Community Design Center; Habitat for Humanity; and Tinkers Workshop, Berkeley, California.

Finally, we wish to acknowledge all of the personal contributions of education, inspiration, support, and love that have helped build a foundation informing our architecture and this book. We wish to thank the Anderson and Tufte families, especially Charles and Margaret Anderson, Kristen Kalbrener, Marge and Paul Reigstad, Esther and Norman Rian, Stanford Anderson and Nancy Royal, Helen Andrewson, Louise Redstone, Dora and Carl Anderson, Olaf and Alma Tufte, and Ruth, Lorraine, and Paul Tufte.